WATER–BASED PAINT FORMULATIONS

WATER–BASED PAINT FORMULATIONS

Volume 3

by

Ernest W. Flick

Library of Congress Catalog Card Number: 75-2939
ISBN: 0-8155-1345-3

Published in the United States of America by
Noyes Publications
Mill Road, Park Ridge, New Jersey 07656

10 9 8 7 6 5 4 3 2 1

Library of Congress Cataloging-in-Publications Data
(Revised for vol. 3)

Flick, Ernest W.
   Water-based paint formulations.

   Vol. 3 published by Noyes Publications.
   Includes index.
   1. Paint--Patents. II. Title.
TP935.F62   1975   667'.63   75-2939
ISBN 0-8155-0571-X (v. 1)
ISBN 0-8155-1345-3 (v. 3)

Transferred to Digital Printing 2009

To

Carol and Bob

and

Janet and Dennis; Karen and David

and

their families

Preface

This collection of 463 up-to-date water-based trade and industrial formulations will be of value to technical and managerial personnel in paint manufacturing companies and firms which supply raw materials or services to these companies, and to those interested in less hazardous, environmentally safer formulations. The book will be useful to both those with extensive experience as well as those new to the field. This book includes new and different formulations than those included in the previously published *Industrial Water-Based Paint Formulations* and *Water-Based Trade Paint Formulations*.

The data consist of selections of manufacturers' suggested formulations made at no cost to, nor influence from, the makers or distributors of these materials. The information given is presented as supplied; the manufacturer should be contacted if there are any questions. Only the most recent data supplied us has been included. Any solvent contained is minimal.

The table of contents is organized in such a way as to serve as a subject index. The formulations described are divided into sections which cover exterior, interior, and exterior and/or interior water-based paints, enamels, and coatings, as indicated below:

    I.   Coatings and Topcoats
    II. Coil Coatings
    III. Enamels
    IV. Enamels—Baking
    V.   Exterior Paints and Related
    VI. Interior Paints and Related
    VII. Lacquers
    VIII. Primers
    IX. Sealers
    X.   Stains
    XI. Texture Paints
    XII. Miscellaneous

Each formula has been placed in the chapter which is most applicable. The reader, if seeking a formula for a specific end use, should check each chapter which could possibly apply.

In addition to the above, there are two other sections which will be helpful to the reader:

  XIII. A chemical trade name section where trade–named raw materials are listed with a chemical description and the supplier's name. The specifications which the raw materials meet are included if applicable.

  XIV.   Main office addresses of the suppliers of trade–named raw materials and/or formulations.

Included in the descriptive information for each formulation, where available, the following properties may be listed: viscosity, solids content, % nonvolatiles, pigment volume concentration, density, pH, spatter, leveling, sag resistance, scrub cycles to failure, contrast ratio, ease of dispersion, fineness of grind, heat stability, freeze–thaw stability, ease of application, gloss foaming, cratering, brightness, opacity, water spotting, adhesion to chalk, brush clean–up, reflectance, and sheen.

Ernest W. Flick

Notice

To the best of our knowledge the information in this publication is accurate; however, the Publisher does not assume any responsibility or liability for the accuracy or completeness of, or consequence arising from, such information. These formulations do not purport to contain detailed manufacturing nor user instruction, and by its range and scope could not possibly do so. Mention of trade names or commercial products does not constitute endorsement or recommendation for use by the Author or Publisher.

Water–based paint raw materials could be toxic, and therefore, due caution should always be exercised in the use of these hazardous materials. Final determination of the suitability of any information or product for use contemplated by any user, and the manner of that use, is the sole responsibility of the user. We strongly recommend that users seek and adhere to a manufacturer's or supplier's current instructions for handling each material they use.

The Author and Publisher have used their best efforts to include only the most recent data available. The reader is cautioned to consult the supplier in case of questions regarding current availability.

Contents and Subject Index

Section I
Coatings and Topcoats

Air Dry or Force Dry Cure Clear Coating-A

Ingredients:	Parts by Weight
Chempol 10-0094	57.11
28% Ammonium Hydroxide	3.00
Water	38.89
5% Cobalt Hydrocure	0.40
5% Manganese Hydrocure	0.40
Activ-8	0.20

Air Dry or Force Dry Cure Clear Coating-B

Ingredients:	Parts by Weight
Chempol 10-0094	57.48
28% Ammonium Hydroxide	3.00
Water	37.42
Hydrocure II	0.40
Hydrocem Calcium	1.20
Hydrocem Zirconium	0.50

Method B yields a somewhat tougher films than Method A.

SOURCE: Freeman Polymers Division: CHEMPOL 10-0094: Formulas

Air Dry or Force Dry Cure Clear Coating-B

Ingredients:	Parts by Weight
Chempol 10-0097	54.05
Ethylene glycol monobutyl ether	3.80
NH4OH (28% min.)	3.40
Water	36.65
5% Cobalt Hydrocure	0.00
5% Manganese Hydrocure	0.00
Activ-8	0.00
Hydrocure II	0.40
Hydrochem Calcium	1.20
Hydrochem Zirconium	0.50

SOURCE: Freeman Polymers Division: CHEMPOL 10-0097: Formula B

Air Dry or Force Dry Clear Coating-A

	Parts by Wt.
Chempol 10-0095	53.30
Ethylene glycol monobutyl ether	3.90
NH4OH (28% min.)	3.00
Water	38.00
5% Cobalt Hydrocure	0.95
5% Manganese Hydrocure	0.55
Activ-8	0.40
Hydrocure II	0.00
Hydrochem Calcium	0.00
Hydrochem Zirconium	0.00

Air Dry or Force Dry Clear Coating-B

	Parts by Wt.
Chempol 10-0095	53.30
Ethylene glycol monobutyl ether	3.90
NH4OH (28% min.)	3.00
Water	37.35
5% Cobalt Hydrocure	0.00
5% Manganese Hydrocure	0.00
Activ-8	0.00
Hydrocure II	0.40
Hydrochem Calcium	1.20
Hydrochem Zirconium	0.50

SOURCE: Freeman Polymers Division: CHEMPOL 10-0095: Formulas

Air or Force Dry Clear Coating-A

	Parts by Wt.
Chempol 10-0097	54.15
Ethylene glycol monobutyl ether	3.80
NH4OH (28% min.)	3.40
Water	36.75
5% Cobalt Hydrocure	0.95
5% Manganese Hydrocure	0.55
Activ-8	0.40
Hydrocure II	0.00
Hydrochem Calcium	0.00
Hydrochem Zirconium	0.00

SOURCE: Freeman Polymers Division: CHEMPOL 10-0097: Formula

Air or Force Dry Clear Coating-A

	Parts by Weight
Chempol 10-1210	54.15
Ethylene glycol monobutyl ether	3.80
NH4OH (28% min.)	3.40
Water	36.75
5% Cobalt Hydrocure	0.95
5% Manganese Hydrocure	0.55
Activ-8	0.40
Hydrocure II	0.00
Hydrocem Calcium	0.00
Hydrocem Zirconium	0.00

Air or Force Dry Clear Coating-B

	Parts by Weight
Chempol 10-1210	54.05
Ethylene glycol monobutyl ether	3.80
NH4OH (28% min.)	3.40
Water	36.65
5% Cobalt Hydrocure	0.00
5% Manganese Hydrocure	0.00
Activ-8	0.00
Hydrocure II	0.40
Hydrocem Calcium	1.20
Hydrocem Zirconium	0.50

SOURCE: Freeman Polymers Division: CHEMPOL 10-1210 Formulation

Alkyd Topcoat Formula

Materials:	Parts by weight
Acrylated alkyd (75% NVM)	259.3
Ammonium hydroxide (28%)	9.9
Deionized water	509.9
Cobalt drier (6%) (1)	1.6
Manganese drier (6%) (1)	1.6
Zirconium drier (12%) (1)	1.6
Anti-skin agent (2)	0.9
Drier accelerator (3)	0.4
Flow agent (4)	0.4
Titanium dioxide (5)	155.6
Water for application viscosity	58.8

Properties:
    Pigment/binder ratio: 0.8/1.0
    pH: 8.0-8.5
    Non-volatile material (NVM), wt %: 35
    Viscosity, #4 Ford Cup, sec: 30-40
    Dry times, ASTM D1640:
       set to touch, min: 10-20
       dry through, hr: 6-7

Compounding Procedure:
1. Solubilize the alkyd with ammonium hydroxide and dilute with water.
2. Add driers, anti-skin agent, drier accelerator and flow agent. Mix thoroughly and adjust pH.
3. Stir in titanium dioxide.
4. Grind resulting paste in a ball mill overnight to a Hegman Grind of 7.
5. Adjust grind base with additional water to application viscosity.
(1) Cyclodex driers
(2) Exkin #2
(3) Activ-8
(4) FC-430
(5) Ti-Pure R-900

Alkyd topcoat performance properties (1,2):
    Air dry time: 1 day:
    60 Gloss, %: 88
    Pencil hardness: F
    Crosshatch adhesion: 5B
    Conical bend, % pass: 100
    Impact resistance, in-lb:
       direct: 60
       reverse: 10
    Water spot resistance, 1 hr: pass, no softening or dulling
(1) Substrate: Bonderite 37, treated cold rolled steel
(2) Film thickness: 1.0 to 1.2 mils

SOURCE: Amoco Chemical Co.: Water-Borne Air Drying Alkyds
      Based on AMOCO IPA and TMA: Formula Table 10 and 11

Aqueous Thermoset Topcoat

Materials	Weight Pounds	Volume Gallons
1. Beetle 65	183.23	17.79
2. Resimene 797	48.79	5.08
3. Methanol	22.44	3.40
4. L-7605	2.24	0.25
5. Butyl Cellosolve	5.21	0.69
6. Isopropyl Alcohol	22.44	3.43
7. Ektasolve EP	14.03	1.85
8. AC-1024	448.88	51.60
9. Tego 1488	4.49	0.54
10. Water	22.44	2.69
11. Chemcor Emulsion AS-35-3	38.47	4.62
12. Methanol	16.83	2.55
13. Nacure 155	44.89	5.50

Formulation Constants:
    Approximate solids, % (weight/volume): 56.5/52.0
    VOC, g/liter: 188
        lbs/gallon: 1.56
    Freeze/Thaw Stability: Protect from Freezing

SOURCE: Rohm and Haas Co.: Formulation KC-24-1

Sealer

Material	Weight Pounds	Volume Gallons
1. E-2955 (37 nv)	528.2	61.6
2. Water	211.3	25.4
3. 14% Ammonia	11.1	1.3

Premix the above and add the following with agitation:

4. Butyl Carbitol	4.2	0.5
5. Butyl Cellosolve	25.4	3.4
6. Isopropanol	29.7	4.5
7. Water (est for visc)	26.4	3.2

Formulation Constants:
    Approximate solids, % (weight/vol): 23.4/21.7
    pH: 7.8
    Viscosity, #2 Zahn Seconds: 19
    VOC, g/liter: 235
        lbs/gallon: 1.96
    Freeze/Thaw Stability: Protect from freezing
    Heat Age Stability: Satisfactory
    Flash Point, Tag closed cup F: 120

SOURCE: Rohm and Haas Co.: Formulation WR-55-1

Black Rust Inhibitive Coating

	Pounds	Gallons
Cargill Water Reducible Epoxy Ester		
73-7331	96.0	11.36
Aqueous Ammonia 28%	5.7	0.76
Triethylamine	1.0	0.17
Ethylene Glycol Monobutyl Ether	9.6	1.28
sec-Butanol	14.4	2.14
Strodex MOK-70	6.0	0.63
Special Black 4	12.5	0.85
Zinc Phosphate J0852	51.3	1.92
Imsil A-10	16.3	0.74
Aerosil R972	2.9	0.16
Byk-020	1.9	0.26
Deionized Water	212.8	25.57

Pebble Mill to a 6 Hegman

	Pounds	Gallons
Cargill Water Reducible Epoxy Ester		
73-7331	192.0	22.72
Aqueous Ammonia 28%	9.6	1.28

Premix the following, then add:

	Pounds	Gallons
Ethylene Glycol Monobutyl Ether	9.6	1.28
sec-Butanol	14.4	2.14
Activ-8	1.0	0.13
5% Cobalt Hydro Cure II	2.5	0.32
Manganese Hydro Cure II	2.1	0.24
Deionized Water	172.8	20.74
Exkin No.2	1.0	0.13
Deionized Water (Hold for Viscosity		
Adjustment)	43.2	5.18

Paint Properties:
   % Nonvolatile (By Weight): 33.28
               (By Volume): 27.07
   Pigment to Binder Ratio: 0.40
   Pigment Volume Concentration: 13.70
   Weight Per Gallon: 8.79
   Theoretical VOC (Pounds/Gallon): 3.08
               (Grams/Liter): 370
   Water:Cosolvent Weight Ratio: 75.00:25.00
   pH: 8.2-8.7
   Viscosity Krebs-Stormer (KU): 65-70

SOURCE: Cargill, Inc.: CARGILL Formulary: Formula P1897-F2

Chrome Yellow Topcoat

Material:	Lbs.	Gals.
Kelsol 3962-B2G-70	90.29	10.45
Water	210.50	25.27
Ammonium Hydroxide	4.41	0.59
Cobalt Hydrocure II	3.40	0.45
Activ-8	1.53	0.18
Manganese Hydrocure II	2.87	0.39
Propasol P	2.42	0.33
Ti-Pure R-902	65.27	1.96
Mapico Yellow 1050A	21.26	0.63
Medium Chrome Yellow X2035	86.77	1.85
Strontium Chromate J-1365	12.64	0.40

Steel ball mill to a 7 Hegman grind.

Letdown:		
Kelsol 3962-B2G-70	168.57	19.51
Water	277.64	33.33
Ammonium Hydroxide	3.66	0.49
Butoxy Ethanol	31.32	4.17

Analysis:
    Pigment Volume Concentration, Percent: 36.66
    Pigment/Binder Ratio: 2.24/1
    Percent Solids, Weight: 39.63
    Percent Solids, Volume: 23.95
    Viscosity @ 25C, #4 Ford, Secs.: 40-50
    Pounds/Gallon: 10.42
    pH: 8.2-8.6
    VOC (excluding water):
        Grams/Liter: 286
        Pounds/Gallon: 2.38

SOURCE: Reichhold Chemicals, Inc.: Waterborne Handbook:
        Formulation 2349-42A

Clear Build Coat Based on Rhoplex CL-103 Emulsion for Air-Dry or Low-Force-Dry Application

Materials	Weight Pounds	Volume Gallons
Rhoplex CL-103 emulsion	704.3	81.7
Dee Fo 3000	1.8	0.3

Premix and add under agitation:

Butyl Carbitol	57.1	7.1
Butyl Propasol	28.5	3.9
Water	20.8	2.5

Add:

14% Ammonia	1.6	0.2

(Adjust viscosity with water to a range of 20 to 26 seconds on a #2 Zahn cup)

Water	36.3	4.3

Formulation Constants:
  Approximate solids, % (wt/vol): 33.8/31.7
  pH: 7.5-8.0
  Viscosity, #2 Zahn cup, seconds: 20 to 26
  VOC, g/liter: 240
       lb/gallon: 2.0
  Coalescent, wt % on polymer solids:
    Butyl Propasol: 10
    Butyl Carbitol: 20
  Heat/age stability (140F/10 days): Passes
  Freeze/thaw stability: Protect from freezing

SOURCE: Rohm and Haas Co.: RHOPLEX CL-103 Acrylic Emulsion:
       Formulation KC-53-4

Clear, Build Coat Based on Rhoplex CL-104 Acrylic Emulsion

Materials:	Weight Pounds	Volume Gallons
Rhoplex CL-104 emulsion	718.1	83.3
Dee Fo 3000	1.7	0.2

Premix and add under agitation:

Butyl Carbitol	55.3	7.0
Butyl Propasol	27.6	3.8
Water	20.8	2.5

Add:

14% Ammonia	1.5	0.2

(adjust viscosity with water to a range of 20 to 26 seconds on a #2 Zahn cup)

Water	25.3	3.0

Formulation Constants:
   Approximate solids, % (wt/vol): 32.7/31.0
   pH: 7.5-8.0
   Viscosity, #2 Zahn cup, seconds: 20 to 26
   VOC, g/liter: 237
        lb/gallon: 1.98
   Coalescent, wt % on polymer solids:
      Butyl Carbitol: 20
      Butyl Propasol: 10
   Freeze/thaw stability: Protect from freezing
   Heat/age stability (140F/10 days): Passes

SOURCE: Rohm and Haas Co.: RHOPLEX CL-104 Acrylic Emulsion:
        Formulation WR-74-6

Clear Coating

	Pounds	Gallons
Cargill Water Reducible Self Crosslinking Resin 73-7390	345.9	40.61
Diethylene Glycol Monobutyl Ether	1.5	0.18
Dimethylethanolamine	12.3	1.67
Ammonium Hydroxide 28%	8.2	1.10
Patcote 577	3.6	0.53
Byk-301	3.7	0.45
Deionized Water	461.9	55.46

Paint Properties:
    % Nonvolatile (By Weight): 29.26
                (By Volume): 27.11
    Weight Per Gallon: 8.37
    Theoretical VOC (Pounds/Gallon): 2.84
                (Grams/Liter): 340
    Water:Cosolvent Weight Ratio: 79.00:21.00
    pH: 8.50
    Viscosity Krebs-Stormer (KU): 66

Typical Cured Film Properties:
    Cure Schedule: 10 Minutes at 325F
    Substrate: Cold Rolled Steel
    Pencil Hardness: Pass HB
    % Crosshatch Adhesion: 100
    Impact (In. Lbs.):
        Direct: 160
        Reverse: 160

SOURCE: Cargill, Inc.: CARGILL Formulary: Formula P1887-82B

Clear Coating for Wood and Concrete

Ingredient	Lbs/100 Gal
Ektasolve EP	72.52
Santicizer 160	23.25
Surfynol 104H	4.65
Diethylamine	9.30
Drew L-475	1.86
Water	71.59
Pliolite 7104 (44% solid)	496.51

Mix well, then add:

Witcobond 234 (Polyurethane)	169.22

    Dry to touch: 15-20 minutes
    60 degree Gloss (2 coats on wood): 80-85%
    Isopropyl alcohol spot test: Pass
    Blushing resistance: Good
    Viscosity: 15-25 seconds, No. 4 Ford Cup
    Age Panel 7 days before testing
    S.G.: 1.02
    Lb/Gal: 8.49
    VOC: 3.41 lb/gallon (402 gram/litre)

SOURCE: The Goodyear Tire & Rubber Co.: Formula #2064-22-90

Clear Satin (-40 gloss at 60) Thermosetting Sealer/Topcoat Based on Rhoplex AC-1024

Materials:	Pounds	Gallons
Part A:		
Add while mixing:		
Beetle 65	179.94	17.47
Methanol	17.63	2.67
Silwet L-77	2.64	0.32
Dehydran 1293	3.52	0.47
Premix and add:		
Water	28.21	3.39
Syloid 169	7.05	0.42
Add to above:		
Rhoplex AC-1024	352.66	40.54
Water	40.91	4.91
Exxate 600	10.58	1.45
Add slowly in order:		
Acrysol RM-825 rheology modifier	17.63	2.03
Chemcor wax emulsion AS-35-3	30.22	3.63
(adjust viscosity with water to a range of 35+-5 seconds on a #2 Zahn cup)		
Water	99.79	11.97
Part B:		
Premix and add slowly:		
Methanol	42.32	6.41
Nacure 155	35.27	4.32

Formulation Constants:
  PVC: 1.14
  Approximate solids, % (wt/vol): 46.1/41.6
  pH, part A: 6.5
  pH, parts A & B combined: 1.4
  Viscosity, part A, #2 Zahn cup, seconds: 35+-5
  Viscosity, parts A & B combined, #2 Zahn cup, seconds: 20
  VOC:
    g/L, theoretical: 203
    g/L, actual, per ASTM D-2369: 350
  Heat-age stability, 140F for 10 days: Passes
  Freeze/thaw stability: Protect from freezing

SOURCE: Rohm and Haas Co.: RHOPLEX AC-1024 Emulsion: Formulation KC-24-2

Coating

Materials:	Parts by weight
WB-138 dispersion (49% NVM)	439.5
Dimethylethanolamine	5.7
Deionized water	284.8
Melamine (1)	53.7
Flow additive (2)	1.5
Titanium dioxide (3)	214.8

Properties:
   Pigment/binder ratio: 0.8/1.0
   Resin/melamine ratio: 80/20
   Water/cosolvent ratio: 95/5
   Non-volatile material (NVM), wt%: 48.9
   Viscosity, Zahn #2, sec: 25
   Weight per gallon, lb: 10.53
   VOC (calculated), lb/gal: 1.4
   pH: 8.15

Compounding procedure for WB-138 coating:
1. Adjust pH of the resin dispersion to 7.5-8.0 with a dimethylethanolamine and deionized water mixture in a 5:95 ratio.
2. Charge the resin dispersion, melamine, flow additive and pigment to a pebble mill. Roll overnight to a Hegman Grind of 7+.
3. Discharge mill and adjust pH to 8.0 and Zahn #2 viscosity to 25 seconds.

(1) Cymel 303
(2) FC-430 (33% NVM in propylene glycol methyl ether acetate)
(3) Ti-Pure R-900

Performance Properties of WB-138 coating(1):Initial:
   CRS (2):
   Film thickness, mils: 1.2
   60 gloss, %: 93
   Pencil hardness (no cut): 3H
   Impact resistance, in-lb: direct: 150
                             reverse: 110
   Conical bend, % pass: 100
   Crosshatch adhesion: 5B
   Double MEK rubs, passed: 100+
   B-37(3):
   Film thickness, mils: 1.3
   60 gloss, %: 91
   Pencil hardness (no cut): 3H
   Impact resistance, in-lb: direct: 160
                             reverse: 120
   Conical bend, % pass: 100
   Crosshatch adhesion: 5B
   Double MEK rubs, passed: 100+
   (1) Cure cycle: 5 minutes flash plus 20 minutes at 177C
   (2) Cold rolled steel
   (3) Bonderite 37 treated cold rolled steel

SOURCE: Amoco Chemical Co.: Water-Borne Bake Polyesters Based on AMOCO IPA and TMA: Formula Table 7 and 8

Curtain Coatable 35 PVC Topcoat Formulation

Materials:	Pounds	Gallons
Acrysol I-62 (22.7%)*	30.0	3.45
Balab 3046A	6.0	0.95
TiO2, Ti-Pure R-960	91.0	2.75
Minex 4	277.0	12.77
Cr2O3 Green	23.0	0.54
Bentone LT	1.0	0.07
Water	65.0	7.80

Grind at high speed, then let down as follows:

Rhoplex AC-1822 (46%)	563.0	63.90
Cymel 373 (85%)	33.9	3.20
Ethylene Glycol Monobutyl Ether (60%)	48.0	6.16
Surfynol 104E**	0.5	----

    Adjust pH to 9.0 with DMAE.

Topcoat Constants:
    PVC: 35%
    Viscosity (#4 Ford Cup): 24 seconds
    Weight Solids: 60%
    Volume Solids: 46%
    TiO2/Binder: 24/76

    * Acrysol I-62 Neutralization:

Acrysol I-62 (50%)	40.0
Water	40.0
DMAE (50%)	8.0

  ** For curtain coater application.

SOURCE: **Rohm and Haas Co.**: Building Products: ACRYSOL I-62:
       Formulation Table II

Curtain Coatable 35 PVC Topcoat Formulation

Materials:	Pounds	Gallons
Acrysol I-62 (50%)	13.4	3.40
Water premix	13.4	
Dimethylaminoethanol/	2.7	
Water, 1/1		
Balab 3046A	5.9	0.94
TiO2, Ti-Pure R-960	89.6	2.71
Minex 4	273.7	12.57
Cr2O3 Green	22.6	0.53
Bentone LT	1.0	0.07
Water	64.0	7.68

Grind at high speed, then letdown as follows:

	Pounds	Gallons
Rhoplex AC-1822 (46%)	554.2	62.90
Cymel 373 (85%)	33.4	3.15
Ethylene Glycol Monobutyl Ether (60%)	47.2	6.06
Surfynol 104E*	0.5	----

Adjust pH to 9.0 with DMAE.

Topcoat Constants:
  PVC: 35%
  Viscosity (#4 Ford Cup): 24 seconds
  Weight Solids: 60%
  Volume Solids: 46%
  TiO2/Binder: 24/76

  * For curtain coater application

SOURCE: Rohm and Haas Co.: Building Products: Acrysol I-62:
        Formulation Table II

Curtain Coatable 35 PVC Topcoat Formulation

Materials:	Pounds	Gallons
Acrysol I-62 (22.7%)*	30.0	3.45
Balab 3046A	6.0	0.95
TiO2, Ti-Pure R-960	91.0	2.75
Minex 4	277.0	12.77
Cr2O3 Green (X-1134CP)	23.0	0.54
Bentone LT	1.0	0.07
Water	65.0	7.80

Grind at high speed then let down as follows:

Rhoplex AC-1822 (46%)	563.0	63.90
Cymel 373 (85%)	33.9	3.20
Butyl Cellosolve (60%)	48.0	6.16
Surfynol 104E**	0.5	-----

Adjust pH to 9.0 with DMAE.

Topcoat Constants:
    PVC: 35%
    Viscosity (#4 Ford Cup): 24 seconds
    Weight Solids: 60%
    Volume Solids: 46%
    TiO2/Binder: 24/76

* Neutralization of Acrysol I-62:	
Acrysol I-62 (50%)	40.0
Water	40.0
DMAE (50%)	8.0

** For curtain coater application

SOURCE: Rohm and Haas Co.: Building Products Board Coatings:
        RHOPLEX AC-1822: Formulation Table III

Experimental Coating: Spray

Materials:
Solubilized resin (1)	680.3
Melamine (2)	26.1
Titanium Dioxide (3)	208.6
Flow Agent (4)	3.9
Water (distilled)	81.1

    Resin/Melamine Ratio: 90/10
    Pigment/Binder Ratio: .8/1.0
    pH: 8.38
    NVM, %: 47.1
    Viscosity, #4 Ford Cup, sec.: 37

1. Solubilized Resin Composition:

Polymer (Resin BAL-389)	700
Diethylene glycol butyl ether	300
Dimethylaminoethanol	29
Water	1000

    NVM = 34.5%   pH = 8.0-8.5

2. Cymel 303
3. Ti-Pure R-900
4. FC-430

Performance Properties:
    CRS (1):
    Cure Cycle 10 min/180F plus 20 min/350F
    Film Thickness, mils: 1.1
    Sward Hardness: 44
    Pencil Hardness: H
    Crosshatch Adhesion, % pass: 100
    Impact Resistance, in-lb:
      Direct: 160
      Reverse: 160
    Conical Bend, % pass: 100

    B-37 (2):
    Cure Cycle 10 min/180F plus 20 min/350F
    Film Thickness, mils: 1.5
    Sward Hardness: 54
    Pencil Hardness: H
    Crosshatch Adhesion, % pass: 100
    Impact Resistance, in-lb.:
      Direct: 160
      Reverse: 160
    Conical Bend, % pass: 100

    1. CRS: Cold rolled steel
    2. B-37: Bonderite 37 treated cold rolled steel

SOURCE: Amoco Chemical Co.: Resin BAL-389, Water Borne Polyester
      for Coil and Spray Applied Flexible Coatings: Coating No.
      6265-124

Flat Topcoat Based on Experimental Emulsion E-2955

Materials:	Weight Pounds	Volume Gallons
Experimental emulsion E-2955	692.9	79.6
Dow 65	0.6	0.1
Water	25.0	3.0

Mix above then add slowly with good agitation:

TS-100	25.6	1.4
14% ammonia to adjust pH to 8.5		

Premix and add slowly under agitation:

Butyl Cellosolve	33.3	4.4
Ektasolve EEH	5.1	0.7
Isopropanol (15%)	38.5	5.9

Add:

Water	40.8	4.9
7% ammonia to adjust pH between 7.5-8.0		

Formulation Constants:
   Approximate solids, % (wt/vol): 32.8/29.2
   pH: 7.5-8.0
   Flash point, closed cup, F: 125
   VOC, g/liter: 156
      lb/gallon: 1.3
   Coalescent, wt % on polymer solids:
      Butyl Cellosolve: 13
      Ektasolve EEH: 2
   Freeze/thaw stability: Protect from freezing
   Heat/age stability (140F/10 days): Passes

SOURCE: Rohm and Haas Co.: Experimental Emulsion E-2955:
        Formulation WR-55-4

Force Dry White High Gloss Top Coat Dip/Spray

	Lbs/100 Gal
Diethylamine	14.12
Ektasolve EP	56.50
Surfynol 104H	5.04
Drew L-475	4.03
DSX 1550	10.00
Water	76.67
Pliolite 7104	66.59

Mix well then add:
TiO2 R-900	238.10

High Speed to 7+ NS

Follow with:
Pliolite 7104	463.08
Ektasolve EP	30.27
Santicizer 160	24.21
Texanol	11.10
10% Ammonium Benzoate	20.18

Force dry: 20 min. at 180 deg F
N.V.V.: 35.6%
T.N.V.: 48%
Pigment/binder ratio: 1:1
W.P.G.: 10.07
Sp. Gr.: 1.21
Gloss 80-85% at 1.5 mil, 60 deg
Visc: 60-65 sec #4 Ford
Adjust visc. as needed with water for dip application
V.O.C.: 2.34 lbs/gal (276 gram/litre)
pH: 9.0-9.5

SOURCE: The Goodyear Tire & Rubber Co.: Formula #2064-38-90

Gloss Black Topcoat for Brush and Roller Application

Ingredients:	Pounds	Gallons
Water	49.5	5.94
Methyl Carbitol	50.0	5.80
Maincote HG-54	609.4	70.99
Texanol	38.0	4.80
Foamaster AP	5.0	0.65
Chrom-Chem (896-9901) Lampblack	18.3	1.76
Water/Acrysol RM-1020, to adjust to 85 KU	73.8	8.85
NH4OH (28% NH3)	3.5	0.42
Sodium Nitrite (15% aqueous solution)	6.6	0.79

Formulation Constants:
    Pigment Volume Content, %: 2.6
    Volume Solids, %: 28.8
    pH: 9.2
    Stormer Viscosity, Kreb Units: 85
    ICI Viscosity, poise: 1.5
    VOC, gm/liter: 235
    Gloss (60/20): 90/75

SOURCE: Rohm and Haas Co.: Maintenance Coatings: MAINCOTE HG-54:
        Formulation GD-54-3

Gloss Black Topcoat for Brush and Roller Application

Ingredients:	Pounds	Gallons
Water	49.5	5.94
Methyl Carbitol	50.0	5.80
Maincote HG-54	609.4	70.99
Texanol	38.0	4.80
Foamaster AP	5.0	0.65
Chrom-Chem (896-9901) Lampblack	18.3	1.76
Water/Acrysol RM-1020, to adjust to 85 KU	73.8	8.85
NH4OH (28% NH3)	3.5	0.42
Sodium Nitrite (15% aqueous solution)	6.6	0.79

Formulation Constants:
    Pigment Volume Content, %: 2.6
    Volume Solids, %: 28.8
    pH: 9.2
    Stormer Viscosity, Krebs Units: 85
    ICI Viscosity, poise: 1.5
    VOC, gm/liter: 235
    Gloss (60/20): 90/75

SOURCE: Rohm and Haas Co.: Maintenance Coatings: MAINCOTE HG-54:
        Formulation GD-54-3

Gloss Green Acrylic/Epoxy Topcoat

Materials:	Pounds	Gallons
Acrylic Component A:		
Grind Preparation:		

Grind the following materials using a high speed dissolver for 20 minutes:

Methyl Carbitol	45.0	5.22
Tamol 165	3.9	0.44
Triton CF-10	0.6	0.07
Patcote 519	0.5	0.07
Chrome oxide green, X-1134	81.4	1.91

Letdown preparation:
Add the following in the order listed and mix thoroughly:

Maincote AE-58	543.4	63.15
NH4OH (28% NH3)	1.2	0.16
Water	64.7	7.76
Grind (from above)	131.4	7.71
Ektasolve EEH	53.1	7.16
Patcote 531	2.0	0.28
Acrysol RM-1020 (20%)	9.6	1.08
Sodium Nitrite (15%)	9.0	1.08

Epoxy Component B:		
Genepoxy 370-H55 (55%)	104.7	11.62

Topcoat Formulation Constants:
    Pigment Volume Content, %: 5.7
    Volume Solids, %: 33.5
    VOC, gm/l: 250
    Gloss, 60/20: 70/30
Component A Formulation Constants:
    pH: 8.5
    Stormer Viscosity, Kreb Units: 85
    ICI Viscosity, poise: 0.9

SOURCE: Rohm and Haas Co.: Maintenance Coatings: MAINCOTE AE-58:
        Formulation G-58-2

Gloss Midtone Tint Base Acrylic/Epoxy Topcoat

Materials:	Pounds	Gallons
Acrylic Component A:		
Grind Preparation:		

Grind the following materials using a high speed dissolver for 20 minutes:

	Pounds	Gallons
Methyl Carbitol	11.0	1.27
Tamol 165	3.9	0.44
NH4OH (28% NH3)	1.0	0.12
Triton CF-10	0.4	0.05
Patcote 519	0.4	0.06
TiPure R-900	54.8	1.60
Water	5.2	0.63

Letdown Preparation:
Add the following in the order listed and mix thoroughly:

	Pounds	Gallons
Maincote AE-58	462.8	53.81
Water	60.0	7.20
NH4OH (28% NH3)	4.7	0.63
Grind (from above)	76.7	4.17
Methyl Carbitol)	27.8	3.23
Ektasolve EEH) premix	45.2	6.10
Patcote 531)	2.0	0.28
Acrysol RM-1020	25.0	2.82
Water	89.8	10.77
Sodium nitrite (15% aqueous solution)	9.0	1.08

Epoxy Component B:

	Pounds	Gallons
Genepoxy 370-H55 (55%)	89.5	9.91

Topcoat Formulation Constants:
    Pigment Volume Content, %: 5.6
    Volume Solids, %: 28.5
    VOC, gm/l: 225
    Gloss, 60/20: 89/79
Component A Formulation Constants:
With 8 oz/gal of Chroma Chem 896-5501

	No	Yes
phthalo green		
pH	8.7	8.4
Stormer Viscosity, Krebs Units	94	78
ICI Viscosity, poise	1.5	1.2

SOURCE: Rohm and Haas Co.: Maintenance Coatings: MAINCOTE AE-58: Formulation M-58-1

Gloss Topcoat

Materials:	Weight Ratio*	Parts Per Hundred (Volume Basis)
Propylene glycol	60.0	6.98
Tamol SG-1	7.9	0.87
Water	5.1	0.62
Triton X-100	2.5	0.28
AMP-95	3.2	0.42
Nopco NDW	1.0	0.13
Titanium dioxide (Zopaque RCL-9)	250.0	7.30
Zinc oxide	25.0	0.54

Grind the above materials in a high-speed mill for 20 minutes and letdown.

Water	30.0	3.60
Rhoplex AC-507 (46.5%)	546.7	62.38
Texanol	25.0	3.16
Propylene glycol	38.0	4.42
Skane M-8	2.0	0.23
Nopco NDW	1.0	0.13
Water/Natrosol 250 MR (2.5%)	74.2	8.94

Formulation Properties:
    Pigment Volume Content, %: 22.5
    Volume Solids, %: 34.8
    Initial viscosity, KU: 70 to 74
    Initial pH: 9.0 to 9.2

* Using weight ratio in pound units will yield approximately 100 gallons of paint; using kilograms, about 833 liters will result.

SOURCE: Rohm and Haas Co.: Maintenance and Marine Coatings: RHOPLEX Acrylic Emulsions for Latex Maintenance Coatings: Formulation G-07-7

Gloss Topcoat Based on Experimental Emulsion E-2955

	Weight Pounds	Volume Gallons
Materials:		
Experimental emulsion E-2955	659.9	76.9
Water	22.0	2.6

Mix above then add slowly with good agitation:

14% ammonia to adjust pH to 8.5

Premix and add slowly under agitation:

Butyl Cellosolve	31.7	4.2
Ektasolve EEH	4.8	0.6
Isopropanol (15%)	36.5	5.5

Add:

Dow Corning #14	24.6	3.7
Michem Emulsion 39235	21.1	2.5
Water	32.5	3.9
7% Ammonia to adjust pH between 7.5-8.0		

Formulation Constants:
   Approximate solids, % (wt/vol): 30.5/27.9
   pH: 7.5-8.0
   Flash point, closed cup, F: 120
   VOC, g/liter: 218
        lb/gallon: 1.85
   Coalescent, wt % on polymer solids:
      Butyl Cellosolve: 13
      Ektasolve EEH: 2
   Freeze/thaw stability: Protect from freezing
   Heat/age stability (140F/10 days): Passes

SOURCE: Rohm and Haas Co.: Experimental Emulsion E-2955:
        Formulation WR-55-3

Gloss White Acrylic/Epoxy Topcoat

Materials:	Pounds	Gallons

Acrylic Component A:
Grind Preparation:
Grind the following materials using a high speed dissolver for
    20 minutes:

	Pounds	Gallons
Methyl Carbitol	38.8	4.50
Tamol 165	13.8	1.57
NH4OH (28% NH3)	1.0	0.12
Triton CF-10	1.6	0.19
Patcote 519	0.4	0.06
TiPure R-900	193.7	5.66

Add the following and continue to grind for 2-3 minutes at lower
    speed:

Water	19.9	2.39

Letdown Preparation:
Add the following in the order listed and mix thoroughly:

Maincote AE-58	493.0	57.25
Water	58.5	7.00
NH4OH (28% NH3)	2.4	0.29
Grind (from above)	269.2	14.49
Ektasolve EEH	48.2	6.50
Patcote 531	2.0	0.28
Water	14.2	1.71
Acrysol RM-1020	8.0	0.90
Sodium Nitrite (15% aqueous solution)	8.8	1.06

Epoxy Component B:

Genepoxy 370-H55	94.8	10.52

Topcoat Formulation Constants:
    Pigment Volume Content, %: 16.5
    Volume Solids, %: 34.0
    VOC, gm/l: 230
    Gloss, 60/20: 80/45
Component A Formulation Constants:
    pH: 8.5
    Stormer Viscosity, Krebs Units: 75
    ICI Viscosity, poise: 0.8

SOURCE: Rohm and Haas Co.: Maintenance Coatings: MAINCOTE AE-58:
    Formulation G-58-1

Gloss White Acrylic/Epoxy Topcoat

Materials:	Pounds	Gallons
Acrylic Component A:		
Grind Preparation:		

Grind the following materials using a high speed dissolver for 20 minutes:

	Pounds	Gallons
Methyl Carbitol	38.8	4.50
Tamol 165	13.8	1.57
NH4OH (28% NH3)	1.0	0.12
Triton CF-10	1.6	0.19
Patcote 519	0.4	0.06
TiPure R-900	193.7	5.66

Add the following and continue to grind for 2-3 minutes at lower speed:

	Pounds	Gallons
Water	19.9	2.39

Letdown Preparation:
Add the following in the order listed and mix thoroughly:

	Pounds	Gallons
Maincote AE-58	461.0	53.62
Water	45.8	5.50
NH4OH (28%/NH3)	1.8	0.22
Grind (from above)	269.2	14.49
Texanol	29.4	3.91
Patcote 531	2.0	0.28
Sodium Nitrite (15% aqueous solution)	9.0	1.08
Acrysol RM-1020	8.8	1.00

Epoxy Component B:	Pounds	Gallons
Epi-Rez WJ-3520 (55%)	125.5	13.81
Water	47.1	5.66
Acrysol RM-1020	5.6	0.62

Topcoat Formulation Constants:
    Pigment Volume Content, %: 16.5
    Volume Solids, %: 34.3
    VOC, gm/l: 196
    Gloss, 60/20: 79/37
Component A Formulation Constants:
    pH: 8.5
    Stormer Viscosity, Kreb Units: 76
    ICI Viscosity, poise: 1.0

SOURCE: Rohm and Haas Co.: Maintenance Coatings: MAINCOTE AE-58: Formulation G-58-4

Gloss White Topcoat for Brush and Roller Application

Ingredients:	Pounds	Gallons
Grind Preparation:		
Grind in Cowles Dissolver 15 minutes. Then let down at slower speed.		
Maincote HG-54	300.0	34.95
Tamol 681	8.4	0.92
NH4OH (28% NH3)	1.0	0.12
Triton CF-10	2.0	0.23
Drew L-405	1.0	0.14
Ti-Pure R-900 HG	195.0	5.70
Allow pigment to wet thoroughly, then add:		
Acrysol RM-1020	4.0	0.45
Letdown Preparation:		
Water	38.8	4.65
Methyl Carbitol	59.0	6.83
Maincote HG-54	332.6	38.75
Texanol	39.4	4.98
Drew Y-250	2.5	0.35
NH4OH (28% NH3)	4.0	0.48
Sodium Nitrite (15% Aqueous Solution)	9.0	1.08
Acrysol RM-1020	14.0	1.63

Formulation Constants:
    Pigment Volume Content, %: 16.3
    Volume Solids, %: 34.5
    pH: 9.1
    Stormer Viscosity, Krebs Units: 90
    ICI Viscosity, poise: 1.8
    VOC, gm/liter: 250
    Gloss (60/20): 75/40

To improve flow, if desirable, dilute this formulation with 1/2 pt. water per gallon.

SOURCE: Rohm and Haas Co.: Maintenance Coatings: MAINCOTE HG-54: Formulation G-54-3

High Build Gloss White Topcoat for Airless Spray Application

Ingredients:	Pounds	Gallons
Grind Preparation:		

Grind in Cowles Dissolver 15 minutes. Then let down at slower
  speed.

	Pounds	Gallons
Maincote HG-54	400.0	46.60
Tamol 681	9.1	1.00
NH4OH (28% NH3)	1.0	0.12
Drew L-405	2.4	0.34
Ti-Pure R-900 HG	127.6	3.70
Allow pigment to wet thoroughly, then add:		
Rheology Modifier QR-708	0.5	0.06

Letdown Preparation:

	Pounds	Gallons
Maincote HG-54	318.5	37.10
Texanol	44.7	5.63
Drew Y-250	3.5	0.46
NH4OH (28% NH3)	4.0	0.48
Sodium Nitrite (15% Aqueous Solution)	8.2	0.98
Water	29.0	3.48
Rheology Modifier QR-708	0.5	0.06

Formulation Constants:
   Pigment Volume Content, %: 10.0
   Volume Solids, %: 37.0
   pH: 9.2
   Stormer Viscosity, Kreb Units: 100
   ICI Viscosity, poise: 0.4
   VOC, gm/liter: 124
   Gloss (60/20): 79/31

SOURCE: Rohm and Haas Co.: Maintenance Coatings: MAINCOTE HG-54:
        Formulation HB-54-1

High Gloss DTM Maintenance Coating

Butyl Carbitol	26 lbs/100 gal
Diethylamine	13
Surfynol 104H	4
Manchem APG	5.5
DSX 1514	5.5
Santicizer 160	4
Butyl Dipropasol	4
Water	79

Mix well; then add:

Pliolite 7104	72

Mix well; then add:

Titanox 2160	210
Kadox 911	9

Sandmill to 7.5 NS, then drop to fully assembled letdown with
   mixing

Let Down Phase:

Pliolite 7104	476
Santicizer 160	29
Butyl Dipropasol	15
10% Ammonium Benzoate Solution	22
Propylene Glycol	30
Drew L-475	2

   Lb/gal: 10.08
   NVV: 38%
   NVW: 48%
   PVC: 18%
   KU Viscosity: 85-95 KU
   Freeze Thaw Stability: Pass 5 cycles
   Oven Stability: Pass 4 weeks, 120 deg F
   Dry to Touch @ 2 mil DFT: 1 hour
   Recoat Time: 4 hours
   Gloss 60 deg @ 2 mil DFT: 85%
   Method of Application: Brush or Spray
   VOC: 236 g/l

SOURCE: The Goodyear Tire & Rubber Co.: Formula 2074-42

High Gloss Maintenance/DTM Coating

Ingredient:	Lbs/100 Gal
Ektasolve EP	27
Diethylamine	13
Surfynol 104H	4
Manchem APG	6
DSX 1550	1
Santicizer 160	4
Texanol	4
Water	90

Mix well, then add:
Pliolite 7104	74

Mix well, then add:
Titanox 2160	216
Kadox 911	9

Sandmill to 7.5 NS, then drop to fully assembled letdown with mixing:

Let Down Phase:	
Pliolite 7104	489
Santicizer 160	30
Texanol	16
10% Ammonium Benzoate	22
Drew L-475	2

Bulk density: 10.07#/gal
Viscosity: 95-110 KU
VOC: 156 g/l (1.30 lb/gallon)
PVC: 15%
NVV: 40%

Gloss 60 deg/20 deg at 2 mil DFT: 89%/66%
Gloss retention:
500 hours QUV: 100%/100%
750 hours QUV: 79%/61%

120 deg F oven stability: Pass one week

Freeze Thaw: Protect from freezing

Age panel 7 days before testing
Allow 4 hours before recoating

SOURCE: The Goodyear Tire & Rubber Co.: Formula #2065-59-21

High Performance, Flexible Coating

Materials:	Parts by weight
WB-389 dispersion (34.5% NVM)	680.3
Deionized water	81.1
Melamine (1)	26.1
Flow additive (2)	3.9
Titanium dioxide (3)	208.6

Properties:
  Pigment/binder ratio: 0.8/1.0
  Resin/melamine ratio: 90/10
  Non-volatile material (NVM), wt%: 47.1
  Viscosity, #4 Ford Cup, sec: 37
  pH: 8.38

Compounding Procedure:
1. Charge the dispersion, melamine, flow additive, water, and pigment to a pebble mill. Roll overnight to a Hegman Grind of 7+.
2. Discharge mill and adjust pH to 8.0-8.5 and #4 Ford cup viscosity to 37 seconds.

(1) Cymel 303
(2) FC-430 (33% NVM in diethylene glycol n-butyl ether)
(3) Ti-Pure R-900

Performance Properties (1):
  CRS (2):
  Film thickness, mils: 1.1
  Sward hardness: 44
  Pencil hardness: H
  Impact resistance, in-lb:
     direct: 160
     reverse: 160
  Conical bond, % pass: 100
  Crosshatch adhesion: 5B

  B-37 (3):
  Film thickness, mils: 1.5
  Sward hardness: 54
  Pencil hardness: H
  Impact resistance, in-lb:
     direct: 160
     reverse: 160
  Conical bond, % pass: 100
  Crosshatch adhesion: 5B
  (1) Cure Cycle: 10 minutes at 82C plus 20 minutes at 177C
  (2) Cold rolled steel
  (3) Bonderite 37 treated cold rolled steel

SOURCE: Amoco Chemical Co.: Water-Borne Bake Polyesters Based on AMOCO IPA and TMA: Formula Table 14 and 15

High Performance Water-Borne Air Dry Coating
Topcoat Formulation

Materials:	Parts by weight
Resin WB-151 (75% NVM)	256.1
Ammonium hydroxide, 28%	11.5
Deionized water	439.7
Cobalt drier, 5% (1)	3.8
Calcium drier, 5% (2)	3.8
Zirconium drier, 12% (3)	1.6
Anti-skin agent (4)	0.9
Flow agent (5)	0.3
Titanium dioxide (6)	173.1
Secondary butanol	20.8
Ethylene glycol monobutylether	13.9
Deionized water	74.5

Properties:
    Non-volatile material (NVM), %: 36.5
    Pigment/binder ratio: 0.9/1.0
    pH: 8.5
    Water/cosolvent ratio (volume): 82/18
    Viscosity, #4 Zahn Cup, sec: 18
    Dry times at 20C, 50% RH:
        set to touch, min: 20
        tack free, min: 52
        dry hard, hr: 3.0

Compounding Procedure:
1. Solubilize the resin with ammonium hydroxide.
2. Dilute the mixture with 75-85% of the water.
3. Add driers, anti-skin agent and flow agent; mix thoroughly.
4. Slowly stir in the titanium dioxide.
5. Grind the mixture overnight to a 7 Hegman Grind in a pebble mill.
6. Adjust with water, ammonia and cosolvent to spraying viscosity and 8.0-8.5 pH.

(1) Cobalt Hydro-Cure II
(2) Calcium Hydro-Cem
(3) Zirconium Hydro-Cem
(4) Exkin #2
(5) Byk 301
(6) Ti-Pure R-900

SOURCE: Amoco Chemical Co.: Water-Borne Air Drying Alkyds
        Based on Amoco IPA and TMA: Table 7

High Solids Water-Borne Clear Coating-B

Formulation:
K-Flex 188	259.5
K-Flex UD-320W	259.5
Cymel 303	257.7
Nacure 5225	15.0
Silwet L-7600	0.7
Water	175.2
Arcosolv PM	32.4

Polyol/Urethane Diol/HMM: 34/31/35
% Solids, calc.: 74.6
% Solids, 20 min/120C: 67.8
Viscosity, 25C, cps: 275

Film Properties:
Substrate: Bonderite 1000 Cold-Rolled Steel
Cure Time: 20 minutes
Film Thickness: 1.25 mil DFT

Cure Temperature, C:	120	150
Pencil Hardness:	F-H	H-2H
Knoop Hardness:	4.2	18.0
Impact Rev., in/lb:	>160	10-20
Dir., in/lb:	>160	60-80

High Solids Water-Borne Clear Coating-C

Formulation:
K-Flex 188	222.6
K-Flex UD-320W	222.6
Cymel 303	361.1
Nacure 5225	15.0
Silwet L-7600	0.6
Water	150.3
Arcosolv PM	27.8

Polyol/Urethane Diol/HMM: 25/27/48
% Solids, calc.: 74.8
% Solids, 20 min/120C: 70.1
Viscosity, 25C, cps: 250

Film Properties:
Substrate: Bonderite 1000 Cold-Rolled Steel
Cure Time: 20 minutes
Film Thickness: 1.25 mil DFT

Cure Temperature, C:	120	150
Pencil Hardness	H-2H	H-2H
Knoop Hardness	11.0	21.0
Impact Rev., in/lb:	>160	0-10
Dir., in/lb:	>160	40-60

SOURCE: King Industries: Formulation UDW-1B and UDW-1C

Industrial Airdry White High Gloss Top Coat

Ingredient:	Lbs/100 Gal
Surfynol 104H	9.71
Butyl Carbitol	29.26
Methyl Carbitol	26.21
Diethylamine	11.65
DSX 1514	15.53
Santicizer 160	3.88
Texanol	3.88
Water	117.48
Rutile TiO2	218.45
L-475 Defoamer	4.85

High speed or Sandmill to 7 N.S., then add:

Pliolite 7104	485.44
Santicizer 160	26.21
Ektasolve EP	26.21
10% Ammonium Benzoate	19.42

Airdry
Visc: 75 KU
pH: 9.0-9.5
Gloss 90% at 60 deg Gardner
Pass: 120 deg F, 4 weeks
Re-coat after 2 hrs airdry
For fast dry, Butyl Carbitol can be replaced with Ektasolve EP
VOC: 2.21 lb/gal or 260 grams/lit.
W.P.G.: 9.98
Sp. Gr.: 1.20
T.N.V.: 47.85
N.V.V.: 35.66

SOURCE: The Goodyear Tire & Rubber Co.: Formula #2064-34-90

Low Solids Clear Topcoat

Materials:	Weight Pounds	Volume Gallons
Rhoplex CL-103 emulsion	484.5	56.2
Dee Fo 3000	1.0	0.1

Premix and add under agitation:

Butyl Carbitol	19.6	2.5
Hexyl Cellosolve	29.5	4.0
Dow Corning #14	19.6	2.9
Water	30.3	3.6

Add:

Michem Emulsion 39235	16.8	2.0
Acrysol RM-825 modifier	15.8	1.8
14% ammonia	0.2	0.02
(adjust viscosity with water to a range of 20 to 26 seconds on a #2 Zahn cup)		
Water	223.3	26.9

Formulation Constants:
  Approximate solids, % (wt/vol): 24.9/23.2
  pH: 7.5-8.0
  Viscosity, #2 Zahn cup, seconds: 20 to 26
  VOC, g/liter: 256
       lb/gallon: 2.1
  Coalescent, wt % on polymer solids:
     Butyl Cellosolve: 15
     Butyl Carbitol: 10
  Rheology modifier, wt % on polymer solids: 2.0
  Heat/age stability (140F/10 days): Passes
  Freeze/thaw stability: Protect from freezing

SOURCE: Rohm and Haas Co.: RHOPLEX CL-103 Acrylic Emulsion:
        Formulation KC-53-1LS

Low-Solids Full-Gloss Topcoat Based on Rhoplex CL-104 Emulsion
For Air-Dry or Low-Force-Dry Application

Materials:	Weight Pounds	Volume Gallons
Rhoplex CL-104 emulsion	513.9	60.1
Dee Fo 3000	0.9	0.1

Premix and add under agitation:

Dow Corning #14	19.7	3.0
Butyl Carbitol	19.7	2.5
Hexyl Cellosolve	29.7	4.0
Water	30.5	3.6

Add:

Michem Emulsion 39235	16.9	2.0
QR-708 Rheology Modifier	1.4	1.4
14% Ammonia	0.5	0.1
(adjust viscosity with water to a range of 38 to 44 seconds on a #2 Zahn cup)		
Water	194.3	23.3

Formulation Constants:
   Approximate solids, % (wt/vol): 25.2/23.7
   pH: 7.0-7.5
   Viscosity, Brookfield LV Spindle #1, cP @ 30 rpm: 175-200
   Viscosity, #2 Zahn cup, seconds: 38-44
   VOC, g/liter: 257
        lb/gallon: 2.14
   Coalescent, wt % on polymer solids:
      Hexyl Cellosolve: 15
      Butyl Carbitol: 10
   Rheology modifier, wt % on polymer solids: 2.0
   Freeze/thaw stability: Protect from freezing
   Heat/age stability (140F/10 days): Passes

SOURCE: Rohm and Haas Co.: RHOPLEX CL-104 Acrylic Emulsion:
        Formulation WR-74-7LS

Red Oxide Maintenance Coating

	Lb/100 Gallon
Butyl Carbitol	21
Santicizer 160	4
Texanol	4
Methyl Carbitol	42
DSX 1514	16
Surfynol 104H	5
Diethylamine	13
Water	57
Drew L-405	0.34

Mix well; then add:

Pliolite 7104	70

Mix well; then add:

Red Oxide 130M	100
Heubach ZBZ	146

Sandmill to 6.5 N.S. than add the following to grind in order shown.

Pliolite 7104	487
Santicizer 160	22
Propyl Cellosolve	22
10% Ammonium Benzoate Solution	21

    Density: 10.30 lb/gal
    VOC: 241 g/l
    Pigment/binder ratio: 1:1 wt
    NVV: 39.61%
    NVM: 48.41%
    KU Viscosity: 70-80 KU
    Freeze Thaw Stability: Pass 5 cycles
    Oven Stability: Pass 4 weeks 120 degrees F
    Dry Time @ 2 mil DFT: 1 hour
    Recoat Time: 4 hours
    Method of Application: Brush or spray
    For Brushing: 2 coats recommended
    Recommended Substrate: Blasted hot or cold rolled steel

SOURCE: The Goodyear Tire & Rubber Co.: Formula 2074-44

Satin (-40 gloss at 60) Thermosetting Sealer/Topcoat
Based on Rhoplex AC-1024 Emulsion
With Early Water and Stain Resistance

Materials:	Pounds	Gallons
Part A:		
Add while mixing:		
Beetle 65	143.24	13.91
Resimene 797	38.14	3.97
Methanol	17.55	2.66
Silwet L-77	2.63	0.31
Dehydran 1293	3.51	0.47
Premix and add:		
Water	28.07	3.37
Syloid 169	7.02	0.42
Add to above:		
Rhoplex AC-1024 emulsion	350.90	40.33
Water	40.71	4.89
Exxate 600	10.53	1.44
Add slowly in order:		
Acrysol RM-825 rheology modifier	17.55	2.02
Chemcor wax emulsion AS-35-3	30.07	3.61
(adjust viscosity with water to a range of 35+-5 seconds on a #2 Zahn cup)		
Water	99.29	11.92
Part B:		
Premix and add slowly:		
Methanol	42.11	6.38
Nacure 155	35.09	4.30

Formulation Constants:
   PVC: 1.26
   Approximate solids, % (wt/vol): 46.0/41.6
   pH part A: 6.5
   pH, parts A & B combined: 1.4
   Viscosity, part A, #2 Zahn cup, seconds: 35+-5
   Viscosity, parts A & B combined, #2 Zahn cup, seconds: 20
   VOC:
      g/L, theoretical: 207
      g/L, actual, per ASTM D-2369: 350
   Heat-age stability, 140F for 10 days: Passes
   Freeze/thaw stability: Protect from freezing

SOURCE: Rohm and Haas Co.: RHOPLEX AC-1024 Emulsion: Formulation
          KC-24-3

Waterbased Charcoal Basecoat with Pearlescent Waterbased Clearcoat Conventional Urethane Clearcoat

Ingredients:	Pounds	Gallons
Chempol 20-4301	661.18	75.13
Nopco NXZ	3.00	0.39
2-butoxyethanol	58.43	7.78
Tinuvin 1130	4.17	0.43
Tinuvin 440	2.09	0.23
Surfynol 104BC	0.80	0.11
Nuocure CK-10	1.39	0.17
Water	52.50	6.30

Add defoamer to 20-4301 slowly and with good agitation. Premix the cosolvent, light stabilizers, surfactant and drier. Add this mixture to the vortex of the emulsion under agitation. Add the water at the end or along with the co-solvent mixture to lower the viscosity. Mix well, then add slowly while continuing to agitate.

Mearlin Fine-Pearl #139V	46.93	1.81
WD-2348 Tint Paste-Black	7.35	0.76
Water	26.70	3.21

Add the following to increase thixotropy to levels necessary to maintain metal orientation. Premix the first two items that follow:

Water	13.05	1.57
Isopropyl alcohol	13.05	1.98
Ammonia	1.00	0.13

Allow paint to stand overnight before adjusting or using.

Properties:
    Total solids, % by weight: 33.41
    Total solids, % by volume: 27.40
    Weight/gallon: 8.92
    PVC: 7.22
    VOC, lbs./gal.: 2.10
    Viscosity, application: 17" #4 Zahn
    pH: 7.4
    Substrate: B-1000
    Cure: 1 wk. A.D.
    DFT: 1.0-1.5 mil

SOURCE: Freeman Polymers Division: CHEMPOL 20-4301-H Formulation

Waterbased Clearcoat

Ingredients:	Pounds	Gallons
Chempol 20-4301	605.09	68.76
Nopco NXZ	2.74	0.36
2-butoxyethanol	45.78	6.10
Dibutyl phthalate	9.14	1.05
Butyl Carbitol	22.84	2.88
Surfynol 104BC	2.74	0.36
Nuocure CK-10	2.56	0.31
Water	72.48	8.70

Add defoamer to 20-4301 slowly with good agitation. Premix the co-solvents, plasticizer, surfactant and drier. Add this mixture to the vortex of the emulsion under agitation. Add the water at the end or along with the co-solvent mixture to lower the viscosity.

Add the following to insure thixotropy to levels necessary to maintain sag control. Premix the first two items that follow.

Water	11.94	1.43
2-butoxyethanol	11.94	1.59
Ammonia	0.91	0.12
Byk 301	0.91	0.11
Water (hold to adjust viscosity)	68.53	8.23

Allow paint to stand overnight before adjusting or using. To be applied over Waterbased Charcoal Basecoat with Pearlescent.

Properties:
    Total solids, % by weight: 27.16
    Total solids, % by volume: 23.95
    Weight/gallon: 8.58
    VOC, lbs./gal.: 2.53
    Viscosity, application: 38" #2 Zahn
    pH: 7.3
    Substrate: B-1000
    Cure: 1 wk. A.D.
    DFT: 1.5 mil
    Gloss, 60: 80
          20: 46
    Pencil hardness: F-H
    Crosshatch adhesion: 100%

SOURCE: Freeman Polymers Divison: CHEMPOL 20-4301-H Formulation

Water-Borne Clear Coating-A

This formulation has been developed to demonstrate the use of K-FLEX UD-320 W polyol as a co-solvent for a low molecular weight polyester resin. The solvent is water and about 15% 2-methoxypropanol is used as a co-solvent. This formulation does not contain any amine to solubilize the resins, therefore it is possible to catalyze this formulation for low temperature cure for applications on paper or wood. For extended shelf life the formulation should be shipped without water.

Formulation:

K-Flex 188	281.7
K-Flex UD-320W	281.7
Cymel 303	195.8
Nacure 5225	14.6
Silwet L-7600	0.7
Water	190.2
Arcosolv PM	35.2

Polyol/Urethane Diol/HMM: 38/35/27
% Solids, calc.: 72.7
% Solids, 20 min/120C: 64.9
Viscosity, 25C, cps: 200

Film Properties:
Substrate: Bonderite 1000 Cold-Rolled Steel
Cure Time: 20 minutes
Film Thickness: 1.25 mil DFT
Cure Temperature, C: 120
Pencil Hardness: <2B
Knoop Hardness: *
Impact Rev., in/lb: *
Dir., in/lb: *
* Not cured
Cure Temperature, C: 150
Pencil Hardness: H-2H
Knoop Hardness: 13.8
Impact Rev., in/lb: 40-60
Dir., in/lb: 100-120

SOURCE: King Industries: Formulation UDW-1A

Waterborne Acrylic Conversion Coating for Wood

Improvements in the water and stain resistance of a waterborne acrylic conversion coating for wood can be achieved by selecting the proper catalyst. Using a more hydrophobic catalyst, such as NACURE 155 (DNNDSA), instead of the more hydrophilic pTSA catalyst can greatly improve the resistance properties of the coating as shown in the following formulation:

Composition:	Weight
Beetle 65	25.20
Resimene 797	6.30

Premix the above and add:

AC 1024	63.00
ASA 35.3	5.36
FC 430	0.14

Characteristics:
    Total Solids: 65.0%
    Resin Solid: 63.0%
    Acrylic/Crosslinker: 50/50
    % Wax: 3% on Resin Solids

Catalyst Levels: 6.3% K-CURE 1040 (4% pTSA on TRS) 8.7% NACURE
    155 reduced to 40% active with methanol (5.5% DNNDSA on TRS)
Substrate: Stained and sealed birch plywood
Cure: IR curing equipment can be oven cured 10'/130F
Application: Spray applied to approximately 3 mils. wet

    IR Cure: Unitube Infrared Heaters
    Peak Emmission Wave Length: 2.95 microns
    Energy Output: 25 Watts/in 2
    Emitter Temperature: 1400F
    Flash Time: None
    Dwell Time: 8"

    All panels cured to 20 MEK rubs, and aged one week.

Stain Test Results:
    Catalyst: K-Cure 1040
    Acid: p-TSA
    16 Hour Immersion: Water: Light
                       Coffee: Moderate
                       Tea: Heavy

    Catalyst: NACURE 155
    Acid: DNNDSA
    16 Hour Immersion: Water: No Effect
                       Coffee: V. Light
                       Tea: Light

SOURCE: King Industries: Formulation WR-21

Water-Borne Coating for Plastics

Materials:	Parts by Weight
Resin WB-151 (75% NVM)	256.1
Ammonium Hydroxide, 28%	11.5
Deionized Water	439.7
Cobalt Drier, 5% (1)	3.8
Calcium Drier, 5% (2)	3.8
Zirconium Drier, 12% (3)	1.6
Anti-Skinning Agent (4)	0.9
Flow Agent (5)	0.3
Titanium Dioxide (6)	173.1
Secondary Butanol	20.8
Ethylene Glycol Monobutyl Ether	13.9
Deionized Water	74.5

Coating Properties:
    Non-Volatile Material (NVM), wt %: 36.2
    pH: 8.37
    Viscosity, #4 Zahn Cup, seconds: 17
    Wt/gal, lbs: 9.58
    Pigment/Binder Ratio: 0.9/1.0
    Water/Cosolvent Ratio (Volume): 82/18
    Dry Times at 23C:
        Set-to-Touch, min: 10
        Tack Free, min: 50
        Dry Hard, hr: >6.0
        Dry Through, hr: >6.0
Coating Compounding Procedure:
1. Solubilize the resin with ammonium hydroxide.
2. Dilute the mixture with water.
3. Add driers, anti-skin agent, and flow agent; mix thoroughly.
4. Stir in the titanium dioxide.
5. Grind the mixture overnight in a pebble mill to a Hegman
   Grind of 7+.
6. Adjust with ammonia, water and cosolvent to a 8.0 to 8.5
   pH and spray viscosity.
(1) Cobalt Hydro-Cure II
(2) Calcium Hydro-Cem
(3) Zirconium Hydro-Cem
(4) Exkin #2
(5) Byk 301
(6) Ti-Pure R-900
Coating Physical Properties:
    Substrate: ABS

7 Day Cure:	30 Day Cure:
Gloss, 60, %: 86.5	Gloss, 60, %: 88.9
Hardness, Sward: 50	Hardness, Sward: 54
Pencil, no cut: 3B	Pencil, no cut: 2B
Crosshatch Adhesion, % pass: 99	Crosshatch Adhesion, % pass: 100

SOURCE: Amoco Chemical Co.: Water-Borne Coating for Plastics:
    Formula Table II and III

Water Soluble Acrylic Clear Coating-A
Modified with K-FLEX UD-320W

This formulation demonstrates the use of K-FLEX UD-320 poly-urethane diol as a modifier for a water soluble acrylic resin. In such a formulation the urethane diol acts as a reactive modifier and raises the application solids and the hardness of the coating.

Formulation:

Acryloid WR-97	277.5
K-Flex UD-320W	0.0
Cymel 303	49.6
Nacure-5225	3.0
Dimethylethanol Amine	12.4
Water	657.5

Acrylic/Urethane Diol/HMM: 80/-/20
% Solids, calc.: 24.3
% Solids, 60 min/110C: 23.4
Viscosity, 25C, cps: 351

Film Properties:
Substrate: Bonderite 1000 (Iron Phosphated CRS)
Cure Schedule: 20 min/150C
Film Thickness, mil DFT: .7
    Pencil Hardness: HB-F
    Knoop Hardness: 7.0
    Impact Rev., in/lb: >160
           Dir., in/lb: >160
Salt Spray, 100 hrs:
    Creepage: 10 mm
    Crosshatch Adhesion, %: 100
    Mandrel 1/8" dia: Pass

Film Thickness, mil DFT: .5
    Pencil Hardness: H-2H
    Knoop Hardness: 9.1
    Impact Rev., in/lb: >160
           Dir., in/lb: >160
Salt Spray, 100 hrs:
    Creepage: ----
    Crosshatch Adhesion, %: 100
    Mandrel 1/8" dia: Pass

SOURCE: King Industries: Formulation UDW-2A

Water Soluble Acrylic Clear Coating-B

Formulation:
Acryloid WR-97	274.8
K-Flex UD-320W	23.8
Cymel 303	69.5
Nacure-5225	3.0
Dimethylethanol Amine	12.2
Water	616.7

    Acrylic/Urethane Diol/HMM: 68/7/25
    % Solids, calc.: 28.2
    % Solids, 60 min/110C: 26.8
    Viscosity, 25C, cps: 601

Film Properties:
    Substrate: Bonderite 1000 (Iron Phosphated CRS)
    Cure Schedule: 20 min/150C

Film Thickness, mil DFT:	.7	.5
Pencil Hardness:	H-2H	H-2H
Knoop Hardness:	14.0	19.0
Impact Rev., in/lb:	120-140	>160
Dir., in/lb:	120-140	>160

Water Soluble Acrylic Clear Coating-C

Formulation:
Acryloid WR-97	278.0
K-Flex UD-320W	54.0
Cymel 303	96.2
Nacure-5225	3.0
Dimethylethanol Amine	12.4
Water	556.4

    Acrylic/Urethane Diol/HMM: 57/15/28
    % Solids, calc.: 33.8
    % Solids, 60 min/110C: 32.4
    Viscosity, 25C, cps: 952

Film Properties:
    Substrate: Bonderite 1000 (Iron Phosphated CRS)
    Cure Schedule: 20 min/150C

Film Thickness, mil DFT:	.7	.5
Pencil Hardness:	H-2H	H-2H
Knoop Hardness:	15.0	17.0
Impact Rev., in/lb:	40-60	>160
Dir., in/lb:	80-100	140-160

SOURCE: King Industries: Formulation UDW-2B and UDW-2C

White High Gloss Railcar Coating

Propyl Carbitol	30
Methyl Carbitol	27
Butyl Dipropasol	4
Water	81
L475 Drew	5
Surfynol 104H	10
DSX-1550	2
T.E.A.	10
Mix well then add:	
Pliolite 7104	67
Mix well then add:	
TiO2	225
Mix well then high speed or sand mill to 7+ N.S.:	
Follow with:	
7104	467
7217	107
Mix well then add:	
Butyl Dipropasol	14
Santicizer 160	27
Ammonium Benzoate (10% solution)	20

Panels sprayed and 72 hrs air dried on B-1000 panels
Impact 80" lb: Passes after 72 hrs air dry.
Film Thickness: 3.5 to 4 mils dry
Salt spray: 300 hrs passes. No creep or blister
              600 hrs few blisters, no creep
Gloss: 3.5 to 4 mils, 80% (60 degree)
Passes: 1/4" Mendral bend 180 deg
Viscosity: 75-80 KU
TNV: 51.1
NVV: 41%
Density: 9.98 lbs/gal
VOC: 200 g/l

SOURCE: The Goodyear Tire & Rubber Co.: Formula 2064-57-91

White Roof Coating

Water	122
Natrosol 250HR	3
AMP 95	3
Tamol 850	6
Foamburst 363	4
Flexbond 471	210
TiPure R-902	100
Titanox AWD	5
Nytal 300	150
Snowflake	300
Troysan Polyphase AF-1	3
Texanol	10
Water	35
Flexbond 471	304
GE 60 Defoamer	1

    Yield, gallons: 102.5
    PVC, %: 45
    Solids, %: 64.7
    Viscosity, KU: 93

Commercial Styrene-Acrylate Comparison: 45% PVC Roof Coating:
Flexbond 471:
    Tensile, cured films: 218 psi
    Ultimate Elongation: 538%
    Water Absorption, 7 days immersion: 12.9%
    MVTR (grams/100 square inches/24 hours - 20 mil film): 6.38
    Mandrel Bend at 0F: Breaks
    Peel Adhesion, Urethane Foam, dry: 6.0 pli
    Peel Adhesion, Urethane Foam, wet: 5.0-5.5 pli
    Peel Adhesion to Aluminum, dry: 4.5 pli
    Peel Adhesion to Aluminum, wet: 5.0 pli

Commercial--15 Tg Styrene-Acrylate:
    Tensile, cured films: 243 psi
    Ultimate Elongation: 560%
    Water Absorption, 7 days immersion: 14.5%
    MVTR (grams/100 square inches/24 hours - 20 mil film): 4.57
    Mandrel Bend at 0F: Breaks
    Peel Adhesion, Urethane Foam, dry: 1.5-2.0 pli
    Peel Adhesion, Urethane Foam, wet: 1.0-1.5 pli
    Peel Adhesion to Aluminum, dry: 5.0 pli
    Peel Adhesion to Aluminum, wet: 5.0 pli

SOURCE: Air Products and Chemicals, Inc.: FLEXBOND 471 Acrylic
        Emulsion: Suggested Formulation F-471-83

White Satin Thermosetting Topcoat

Materials:	Pounds	Gallons
Part A:		
Add while mixing:		
Beetle 65	140.95	13.68
Resimene 797	37.53	3.91
Methanol	17.26	2.62
Silwet L-77	2.59	0.31
Dehydran 1293	3.45	0.46
Ti-Pure R-900	64.33	1.93
Premix and add:		
Water	27.62	3.32
Syloid 169	6.91	0.41
Add to above:		
Rhoplex AC-1024 emulsion	345.29	39.69
Water	70.23	8.43
Exxate 600	10.36	1.42
Add slowly in order:		
Acrysol RM-825 rheology modifier	17.26	1.99
Chemcor wax emulsion AS-35-3	29.59	3.55

   (adjust viscosity with water to a range of 35+-5 seconds on
    a #2 Zahn cup)

Water	64.76	7.77
Part B:		
Premix and add slowly:		
Methanol	41.53	6.28
Nacure 155	34.53	4.23

Formulation Constants:
    PVC: 6.75
    Approximate solids, % (wt/vol): 46.9/42.9
    pH, part A: 6.5
    pH, parts A & B combined: 1.4
    Viscosity, part A, #2 Zahn cup, seconds: 35+-5
    Viscosity, parts A & B combined, #2 Zahn cup, seconds: 20
    VOC:
        g/L, theoretical: 199
        g/L, actual, per ASTM D-2369: 350
    Heat-age stability, 140F for 10 days: Passes
    Freeze/thaw stability: Protect from freezing

SOURCE: Rohm and Haas Co.: RHOPLEX AC-1024 Emulsion: Formulation
        KC-24-4

25 PVC Topcoat

Materials:	Pounds	Gallons
Cowles Grind:		
Acrysol I-62 (22.7%)*	25.4	2.93
Balab 3056A	4.5	0.61
TiPure R-960	150.0	4.56
Minex 4	138.8	6.40
Bentone LT	1.0	0.07
Water	39.2	4.70
Letdown:		
Rhoplex AC-2045 (45%)	625.1	71.01
Monomeric Melamine (Cymel 303 or Resimene 745)	31.3	3.13
Ethylene Glycol Monobutyl Ether (60% in water)	52.0	6.66
Surfynol 104E	0.5	0.06

Formulation Constants:
    pH: adjust to 8.3-8.8 with DMAE (50% in H20)
    Viscosity (#4 Ford Cup), sec.: 21
    % PVC: 25
    % Weight Solids: 57
    % Volume Solids: 47
    Acrylic/Melamine (solids ratio): 90/10
    * Acrysol I-62 Neutralization:

Acrysol I-62 (50%)	40.0
Water	40.0
DMAE (50%)	8.0

Formulation #TC-2045-6

Flatted Clear Topcoat

Materials:	Pounds	Gallons
Rhoplex AC-2045 (45.0%)	729.1	82.83
Monomeric Melamine (Cymel 303 or Resimene 745)	36.5	3.65
Ethylene Glycol Monobutyl Ether	47.4	6.30
Ethylene Glycol	25.5	2.74
Surfynol 104E	7.3	0.90
Balab 3056A	1.8	0.25
Flatting Agent (Syloid 74 or OK-412)	36.5	2.21
Water	9.5	1.14

Formulation Constants:
    pH: adjust to 8.3-8.8 with DMAE (50% in H20)
    Viscosity (#4 Ford Cup), sec.: 35
    % Weight Solids: 45
    % Volume Solids: 43
    Acrylic/Melamine (solids ratio): 90/10

SOURCE: Rohm and Haas Co.: Building Products Hardboard Coatings:
        RHOPLEX AC-2045: Formulation #CL-2045-2

35 PVC Curtain Coatable Topcoat

Materials:	Pounds	Gallons
Cowles Grind:		
Acrysol I-62 (22.7%)*	29.2	3.36
Balab 3056A	5.8	0.92
TiPure R-960	87.8	2.67
Minex 4	269.4	12.42
Chrome Oxide Green	22.6	0.53
Bentone LT	1.0	0.07
Water	63.3	7.59
Letdown:		
Rhoplex AC-2045 (45%)	559.8	63.59
Monomeric Melamine (Cymel 303 or Resimene 745)	28.0	2.80
Ethylene Glycol Monobutyl Ether (60% in water)	46.8	5.99
Surfynol 104E	0.5	0.06

Formulation Constants:
  pH: adjust to 8.3-8.8 with DMAE (50% in H20)
  Viscosity (#4 Ford Cup), sec.: 21
  % PVC: 35
  % Weight Solids: 52
  % Volume Solids: 48
  Acrylic/Melamine (solids ratio): 90/10

  * Acrysol I-62 Neutralization:
    Acrysol I-62 (50%)   40.0
    Water   40.0
    DMAE (50%)   8.0

SOURCE: Rohm and Haas Co.: Building Products Hardboard Coatings: RHOPLEX AC-2045: Formulation #TC-2045-9

35 PVC Hardboard Topcoat Formulation

Materials:			Pounds	Gallons
Grind (20 minutes)				
Add in order:				
Water			65.0	7.80
Bentone LT			1.0	0.07
Acrysol I-62 (50%)	)		13.6	
Water	)	Premix	13.6	3.45*
Dimethylaminoethanol/Water, 1/1	)		2.8	
Foamaster VL			1.4	0.19
Foamaster TCX			0.6	0.08
Ti-Pure R-960			91.0	2.75
Minex 4			277.0	12.77
Cr2O3 Green			23.0	0.54
Letdown:				
Rhoplex AC-1230 (47%)			612.3	71.50
Ethylene glycol monobutyl ether (60% in				
water)			48.0	6.16
Surfynol 104E			0.5	0.06

Dimethylaminoethanol/Water, 1/1: Adjust pH to 9.5
Thickener or water as needed: Adjust viscosity to 25 seconds,
#4 Ford Cup

Formulation Constants:
    Pigment volume content: 35%
    TiO2/Binder ratio, by wt.: 24/76
    Solids content: 60% (weight)
                    48% (volume)

* 3.45 gallons of the premix.

SOURCE: Rohm and Haas Co.: Building Products: RHOPLEX AC-1230:
        Formulation Table IV

Section II
Coil Coatings

Gloss Coil Coating--Black Deeptone

Sand Mill Grind (15 minutes):		Pounds	Gallons
Acrysol I-62 (50.0%)	)	1.5	0.17
Water	) Premix	1.5	0.18
Dimethylaminoethanol/Water, 1/1)		0.3	0.03
Resimene 745		36.8	3.68
Ethylene Glycol		27.9	3.00
Rhoplex AC-1822 (46.5%)		195.5	22.21
Balab 3056A		7.3	0.98
Raven 420		27.7	1.85
Water		10.8	1.30

Adjust grind to pH 8.7 with Dimethylaminoethanol/water, 1/1, if
    necessary.

	Pounds	Gallons
Rhoplex AC-1822 (46.5%)	513.2	58.29
Ethylene glycol monobutyl ether/Water,		
80/20	56.2	7.39
Surfynol 104E	7.4	0.92

    Viscosity, #2 Zahn Cup: Adjust to 30 seconds with water
    Volume Solids: 39.8%
    Weight Solids: 44.7%
    P.V.C.: 4.7
    P/B: 7/93
    Acrylic/Melamine: 90/10
    V.O.C.: 0.22 kg/l (1.84 pounds/gallon) minus water
          0.25 kg/l (2.08 pounds/gallon) of coating solids
          applied
    Gloss, 60/20: 80/49
    T-Bend Microscope: 2T
        Tape: 0T
    Pencil Hardness: F

SOURCE: Rohm and Haas Co.: Building Products: ACRYSOL I-62:
      Formulation Table IV

Gloss Coil Coating—-Dark Brown Deeptone

Sand Mill Grind (15 minutes):		Pounds	Gallons
Acrysol I-62 (50.0%)	)	8.2	0.94
Water	) Premix	8.2	0.99
Dimethylaminoethanol/Water, 1/1	)	1.7	0.20
Resimene 745		35.2	3.52
Ethylene Glycol		27.6	2.97
Rhoplex AC-1822		187.1	21.26
Balab 3056A		7.0	0.95
Brown Iron Oxide 444		178.0	4.45
Raven 420		9.4	0.63

Adjust grind to pH 8.7 with Dimethylaminoethanol/water, 1/1,
   if necessary.

Letdown:	Pounds	Gallons
Rhoplex AC-1822	493.9	56.12
Ethylene Glycol Monobutyl Ether/Water, 80/20	54.0	7.10
Surfynol 104E	7.0	0.87

   Viscosity, #2 Zahn Cup: Adjust to 30 seconds with water
   Volume Solids: 42.5%
   Weight Solids: 54%
   P.V.C.: 12%
   P/B: 35/65
   Acrylic/Melamine: 90/10
   V.O.C.: 0.21 kg/l (1.75 pounds/gallon) minus water
           0.25 kg/l (2.06 pounds/gallon) of coating solids
           applied
   Gloss, 60/20: 77/40
   T-Bend Microscope: 2T
           Tape: 0T
   Pencil Hardness: F

SOURCE: Rohm and Haas Co.: Building Products: ACRYSOL I-62:
       Formulation Table III

Phthalo Green Coil Coating
Green Semi-Gloss

Materials:	Pounds	Gallons
Cowles Grind (15 minutes):		
Rhoplex AC-1230 (47%)	252.9	28.3
Acrysol I-62 (25% total solids, pH 9.2		
with DMAE)	44.5	5.0
DMAE (50% in water)	3.0	0.3
Syloid 72	20.7	1.2
Ti-Pure R-960	227.1	6.8
Green Pebble Mill Grind (from below)	20.2	2.1

Grind the above at high speed, then letdown as follows:

Letdown:		
Rhoplex AC-1230 (47%)	458.3	51.4
Foamaster TCX	1.6	0.2
Diethylene glycol monomethyl ether		
(60% in water)	29.4	3.5
Thickener or water as needed	10.0	1.2

Green Pebble Mill Grind:		
GT-674 Phthalo Green	242.5	14.55
Acrysol I-62 (25% total solids,		
pH 9.2 with DMAE)	48.5	5.34
Triton X-114	9.8	1.07
Water	654.7	78.57
Foamaster TCX	3.7	0.47

Formulation Constants:
    Solids content, wt %: 59
    Pigment volume content, %: 19
    TiO2/Binder ratio: 40/60
    TiO2/Phthalo green/Silica ratio: 89/2/9

SOURCE: Rohm and Haas Co.: Building Products: RHOPLEX AC-1230:
    Formulation SGT-1230-1

Water-Reducible Acrylic Coil Coating

Ingredients:	Pounds	Gallons
Chempol 10-0509	149.13	17.34
Dimethylethanolamine	10.41	1.41
Byk P-104S	3.55	0.45
2-butoxyethanol	11.56	1.54
Butyl Carbitol	11.57	1.46
Water	104.07	12.49
Ti-Pure R-900	155.34	4.53

Weigh in first five ingredients and mix well. Add water slowly with agitation. Mix well, then add pigment slowly with continued agitation.

Sandmill to #7 Hegman, keeping grind temperature below 120F with water jacket.

Let down with:

Chempol 10-0509	74.07	8.61
Dimethylethanolamine	3.47	0.47
Cymel 303	47.71	4.77
2-butoxyethanol	5.78	0.77
Water	326.78	39.23
Water (hold to adjust viscosity)	57.73	6.93

In the "let down," premix the first four ingredients, mix well, and add the water with continued agitation. After thoroughly mixing the letdown portion, add it slowly to the grind with agitation. Mix well, and allow the paint to set overnight.

Check the pH and adjust with dimethylethanolamine if necessary. Use the "adjust viscosity" water to bring the viscosity of the coating to spec at room temperature.

Properties:
    Total solids, % by weight: 36.89%
    Total solids, % by volume: 25.14%
    Weight/gallon: 9.61 lbs/gal
    PVC: 18.01
    Pigment/binder ratio: 0.8
    VOC: 2.97 lbs/gal
    Viscosity: 36" #4 Zahn
    pH: 8.3
    Substrate: Aluminum
    DFT: 0.5-0.7 mil
    Cure: 60" @ 500F
    PMT: 420F-435F
    Gloss, 60: 91
        , 20: 71
    Pencil hardness: H-2H
    Crosshatch adhesion: 100% retention
    Impact resistance, reverse: Pass 24 inch pounds
    T-bend: 2T-3T

SOURCE: Freeman Polymers Divison: CHEMPOL 10-0509-A Formulation

White Coil Coating: All TiO2

Materials:	Weight Ratio*	Parts per hundred (Vol. Basis)
Rhoplex AC-658 (47%)	175.1	19.6
Acrysol I-98 (25%, pH to 9.2 with DMAE)	28.5	3.2
Foamaster VL	0.3	0.03
Foamaster TCX	1.6	0.20
Ti-Pure R-960	267.7	8.1

Grind the above at high speed, then let down as follows:

Rhoplex AC-658 (47%)	524.3	58.8
Diethylene glycol monomethyl ether (60% in water)	55.0	6.6
DMAE (1/1 with water)	6.6	0.8
Cellosize QP-40 (5% in water)	22.4	2.7

Formulation Constants:
  Pigment volume content: 19
  TiO2/binder ratio, by weight: 45/55
  pH: 9.7 to 10.0
  Viscosity, sec. (#4 Ford Cup): 30
  Solids Content:
    by weight: 56.5
    by volume: 46

  * Using weight ratio in pound units will yield approximately
    100 gallons of paint; using kilograms, about 833 liters
    will result.

SOURCE: Rohm and Haas Co.: Building Products: RHOPLEX AC-658:
        Formulation: All TiO2

White Coil Coatings: Low Gloss

	Weight Ratio*	Parts per hundred (Vol. Basis)
Materials:		
Rhoplex AC-658 (47%)	220.4	24.7
Acrysol I-98 (25%, pH to 9.2 with DMAE)	29.3	3.2
Foamaster VL	0.3	0.03
Foamaster TCX	1.6	0.20
Water	11.1	1.3
Ti-Pure R-960	223.4	6.8
Syloid 72	44.5	2.5

Grind the above at high speed, then let down as follows:

Rhoplex AC-658 (47%)	481.2	53.9
Diethylene glycol monomethyl ether (60% in water)	54.6	6.6
DMAE (1/1 with water)	6.6	0.8

Formulation Constants:
   Pigment volume content: 22
   TiO2/binder ratio, by weight: 40/60
   pH: 9.7 to 10.0
   Viscosity, sec. (#4 Ford Cup): 30
   Solids Content:
      by weight: 53
      by volume: 46

* Using weight ratio in pound units will yield approximately
  100 gallons of paint; using kilograms, about 833 liters
  will result.

SOURCE: Rohm and Haas Co.: Building Products: RHOPLEX AC-658:
        Formulation: Low Gloss

Section III
Enamels

Acrylic Semi-Gloss White Enamel

Dispersion:	Pounds	Gallons
Dispersant (1)	11	1.3
Defoamer (2)	2	.27
Propylene glycol	63	7.3
High gloss titanium dioxide (3)	270	7.9

Reduction:		
Propylene glycol	30	3.5
Acrylic latex (4)	556	62.7
Vancide TH	1	0.1
Water)	23	2.8
Coalescent (5)) Premix	12	1.5
Surfactant (6))	2	0.23
Defoamer	3	0.4
3% Van Gel (Water/propylene glycol 75/25 gel)	103	12.0

1) Tamol 731 or equivalent
2) Nopco NDW or equivalent
3) TiPure R-900 or equivalent
4) Rhoplex AC-490 or equivalent
5) Texanol or equivalent
6) Triton GR-7 or equivalent

Paint Properties:
    Consistency, KU: 77
    Pigment volume concentration, %: 23
    Solids by weight, %: 49
    Solids by volume, %: 35
    60 Gloss, 1 week: 65-70

This formulation exhibits the substitution of VAN GEL for bacteria-sensitive cellulosic thickener in an acrylic semi-gloss enamel.

Manufacturing Instructions:
    (Equipment - high shear dispersion type)

Paint Preparation:
1. Charge dispersant, defoamer and propylene glycol to suitable container.
2. Mix at low speed until uniform.
3. Add and disperse titanium dioxide, increasing the mixing speed as necessary.
4. Reduce speed and add the following ingredients, mixing until uniform after each addition: propylene glycol, latex, coalescent premix, VANCIDE TH, defoamer and VAN GEL dispersion.

Note: The use of VAN GEL dispersion prevents air entrapment and optimizes gloss. The coalescent is emulsified in premix to minimize effect on latex upon incorporation.

SOURCE: R.T. Vanderbilt Co., Inc.: Development Formulation SG-201

Air-Dry Water-Reducible Enamel

Ingredients:	Weight %
Grind:	
Titanium dioxide, R-900	16.07
Polymon blue pigment	0.85
Resin WA-17-2T, 75% N.V.	27.57
Ammonium hydroxide, 28%	1.26
Ektasolve EB solvent	4.34
Cycloden cobalt, 6% Co	0.20
Cyclodex manganese, 6% Mn	0.10
Let Down:	
Water	49.61

Constants:
  Nonvolatiles, wt %: 37.6
  Pigment/binder, wt %: 45/55
  Water/organics, wt %: 82/18
  Viscosity by No. 4 Ford Cup, sec: 60
  pH: 8.9

Drying Properties (One-Mil Dry Film):
  Set to touch, min (b): 30
  Tack free, min (c): 45
  One-hour water spot, after 6 hours: Slight effect, full
                                      recovery

Coating Properties, After 7 Days:
  Pencil hardness: B-HB
  Impact resistance, in.-lb:
    Direct: 40
    Reverse: 4
  60 Gloss: 83
  Salt spray resistance, 100 hours: No creepage, no loss of
                                    gloss

Stability After 18 Months at Ambient Temperature:
  pH: 7.1
  Viscosity by No. 4 Ford Cup, sec: 34
  Set-to-touch, min (b): 30
  Tack free, min (c): 45
  One-hour water spot, after 6 hours: Slight effect, full
                                      recovery

b) ASTM D1640.5.1
c) ASTM D1640.5.3.3.1
d) Unprimed, zinc phosphatized, cold rolled steel panels coated
   with a 1-mil film were used in these evaluations.

SOURCE: Eastman Chemical Products, Inc.: Air-Drying Water-
        Reducible Alkyd Enamel Based on TMPD Glycol

Air-Dry Water-Reducible Enamel

Grind:	Weight
Ti-Pure R-900 pigment	240.1
WA-17-6C (75% N.V.) alkyd	251.5
Ammonium hydroxide (28%)	18.6
Cobalt Cyclodex drier (6%)	3.8
Manganese Cyclodex drier (6%)	3.8
Activ-8 additive	1.2
BYK 301 additive	2.3
n-Butanol	23.8
Let Down:	
Water	454.9

Physical Constants:
    Nonvolatiles, wt %: 42.9
    Pigment/Binder, wt %: 56/44
    Water/Organics, wt %: 84/16
    Viscosity, No. 4 Ford Cup, seconds: 36
    pH: 8.8
Drying Properties (One-Mil Dry Film):
    Set to touch, (b) min: 20
    Tack Free, (c) min: 35
    1-Hour Water Spot, After 6 Hours: Slight effect, full
                                          recovery
Properties After 1 Week Dry (d):
    Pencil Hardness: HB
    Impact Resistance, Direct: 28
                      Reverse: <4
    Gloss, % 60: 90
           20: 70
    Salt-Spray Resistance, 100 hr, creepage from Scribe, in.: 3/16
    Carbon-Arc Weather-Ometer, 500 hr:
        % Gloss Retention, 60: 66
                          20: 44
Properties After 3 Weeks Dry:
    Salt-Spray Resistance, Creepage from Scribe, in.:
        @ 100 hr: None
        @ 200 hr: 1/16
        @ 300 hr: 9/16
Solution Stability:
    After Four Weeks at 120F:
        pH: 7.2
        Viscosity, No. 4 Ford Cup, seconds: 155
        Dry Time, Set to Touch, min: 30
                    Tack Free, min: 45
    Gloss: Excellent   Appearance: Excellent

b) ASTM D 1640.5.1
c) ASTM D 1640.5.3.3.1
d) Untreated cold-rolled steel panels coated with a 1-mil film
   were used in these evaluations.

SOURCE: Eastman Chemical Products, Inc.: Enamel Prepared from
        Polyester Resin WA-17-6C

Black Acrylic Emulsion Enamel

Ingredients:	Pounds	Gallons
Coroc A-2678-M	47.00	5.34
2-butoxyethanol	7.91	1.05
Surfynol 104BC	8.25	1.10
Surfynol DF-75	1.35	0.16
Water	31.05	3.73
Raven #14 powder	10.26	0.70

Pre-mix first four ingredients, then add water with agitation. Load mixture into ball mill followed by pigment.

Steel ball grind to a #7 Hegman. Adjust grind and/or wash mill using water from below, if necessary.

Let down with:		
Chempol 20-4301	587.57	66.77
2-butoxyethanol	32.79	4.37
Butyl Carbitol	20.97	2.64
Dibutyl Phthalate	7.24	0.83
Nuocure CK-10	2.06	0.25
Michem 39235	30.96	3.72
Water (hold to adjust viscosity)	77.83	9.34

Add the grind to the emulsion. Add it slowly and with good agitation. Pre-mix co-solvents, plasticizers and drier. Add this mixture to the vortex of the agitating emulsion <u>VERY SLOWLY</u>. If the mixture becomes heavy, add a small amount of water from the "adjust viscosity." Check mixture for kick-out or seeding. Add the wax dispersion to the "let down" with agitation. Mix well, and allow the paint to set overnight.

Check the pH and adjust with ammonia if necessary. Avoid pH values above 7.6, as this will cause the viscosity of the paint to increase significantly. Use the "adjust viscosity" water to bring the viscosity of the coating to spec at room temperature.

Properties:
  Total solids, % by weight: 31.20%
  Total solids, % by volume: 27.60%
  Weight/gallon: 8.65 lbs./gal.
  PVC: 2.55
  Pigment/binder ratio: 0.04
  VOC: 2.17 lbs./gal.
  Viscosity: 49" #2 Zahn
  pH: 7.5
  Substrate: B-1000 CRS
  Cure: 1 week air dry
  Set to touch: 10'
  Dust free: 15'
  Tack free: 45'-50'

SOURCE: Freeman Polymers Division: CHEMPOL 20-4301-F Formula

Black Air Dry Enamel

	Pounds	Gallons
Cargill Water Reducible Epoxy Ester		
73-7331	97.8	11.58
Aqueous Ammonia 28%	5.9	0.78
Ethylene Glycol Monobutyl Ether	9.8	1.30
sec-Butanol	14.7	2.18
Byk-301	2.0	0.24
Special Black 4A	12.2	0.83
Aerosil R972	4.9	0.27
Deionized Water	190.8	22.90

Pebble Mill to a 7.5 Hegman

Cargill Water Reducible Epoxy Ester		
73-7331	181.0	21.43
Aqueous Ammonia 28%	10.8	1.43

Premix the following, then add:

sec-Butanol	14.7	2.18
Ethylene Glycol Monobutyl Ether	9.8	1.30
Activ-8	1.0	0.12
5% Cobalt Hydro Cure II	2.5	0.33
Manganese Hydro Cure II	2.2	0.25
Deionized Water	273.9	32.88

Paint Properties:
    % Nonvolatile (By Weight): 25.97
                 (By Volume): 23.47
    Pigment to Binder Ratio: 0.09
    Theoretical VOC (Pounds/Gallon): 3.24
                 (Grams/Liter): 388
    Water:Cosolvent Weight Ratio: 77.80:22.20
    pH: 8.2-8.7
    Viscosity Krebs-Stormer (KU): 68-73

Typical Cured Film Properties:
    Cure Schedule: 7 Days Air Dry
    Substrate: Cold Rolled Steel
    Gloss (60): 83
    Pencil Hardness: Pass B
    Dry Time (Minutes): 65
    Humidity (Hours): 120
        Blisters: None
    Salt Spray (Hours): 96
        Scribe Creep (mm): 5
        % Wet Tape Adhesion: 98

SOURCE: Cargill, Inc.: CARGILL Formulary: Formulation P1897-F1

Black Air Dry Enamel

	Pounds	Gallons
Cargill Water Reducible Styrenated Alkyd		
Copolymer 74-7422	90.8	10.75
Aqueous Ammonia 28%	3.6	0.48
Propylene Glycol Monobutyl Ether	9.1	1.24
Byk-301	1.8	0.22
Special Black 4	10.2	0.69

Sand Mill to a 7 Hegman

	Pounds	Gallons
Cargill Water Reducible Strenated Alkyd		
Copolymer 74-7422	181.6	21.49
Aqueous Ammonia 28%	5.9	0.79

Premix the following, then add:

	Pounds	Gallons
Propylene Glycol Monobutyl Ether	18.2	2.48
Intercar 6% Cobalt	1.9	0.25
Intercar 4% Calcium	2.9	0.38
Activ-8	0.9	0.11
Deionized Water	455.4	54.67
Propylene Glycol Monobutyl Ether	10.4	1.43
Deionized Water (Hold for Viscosity		
Adjustment)	41.8	5.02

Paint Properties:
    % Nonvolatile (By Weight): 24.51
                (By Volume): 21.83
    Pigment to Binder Ratio: 0.05
    Pigment Volume Concentration: 3.30
    Weight Per Gallon: 8.35
    Theoretical VOC (Pounds/Gallon): 3.19
                (Grams/Liter): 382
    Water:Cosolvent Weight Ratio: 80.00:20.00
    pH: 8.0
    Viscosity #4 Ford Cup (Seconds): 44

Typical Cured Film Properties:
    Cure Schedule: 7 Days Air Dry
    Substrate: Cold Rolled Steel
    Gloss (60/20): 91/77
    Pencil Hardness: Pass 2B
    % Crosshatch Adhesion: 80
    Impact (In. Lbs.):
        Direct: 10
        Reverse: 5

SOURCE: Cargill, Inc.: CARGILL Formulary: Formula P1857-207A

Black Air Dry Enamel

	Pounds	Gallons
Cargill Water Reducible Styrenated Alkyd		
Copolymer 74-7425	125.6	14.52
Ethylene Glycol Monobutyl Ether	13.2	1.76
Byk-301	2.6	0.33
Aqueous Ammonia 28%	7.9	1.06
Special Black 4A	17.2	1.16
Deionized Water	330.5	39.68

Pebble Mill to a 7.5 Hegman

Cargill Water Reducible Styrenated Alkyd		
Copolymer 74-7425	244.6	28.28

Premix the following, then add:

Ethylene Glycol Monobutyl Ether	13.2	1.76
sec-Butanol	6.6	0.98
5% Cobalt Hydro Cure II	1.7	0.22
8% Manganese Hydro Cure II	1.3	0.15
Activ-8	0.8	0.10
Aqueous Ammonia 28%	11.9	1.59
Exkin No. 2	1.3	0.17
Aqueous Ammonia 28% (adjust to pH 8.0-8.5)	2.6	0.30
Deionized Water (Hold for Viscosity		
Adjustment)	66.1	7.94

Paint Properties:
   % Nonvolatile (By Weight): 33.0
               (By Volume): 29.6
   Pigment to Binder Ratio: 0.07
   Weight Per Gallon: 8.47
   Theoretical VOC (Pounds/Gallon): 3.00
               (Grams/Liter): 360
   pH: 8.0-8.5
   Viscosity Krebs-Stormer (KU): 67-72

Typical Cured Film Properties:
   Cure Schedule: 7 Days Air Dry
   Substrate: Cold Rolled Steel
   Gloss (60): 87-92
   Pencil Hardness (1 Day): Pass 6B
                (7 Days): Pass 3B
   Dry Time 500g Zapon (Hours): 1

SOURCE: Cargill, Inc.: CARGILL Formulary: Formula P1863-D2

Black Air Dry Enamel

	Pounds	Gallons
Cargill Water Reducible Acrylic Modified Alkyd 74-7432	53.0	6.23
Byk-301	1.9	0.24
Aqueous Ammonia 28%	2.0	0.27
Special Black 4	9.7	0.66
Deionized Water	68.3	8.20

Sand Mill to a 7 Hegman

Cargill Water Reducible Acrylic Modified Alkyd 74-7432	195.2	22.97
Aqueous Ammonia 28%	11.9	1.59

Premix the following, then add:

Intercar 6% Cobalt	2.9	0.38
Intercar 6% Manganese	2.9	0.40
Activ-8	1.0	0.12
Propylene Glycol Monopropyl Ether	40.2	5.46
Diacetone Alcohol	3.0	0.38

Exkin No. 2	2.6	0.34
Deionized Water	439.4	52.76

Paint Properties:
    % Nonvolatile (By Weight): 22.42
                 (By Volume): 19.85
    Pigment to Binder Ratio: 0.06
    Pigment Volume Concentration: 3.54
    Weight Per Gallon: 8.34
    Theoretical VOC (Pounds/Gallon): 3.42
                    (Grams/Liter): 410
    Water:Cosolvent Weight Ratio: 80.00:20.00
    pH: 8.40
    Viscosity #4 Ford Cup (Seconds): 61

Typical Cured Film Properties:
    Cure Schedule: 7 Days Air Dry
    Substrate: Cold Rolled Steel
    Gloss (60/20): 85/60
    Pencil Hardness: Pass B
    % Crosshatch Adhesion: 100
    Impact (In. Lbs.):
        Direct: 70
        Reverse: <5

SOURCE: Cargill, Inc.: CARGILL Formulary: Formula P1848-161A

Black Air Dry Enamel

	Pounds	Gallons
Cargill Water Reducible Acrylic Modified Alkyd 74-7432	86.0	10.12
Aqueous Ammonia 28%	4.2	0.57
sec-Butanol	15.4	2.28
Special Black 4	10.6	0.71
Aerosil R972	3.6	0.20
Byk-020	1.8	0.24
Surfynol 104E	1.4	0.17
Deionized Water	96.6	11.60

Ball Mill to a 7 Hegman

	Pounds	Gallons
Cargill Water Reducible Acrylic Modified Alkyd 74-7432	196.5	23.12
Aqueous Ammonia 28%	9.9	1.32

Premix the following, then add:

	Pounds	Gallons
sec-Butanol	2.8	0.42
Intercar 6% Cobalt	1.8	0.23
Intercar 6% Zirconium	2.5	0.34
Intercar 4% Calcium	1.4	0.18
Activ-8	0.8	0.10

	Pounds	Gallons
Exkin No. 2	0.8	0.10
Deionized Water	248.6	29.85

	Pounds	Gallons
Deionized Water (Hold for Viscosity Adjustment)	153.7	18.45

Paint Properties:
    % Nonvolatile (By Weight): 25.68
                  (By Volume): 22.69
    Pigment to Binder Ratio: 0.07
    Pigment Volume Concentration: 4.19
    Weight Per Gallon: 8.38
    Theoretical VOC (Pounds/Gallon): 2.93
                    (Grams/Liter): 351
    Water:Cosolvent Weight Ratio: 81.72:18.28
    pH: 8.4
    Viscosity Krebs-Stormer (KU): 68

Typical Cured Film Properties:
    Cure Schedule: 7 Days Air Dry
    Substrate: Cold Rolled Steel
    Gloss (60/20): 79/38
    Pencil Hardness: Pass 2B
    % Crosshatch Adhesion: 90

SOURCE: Cargill, Inc.: CARGILL Formulary: Formula P1843-125A

Black Air Dry Enamel

	Pounds	Gallons
Cargill Water Reducible Chain Stopped Alkyd 74-7470	80.3	9.24
Aqueous Ammonia 28%	4.2	0.57
sec-Butanol	26.4	3.93
Special Black 4	10.6	0.71
Aerosil R972	3.6	0.20
Byk-020	1.8	0.24
Surfynol 104E	1.4	0.17
Deionized Water	96.8	11.62

Ball Mill to a 7 Hegman

	Pounds	Gallons
Cargill Water Reducible Chain Stopped Alkyd 74-7470	183.6	21.11
Aqueous Ammonia 28%	9.9	1.32

Premix the following, then add:

	Pounds	Gallons
sec-Butanol	10.7	1.59
Intercar 6% Cobalt	1.8	0.23
Intercar 6% Zirconium	2.5	0.34
Intercar 4% Calcium	1.4	0.18
Activ-8	0.8	0.10
Exkin No. 2	0.8	0.10
Deionized Water	402.9	48.35

Paint Properties:
    % Nonvolatile (By Weight): 25.68
                  (By Volume): 22.20
    Pigment to Binder Ratio: 0.07
    Pigment Volume Concentration: 4.29
    Weight Per Gallon: 8.39
    Theoretical VOC (Pounds/Gallon): 2.94
                    (Grams/Liter): 352
    Water:Cosolvent Weight Ratio: 81.72:18.28
    pH: 8.5
    Viscosity Krebs-Stormer (KU): 65

Typical Cured Film Properties:
    Cure Schedule: 7 Days Air Dry
    Substrate: Cold Rolled Steel
    Gloss (60/20): 76/35
    Pencil Hardness: Pass 3B
    % Crosshatch Adhsion: 70
    Impact (in. lbs.):
      Direct: 40
      Reverse: >5

SOURCE: Cargill, Inc.: CARGILL Formulary: Formula P1843-125D

Black Air Dry Enamel

	Pounds	Gallons
Cargill Water Reducible Fatty Acid Alkyd 74-7472	80.1	9.31
Aqueous Ammonia 28%	4.2	0.56
sec-Butanol	26.3	3.91
Special Black 4	10.5	0.71
Aerosil R972	3.6	0.20
Byk-020	1.8	0.24
Surfynol 104E	1.4	0.17
Deionized Water	96.4	11.58

Ball Mill to a 7 Hegman

Cargill Water Reducible Fatty Acid Alkyd 74-7472	183.0	21.28
Aqueous Ammonia 28%	9.9	1.32

Premix the following, then add:

sec-Butanol	10.6	1.58
Intercar 6% Cobalt	1.8	0.23
Intercar 6% Zirconium	2.4	0.34
Intercar 4% Calcium	1.4	0.18
Activ-8	0.8	0.10
Exkin No. 2	0.8	0.10
Deionized Water	401.4	48.19

Paint Properties:
    % Nonvolatile (By Weight): 25.68
                (By Volume): 22.48
    Pigment to Binder Ratio: 0.07
    Pigment Volume Concentration: 4.23
    Weight Per Gallon: 8.36
    Theoretical VOC (Pounds/Gallon): 2.91
                (Grams/Liter): 349
    Water:Cosolvent Weight Ratio: 81.72:18.28
    pH: 8.3
    Viscosity Krebs-Stormer (KU): 74

Typical Cured Film Properties:
    Cure Schedule: 7 Days Air Dry
    Substrate: Cold Rolled Steel
    Gloss (60/20): 86/70
    Pencil Hardness: Pass 2B
    % Crosshatch Adhesion: 90
    Impact (In. Lbs.):
        Direct: 30
        Reverse: <5

SOURCE: Cargill, Inc.: CARGILL Formulary: Formula P1843-125E

Black Air Dry Enamel

	Pounds	Gallons
Cargill Water Reducible Chain Stopped Alkyd 74-7474	85.5	9.71
Aqueous Ammonia 28%	4.5	0.60
sec-Butanol	28.1	4.18
Special Black 4	11.3	0.76
Aerosil R972	3.8	0.21
Byk-020	1.9	0.26
Surfynol 104 E	1.5	0.18
Deionized Water	103.0	12.36

Ball Mill to a 7 Hegman

Cargill Water Reducible Chain Stopped Alkyd 74-7474	195.4	22.20
Aqueous Ammonia 28%	10.5	1.41

Premix the following, then add:

sec-Butanol	11.4	1.69
Intercar 6% Cobalt	1.9	0.25
Intercar 6% Zirconium	2.6	0.36
Intercar 4% Calcium	1.5	0.19
Activ-8	0.8	0.10
Exkin No. 2	0.8	0.10
Deionized Water	378.4	45.44

Paint Properties:
    % Nonvolatile (By Weight): 27.21
                 (By Volume): 23.36
    Pigment to Binder Ratio: 0.07
    Pigment Volume Concentration: 4.34
    Weight Per Gallon: 8.43
    Theoretical VOC (Pounds/Gallon): 2.97
                    (Grams/Liter): 356
    Water:Cosolvent Weight Ratio: 80.22:19.78
    pH: 8.2
    Viscosity Krebs-Stormer (KU): 72
Typical Cured Film Properties:
    Cure Schedule: 7 Days Air Dry
    Substrate: Cold Rolled Steel
    Gloss (60/20): 90/71
    Pencil Hardness: Pass 3B
    % Crosshatch Adhesion: 50
    Impact (in. lbs.):
       Direct: 20
       Reverse: <5
    Dry Time:
       Set to Touch (Minutes): 37
       500g Zapon (Hours): 4 1/2

SOURCE: Cargill, Inc.: CARGILL Formulary: Formula P1843-125F

Black Air Dry Enamel

	Pounds	Gallons
Cargill Water Reducible Fatty Acid Alkyd		
74-7476	80.1	9.31
Aqueous Ammonia 28%	4.2	0.56
sec-Butanol	26.3	3.91
Special Black 4	10.5	0.71
Aerosil R972	3.6	0.20
Byk-020	1.8	0.24
Surfynol 104 E	1.4	0.17
Deionized Water	96.4	11.58

Ball Mill to a 7 Hegman

Cargill Water Reducible Fatty Acid Alkyd		
74-7476	183.0	21.28
Aqueous Ammonia 28%	9.9	1.32

Premix the following, then add:

sec-Butanol	10.6	1.58
Intercar 6% Cobalt	1.8	0.23
Intercar 6% Zirconium	2.4	0.34
Intercar 4% Calcium	1.4	0.18
Activ-8	0.8	0.10
Exkin No. 2	0.8	0.10
Deionized Water	401.4	48.19

Paint Properties:
   % Nonvolatile (By Weight): 25.68
                 (By Volume): 22.36
   Pigment to Binder Ratio: 0.07
   Pigment Volume Concentration: 4.25
   Weight Per Gallon: 8.36
   Theoretical VOC (Pounds/Gallon): 2.91
                 (Grams/Liter): 349
   Water:Cosolvent Weight Ratio: 81.72:18.28
   pH: 8.3
   Viscosity Krebs-Stormer (KU): 65

Typical Cured Film Properties:
   Cure Schedule: 7 Days Air Dry
   Substrate: Cold Rolled Steel
   Gloss (60/20): 87/72
   Pencil Hardness: Pass 2B
   % Crosshatch Adhesion: 95
   Impact (In. Lbs.): Direct: 30
                 Reverse: <5
   Dry Time: Set to Touch (Minutes): 25
             500g Zapon (Hours): >8

SOURCE: Cargill, Inc.: CARGILL Formulary: Formula P1843-125G

Black Air Dry Enamel

	Pounds	Gallons
Cargill Water Reducible Phenolic Modified Alkyd 74-7478	84.9	9.87
Aqueous Ammonia 28%	4.5	0.60
sec-Butanol	27.9	4.15
Special Black 4	11.2	0.75
Aerosil R972	3.8	0.21
Byk-020	1.9	0.26
Surfynol 104 E	1.5	0.18
Deionized Water	102.2	12.27

Ball Mill to a 7 Hegman

	Pounds	Gallons
Cargill Water Reducible Phenolic Modified Alkyd 74-7478	193.9	22.55
Aqueous Ammonia 28%	10.5	1.40

Premix the following, then add:

	Pounds	Gallons
sec-Butanol	11.3	1.67
Intercar 6% Cobalt	1.9	0.25
Intercar 6% Zirconium	2.6	0.36
Intercar 4% Calcium	1.5	0.19
Activ-8	0.8	0.10
Exkin No. 2	0.8	0.10
Deionized Water	375.6	45.09

Paint Properties:
   % Nonvolatile (By Weight): 27.21
                (By Volume): 24.09
   Pigment to Binder Ratio: 0.07
   Pigment Volume Concentration: 4.18
   Weight Per Gallon: 8.37
   Theoretical VOC (Pounds/Gallon): 2.91
                   (Grams/Liter): 349
   Water:Cosolvent Weight Ratio: 80.22:19.78
   pH: 8.7
   Viscosity Krebs-Stormer (KU): 70
Typical Cured Film Properties:
   Cure Schedule: 7 Days Air Dry
   Substrate: Cold Rolled Steel
   Gloss (60/20): 87/75
   Pencil Hardness: Pass 3B
   % Crosshatch Adhesion: 98
   Impact (in. Lbs.):
      Direct: 40
      Reverse: <5
   Dry Time:
      Set to Touch (Hours): 3/4
      550g Zapon (Hours): >8

SOURCE: Cargill, Inc.: CARGILL Formulary: Formula P1843-125H

Black Enamel

Material:	Lbs.	Gals.
Water	100.0	12.00
Foamaster VL	2.0	0.25
Cellosize QP 4400	0.5	0.04
Methyl Carbitol	8.5	1.00
Disperse-Ayd W22	4.4	0.50
Aerosol OT-75	2.3	0.25
Arolon 820-W-49	105.6	12.00
AMP 95	1.9	0.25
Nalzin 2	50.0	1.52
Raven 1020 Beads	15.0	1.00

Pebble mill grind the above to a 7 Hegman.

Letdown:		
Arolon 820-W-49	572.0	65.00
Methyl Carbitol	25.0	3.00
Water	20.3	2.44
Rheolate 278	4.4	0.50
Ammonium Hydroxide	1.9	0.25

Analysis:
   Pigment Volume Concentration, Percent: 6.5
   Pigment/Binder Ratio: 0.20/1.0
   Percent Solids, Weight: 44.3
   Percent Solids, Volume: 39.0
   Viscosity @ 25C, KU: 77
   Pounds/Gallon: 9.14
   pH: 8.5-9.0
   VOC (excluding water):
      Grams/Liter: 88
      Pounds/Gallon: 0.7

SOURCE: Reichhold Chemicals, Inc.: Waterborne Handbook:
        Formulation 2343-87

Black Gloss Chromate Enamel

Material:	Lbs.	Gals.
Kelsol 3918-B2G-70	123.60	14.49
Ammonium Hydoxide	4.30	0.58
Raven 1255	20.60	1.41
Strontium Chromate J1365	25.75	0.83
Igepal CTA-639	1.85	0.21
Water	248.32	29.81

Steel ball mill to a 7 Hegman grind.

Letdown:		
Kelsol 3918-B2G-70	128.96	15.12
Ammonium Hydroxide	7.00	0.94
Water	283.25	34.00
Cobalt Hydrocure II	5.29	0.70
Activ-8	1.26	0.16
Butoxy Ethanol	13.80	1.84

Analysis:
  Pigment Volume Concentration, Percent: 10.48
  Pigment/Binder Ratio: .26/1
  Percent Solids, Weight: 26.2
  Percent Solids, Volume: 21.9
  Viscosity @ 25C, KU: 35-45
  pH: 8.2-8.5
  Pounds/Gallon: 8.64
  VOC (excluding water):
      Grams/Liter: 317.2
      Pounds/Gallon: 2.64

SOURCE: Reichhold Chemicals, Inc.: Waterborne Handbook:
        Formulation 2403-19

Black Gloss Enamel Flowcoat or Dip

Material:	Lbs.	Gals.
Water	233.69	28.05
Ammonium Hydroxide	7.00	.93
Kelsol 3902-BG4-75	116.90	13.36

Premix and add to above mixture:

Cobalt Hydrocure II	4.69	.61
Activ-8	2.33	.29
Butoxy Ethanol	4.90	.66

Premix drier and solvent and add to above mixture:

12% Manganese Cem-All	4.90	.66
Butoxy Ethanol	5.83	.69

Then add in order:

Cab-O-Sil M5	4.68	.26
Strontium Chromate J1365	23.58	.76
Raven 1020	17.20	1.18
Foamkill FBF	3.14	.44
Sand mill to 7 Hegman grind.		

Letdown:

Water	182.43	21.90
Ammonium Hydroxide	6.26	.83
Kelsol 3902-BG4-75	257.16	29.38

Analysis:
    Pigment Volume Concentration, Percent: 10.3
    Pigment/Binder Ratio: .16/1
    Percent Solids, Weight: 38.0
    Percent Solids, Volume: 32.6
    Pounds/Gallon: 8.75
    pH: 8.2-8.5
    VOC (excluding water):
        Grams/Liter: 264
        Pounds/Gallon: 2.2

SOURCE: Reichhold Chemicals, Inc.: Waterborne Handbook:
        Formulation 1648-40

Black Gloss Non-Chromate Enamel

Material:	Lbs.	Gals.
Kelsol 3919-B2G-70	169.11	19.55
Ammonium Hydroxide	9.64	1.29
Cobalt Hydrocure II	5.79	0.76
Activ-8	0.96	0.12
Butoxy Ethanol	22.91	3.05
n-Butanol	10.33	1.53
Heucophos ZPO	56.55	1.83
Alcophor 827	6.99	0.34
Raven 1255	9.76	0.67
Water	252.23	30.28

Steel ball mill to a 7 Hegman grind.

Letdown:		
Kelsol 3919-B2G-70	106.14	12.27
Ammonium Hydroxide	1.31	0.17
Water	234.41	28.14

Analysis:
    Pigment Volume Concentration, Percent: 9.63
    Pigment/Binder Ratio: .33/1
    Percent Solids, Weight: 30.3
    Percent Solids, Volume: 23.7
    Viscosity @ 25C, #4 Ford, Secs.: 35-45
    pH: 8.2-8.5
    Pounds/Gallon: 8.86
    VOC (excluding water):
        Grams/Liter: 350.0
        Pounds/Gallon: 2.9

SOURCE: Reichhold Chemicals, Inc.: Waterborne Handbook:
        Formulation 2403-16

Black High Gloss Brushing Enamel

Material:	Lbs.	Gals.
Water	86.0	10.32
Triethylamine	7.0	1.16
Butyl Carbitol	12.0	1.52
Kelsol 3970-G4-75	98.0	11.00
Raven 1020	24.0	1.64
Foamkill 639Q	0.5	0.07

Steel ball mill to a 7 Hegman grind.

Letdown:		
Kelsol 3970-G4-75	269.0	30.19
Ammonium Hydroxide	8.5	1.13
Propasol P	20.0	2.72

Premix:		
Cobalt Hydrocure II	6.0	0.79
Activ-8	1.0	0.13
Water	327.0	39.26

Analysis:
   Pigment Volume Concentration, Percent: 5.3
   Pigment/Binder Ratio: .09/1
   Percent Solids, Weight: 34.8
   Percent Solids, Volume: 31.0
   Viscosity @ 25C, KU: 90-95
   Pounds/Gallon: 8.60
   pH: 8.2-8.6
   VOC (excluding water):
      Grams/Liter: 326
      Pounds/Gallon: 2.72

SOURCE: Reichhold Chemicals, Inc.: Waterborne Handbook:
        Formulation 2025-73-1A

Black High Gloss Enamel

Material:	Lbs.	Gals.
Arolon 850-W-45	86.6	10.00
Water	83.3	10.00
Methyl Carbitol	8.5	1.00
Disperse Ayd W22	4.4	0.50
Aerosol OT-75	2.3	0.25
Dapro 1181	1.8	0.25
Raven 1020	15.0	1.00
Premix the above, then pebble mill grind to a 7 Hegman.		
Arolon 850-W-45	529.2	61.25
Butoxy Ethanol	39.4	5.25
Methyl Carbitol	29.7	3.50
KP 140	4.2	0.50
Surfynol 104 BC	1.9	0.25
15% Solution FC430	2.1	0.25
Ucar SCT 200	4.3	0.50
Water	43.7	5.25
Ammonium Hydroxide	1.9	0.25

Analysis:
    Pigment Volume Concentration, Percent: 3.0
    Pigment/Binder Ratio: 0.05/1.00
    Percent Solids, Weight: 36.5
    Percent Solids, Volume: 33.0
    Viscosity @ 25C, #4 Ford, Secs.: 94
    Gloss (60): 94
    Pounds/Gallon: 8.6
    pH: 8.0-8.5
    VOC (excluding water):
        Grams/Liter: 233
        Pounds/Gallon: 1.94

SOURCE: Reichhold Chemicals, Inc.: Waterborne Handbook:
        Formulation 2210-1

Blue High Gloss Enamel

Material:	Lbs.	Gals.
Kelsol 3961-B2G-75	43.95	5.13
Ammonium Hydroxide	2.91	0.38
n-Butanol	3.59	0.53
Butoxy Ethanol	3.89	0.52
Water	71.75	8.61
Heucophthal Blue RF BT 698D	4.87	0.36
Raven 1020	2.91	0.19
Y.I.O. 2288D	2.91	0.08
Strontium Chromate J-1365	17.94	0.56
Ti-Pure R-902	35.87	1.04
Pebble mill to 6-7 Hegman grind.		

Letdown:		
Kelsol 3961-B2G-75	242.15	28.29
sec-Butanol	5.84	0.87
n-Butanol	7.71	1.14
Ammonium Hydroxide	7.71	1.02
Water	385.65	46.30

Premix:		
Activ-8	1.17	0.14
Cobalt Hydrocure II	6.62	0.86
Butoxy Ethanol	0.98	0.13
Methyl Ethyl Ketone	26.91	4.01

Note: Water may be used to replace the MEK for a lower CARB
      value.

Analysis:
    Pigment Volume Concentration, Percent: 8.18
    Pigment/Binder Ratio: .3/1
    Percent Solids, Weight: 31.87
    Percent Solids, Volume: 27.25
    Viscosity @ 25C, KU: 75-80
    Pounds/Gallon: 8.74
    pH: 8.2-8.5
    VOC (excluding water):
        Grams/Liter: 351
        Pounds/Gallon: 2.93

SOURCE: Reichhold Chemicals, Inc.: Waterborne Handbook:
        Formulation 2116-51-2

Black Low Gloss Enamel

Material:	Lbs.	Gals.
Kelsol 3963-B2G-70	141.33	16.53
Cobalt Hydrocure II	4.16	0.55
Activ-8	0.72	0.09
n-Butanol	1.15	0.17
Butoxy Ethanol	1.20	0.16
Propoxy Propanol	1.18	0.16
Ammonium Hydroxide	6.95	0.93
Raven 1255 Beads	11.52	0.79
Novacite L207A	34.70	1.57
Atomite	34.88	1.55
Syloid 83	8.18	0.49
Patcote 519	1.17	0.17
Water	363.10	43.59

Steel ball mill to 7 Hegman grind.

Letdown:		
Kelsol 3963-B2G-70	57.97	6.78
Ammonium Hydroxide	2.91	0.39
n-Butanol	1.69	0.25
Patcote 519	0.76	0.11
Water	214.25	25.72

Note: Premix the cobalt drier, Activ-8, and n-butanol before
     adding to the resin.

Analysis:
    Pigment Volume Concentration, Percent: 22.78
    Pigment/Binder Ratio: 0.64/1
    Percent Solids, Weight: 26.42
    Percent Solids, Volume: 19.85
    Viscosity @ 25C, #4 Ford, Secs.: 35-45
    Pounds/Gallon: 8.88
    pH: 8.3-8.6
    VOC (excluding water):
        Grams/Liter: 271.2
        Pounds/Gallon: 2.26

SOURCE: Reichhold Chemicals, Inc.: Waterborne Handbook:
        Formulation 2344-75

Blue Acrylic Emulsion Enamel

Ingredients:	Pounds	Gallons
Coroc A-2678-M	22.22	2.52
2-butoxyethanol	7.48	1.00
Surfynol 104BC	7.81	1.04
Surfynol DF-75	1.28	0.15
Water	7.48	0.90
TiPure R-900	99.68	2.90

Pre-mix first four ingredients, then add water with agitation. Add pigment slowly with agitation.

Cowles to a #7 Hegman with a high speed dispersator. Adjust grind viscosity with water from below if necessary.

Let down with:

Chempol 20-4301	586.40	66.64
2-butoxyethanol	31.00	4.13
Butyl Carbitol	19.82	2.50
Dibutyl Phthalate	6.85	0.79
Nuocure CK-10	1.95	0.24
Michem 39235	29.27	3.51
UCD 4820Q	19.94	2.23
Water (hold to adjust viscosity)	95.38	11.45

Pre-mix co-solvents, plasticizers and drier. Add this mixture to the vortex of the agitating emulsion <u>VERY SLOWLY.</u> If the mixture becomes heavy, add a small amount of water from the "adjust viscosity." Check mixture for kick-out or seeding. Add the wax dispersion to the "let down" with agitation. After mixing well, add the grind to the "let down" with good agitation. Finally, add the blue tint paste slowly to the blended paint and mix until well dispersed. Allow the paint to set overnight.

Check the pH and adjust with ammonia if necessary. Avoid pH values above 7.6, as this will cause the viscosity of the paint to increase significantly. Use the "adjust viscosity" water to bring the viscosity of the coating to spec at room temperature.

Properties:
   Total solids, % by weight: 37.50%
   Total solids, % by volume: 28.85%
   Weight/gallon: 9.37 lbs./gal.
   PVC: 11.10
   Pigment/binder ratio: 0.42
   VOC: 1.97 lbs./gal.
   Viscosity: 23" #2 Zahn
   pH: 7.5
   Substrate: B-1000 CRS
   Cure: 1 week air dry
   Set to touch: 17'
   Dust free: 20'
   Tack free: 75'-90'
   Dry through: 75'-90'

SOURCE: Freeman Polymers Division: Chempol 20-4301-E Formulation

Blue Enamel

Materials

Resin WB-408 (80% NVM)	255
Triethylamine	18
Amino crosslinker, Cymel (1)	36
Manganese naphthenate, 6%	2
Water	443
Phthalocyanine blue (2)	40
Titanium dioxide (3)	8

Properties:
   Pigment/binder ratio: 0.2/1.0
   Melamine/alkyd ratio: 15/85
   Non-volatile (NVM), wt %: 39
   Viscosity, #4 Ford cup, sec: 35-45
   pH: 7.5-8.0
   Stability, 15 days at 49C (4): pass
(1) Cymel 301
(2) Monastral Blue BT
(3) Ti-Pure R-900
(4) Less than 5 seconds viscosity change or 0.5 pH change for
    pass
    The pigmented enamel is compounded using a pebble mill.
The resin, amine solubilizer, amino crosslinker, pigment, and
80% of the water are added to the mill. The enamel is dispersed
to a minimum of #7 Hegman grind. The enamel is then adjusted
to the correct pH and viscosity using additional amine and
water respectively.

Performance properties of blue enamel(1):
Substrate: B-37(3):
   Pencil hardness: H
   Sward hardness: 46
   Crosshatch adhesion: 5B
   1/8-inch conical bend, % pass: 100
   Impact resistance, in-lb:
     direct: 80
     reverse: 80
   60 gloss, %: 90+
   Mar resistance: good
   Salt spray resistance, 250 hr: pass
Substrate: CRS(4):
   Pencil hardness: H
   Sward hardness: 48
   Crosshatch adhesion: 5B
   1/8-inch conical bend, % pass: 100
   Impact resistance, in-lb:
     direct: 80
     reverse: 80
   60 gloss, %: 90+
   Mar resistance: good
(1)Cure cycle 20 minutes at 177C
(3)Bonderite 37 treated cold rolled steel
(4)Less than 5 seconds viscosity change or 0.5 pH for pass
SOURCE: Amoco Chemical Co.: Water-Borne Bake Alkyds Based on
      Amoco TMA and IPA: Table 5 & 7 Formula

Blue High Gloss Enamel

Material:	Lbs.	Gals.
Water	258.20	30.98
Ammonium Hydroxide	12.91	1.72
Kelsol 3970-G4-75	129.10	14.49

Premix:		
Activ-8	1.29	0.17
Cobalt Hydrocure II	2.58	0.34
Manganese Hydrocure	5.16	0.68
Butoxy Ethanol	12.91	1.72
Heliogen Blue L6975F	40.02	3.02
Titanox 2160	40.02	1.12

Steel ball mill to 6+ Hegman grind.

Letdown:		
Water	154.92	18.59
Butoxy Ethanol	6.46	0.86
Ammonium Hydroxide	6.46	0.86
Kelsol 3970-G4-75	225.93	25.36

Analysis:
    Pigment Volume Concentration, Percent: 12.8
    Pigment/Binder Ratio: .3/1
    Percent Solids, Weight: 38.7
    Percent Solids, Volume: 32.3
    Viscosity @ 25C, #4 Ford, Secs.: 38-42
    Pounds/Gallon: 8.97
    pH: 8.2-8.5
    VOC (excluding water):
        Grams/Liter: 278
        Pounds/Gallon: 2.32

SOURCE: Reichhold Chemicals, Inc.: Waterborne Handbook:
        Formulation 2079-42-2

Blue Semi-Gloss Enamel

Material:	Lbs.	Gals.
Kelsol 3980-G4-75	103.70	11.52
Surfynol 104E	2.07	0.25
Defoamer L-475	2.07	0.28
Ethylene Glycol Monobutyl Ether	10.37	1.38
Dimethylethanol Amine (DMEA)	5.19	0.70
Water	82.96	9.96
Ti-Pure R-902	207.41	6.07
A-4434 Tacoma Blue	2.07	0.17
Talc 399	62.22	2.77

Pebble mill or sand mill to a 6 1/2 Hegman grind.

Letdown:		
Kelsol 3980-G4-75	155.56	17.26
Cymel 327	72.07	7.35
Ethylene Glycol Monobutyl Ether	10.37	1.38
DMEA	10.37	1.40
Propoxy Propanol	20.74	2.82
Water	305.41	36.67

Analysis:
    Pigment Volume Concentration, Percent: 25.28
    Pigment/Binder Ratio: 1.04/1
    Percent Solids, Weight: 50.55
    Percent Solids, Volume: 35.62
    Viscosity @ 25C, KU: 70-75
    Pounds/Gallon: 10.53
    pH: 8.2-8.6
    Resin/Melamine Ratio: 72/75
    % Reduction to 30 Secs., #2 Zahn, with water: 15
    VOC (excluding water):
        Grams/Liter: 294
        Pounds/Gallon: 2.45

SOURCE: Reichhold Chemicals, Inc.: Waterborne Handbook:
        Formulation 2185-66

Clear Enamel

Material:	Pounds	Gallons
Part A:		
Epi-Rez WD-510 Epoxy Resin	79.0	8.18
2-Propoxyethanol (2)	7.6	1.00
Water	90.1	10.82
Part B:		
CMD 9790 Acrylic Resin	653.0	75.50
2-Propoxyethanol (2)	22.7	3.00
Mix well under low agitation and add:		
Water	12.5	1.50

```
Composite Blend:          Viscosity
Part A          12 seconds #4 Ford Cup      176.7      20.00
Part B          114 KU                       688.2      80.00
Parts A and B   94 KU
```

(2) Ektasolve EP

Typical Coating Properties:
    Total Weight Solids, percent: 40.8
    Total Volume Solids, percent: 38.0
    Volatile Organic Compounds (VOC):
        Pounds/Gallon: 2.33
        Grams/Liter: 280

Resin Composition (Solids):
    Acrylic Resin: 77.6%
    Epoxy Resin: 22.4%

SOURCE: Shell Chemical Co.: 24-258 Clear Enamel

Clear Enamel (for use with predispersed colorants)

Materials:	Pounds	Gallons
Grind Portion:		
Kelsol 3905	240.0	27.93
Aquacat	1.9	0.24
Magnacat	3.8	0.49
Patcote 519	0.9	0.12
Patcote 577	0.9	0.12
28% Ammonia	5.1	0.61
Water	321.7	38.76

Mix well (% neutralization=58%, pH 5.8 to 6.0)

Add in order, with agitation:		
Rhoplex WL-71	186.7	21.71
Water	32.2	3.86
Butyl Carbitol	20.0	2.52
14% Ammonia (to pH 8.4)	----	----
Water	21.4	2.58

Note: Reduce to spray viscosity with 87/13 water/Butyl Cellosolve

Formulation Constants:
    Pounds/gallon: 8.4
    Non-volatiles (Wt.): 31.3%
                  (Vol.): 28.2%
    Rhoplex WL-71/Kelsol 3905: 30/70
    pH: 8.4
    Viscosity: 300-500 cps.
    VOC: 1.97 lbs./gal.

Note: Colors are obtained by adding pigment dispersions (Colanyl
      series) to this clear enamel with good mixing.

Code No.	Description	Pigment/Binder	Lb./100 Gal.
16-2010	Green 8G	13/87	70
18-1004	Black PR-A	5/95	40
11-1109	Yellow OT	13/87	70

SOURCE: Rohm and Haas Co.: Industrial Coatings: RHOPLEX WL-71:
        Formulation WL-71-5

Gloss White Enamel

Material:		Pounds	Gallons
Part A:			
Epi-Rez WD-510 Epoxy Resin		68.0	7.07
2-Propoxyethanol (2)		22.2	2.93
Water		83.3	10.00
Part B:			
Water		73.3	8.80
Potassium Tripolyphosphate		2.0	0.12
Mix well at low speed, then add:			
Defoamer (4)		2.0	0.26
Surfactant (5)		2.0	0.23
Water Dispersible Lecithin (6)		4.5	0.52
Rutile Titanium Dioxide (7)		250.0	7.33
High speed disperse for 5-10 minutes at 2000 rpm, then add:			
CMD 9790 Acrylic Resin		108.0	12.50
High speed disperse to a texture of 9-10 P.S.C., then add:			
Defoamer		2.0	0.26
2-Propoxyethanol		22.7	3.00
CMD 9790 Acrylic Resin		404.0	46.98

Composite Blend:	Viscosity		
Part A	12 seconds #4 Ford Cup	173.5	20.00
Part B	120 KU	870.5	80.00
Parts A and B	90 KU		

2) Ektasolve EP
4) Colloid 640
5) Triton X-100
6) Alcolec 439-C
7) Titanox 2020 or Tronox CR800

Typical Coating Properties:
Total Weight Solids, percent: 51.0
Total Volume Solids, percent: 38.0
PVC, percent: 19.3

Resin Composition:
Epoxy Resin, percent: 24.0
Acrylic Resin, percent: 76.0

Reduction to Spray Viscosity:
by Volume:
Parts A and B   46 parts
Water   1 part

SOURCE: Shell Chemical Co.: 24-309 Gloss White Enamel Using
CMD 9790 Acrylic Resin

Gloss White Enamel
(Maximum Acid Resistance)

Material:	Pounds	Gallons
Part A:		
Epi-Rez WJ-5522 Epoxy Resin	501.2	56.00
High Flash Aromatic Hydrocarbon (2)	22.3	3.00
2-Propoxyethanol (3)	22.7	3.00
Water	38.9	4.67
Part B:		
CMD J60-8290 Curing Agent	81.0	9.17
Defoamer (4)	4.0	0.56
Rutile Titanium Dioxide (5)	250.0	7.74
Water	33.3	4.00
High speed disperse to a texture of 10 P.S.C.		
Water	96.1	11.53
Glacial Acetic Acid	2.9	0.33

2) Hi-Sol 15 or Enco Aromatic 150
3) Ektasolve EP
4) Patcote 847
5) R-960 High Gloss, organic treatment

Composite Blend:	Viscosity		
Part A	97 K.U.	585.1	66.67
Part B	75 K.U.	467.3	33.33

Reduction for Conventional Spray:	by volume
Parts A and B	25 parts
Water	1 part

Typical Coating Properties:
    Total Weight Solids, percent: 54.6
    Total Volume Solids, percent: 41.1
    PVC, percent: 18.8
    Volatile Organic Compounds (V.O.C.):
        Pounds/Gallon: 2.07
        Grams/Liter: 249

    Induction Time, minutes: 30

Resin Composition (based on solids):	
Epoxy Resin, percent	85.0
Curing Agent, percent	15.0

SOURCE: Shell Chemical Co.: 24-296 Gloss White Enamel

Clear Enamel
(Maximum Acid Resistance)

Material:	Pounds	Gallons
Part A:		
Epi-Rez WJ-5522 Epoxy Resin	540.8	60.42
High Flash Aromatic Hydrocarbon (2)	32.5	4.37
2-Propoxyethanol (3)	33.2	4.37
Water	48.6	5.84
Part B:		
CMD J60-8290 Curing Agent	87.6	9.92
Defoamer (4)	4.0	0.56
Water	121.0	14.52

Composite Blend:	Viscosity		
Part A	93 K.U.	655.1	75.00
Part B	61 K.U.	212.6	25.00
Parts A & B	75 K.U.		

2) Hi-Sol 15 or Enco Aromatic 150
3) Ektasolve EP
4) Patcote 847 Defoamer

Typical Properties:
    Total Weight Solids, percent: 40.3
    Total Volume Solids, percent: 36.1
    Volatile Organic Compounds:
        Pounds/Gallon: 2.57
        Grams/Liter: 308

    Induction Time, minutes: 30

Resin Composition (based on solids):
Epoxy Resin, percent:	85.0
Curing Agent, percent	15.0

Reduction for Conventional Spray (by volume)
    None Required

SOURCE: Shell Chemical Co.: 24-294 Clear Enamel

Clear Enamel
(Maximum Solvent and Abrasion Resistance)

Material:	Pounds	Gallons
Part A:		
Epi-Rez WJ-5522 Epoxy Resin	477.2	53.31
High Flash Aromatic Hydrocarbon (2)	32.5	4.37
2-Propoxyethanol (3)	33.2	4.37
Water	107.9	12.95
Part B:		
CMD J60-8290 Curing Agent	145.8	16.51
Defoamer (4)	4.0	0.56
Water	66.1	7.93

2) Hi-Sol 15 or Enco Aromatic 150
3) Ektasolve EP
4) Patcote 847

Composite Blend:		
Part A	650.8	75.00
Part B	215.9	25.00

Typical Coating Properties:
    Viscosity, Part A: 68 KU
            Part B: 64 KU
            Part A + B: 75 KU
    Total Weight Solids, percent: 40.4
    Total Volume Solids, percent: 36.0
    Volatile Organic Compounds (VOC):
        Pounds/Gallon: 2.84
        Grams/Liter: 340
    Induction Time, minutes: 30

Resin Composition (based on solids):
    Epoxy Resin, percent: 75.0
    Curing Agent, percent: 25.0

Reduction for Conventional Spray (by volume):
    None required

SOURCE: Shell Chemical Co.: 24-295 Clear Enamel

Economical Flat Black Air Dry Enamel

	Pounds	Gallons
Cargill Water Reducible Styrenated Alkyd		
Copolymer 74-7425	111.3	12.87
ASP 170 Aluminum Silicate	134.8	6.27
Special Black 4	7.7	0.52
Aerosil R972	5.7	0.31
Triethylamine	6.7	1.11
M-P-A 1075	2.8	0.40
Byk-080	1.1	0.12
sec-Butanol	12.0	1.79
Deionized Water	226.4	27.18

Sand Mill to a 5 Hegman

Premix the following, then add:		
sec-Butanol	1.5	0.22
5% Cobalt Hydro Cure	0.9	0.12
Manganese Hydro Cure II	0.4	0.05
Activ-8	0.2	0.03
Triethylamine	9.3	1.54
Exkin No. 2	2.6	0.34
Byk-080	1.1	0.12
sec-Butanol	52.3	7.77
Acrysol ASE-60	31.2	3.42
Deionized Water	31.2	3.75
Deionized Water	267.0	32.07

Paint Properties:
   % Nonvolatile (By Weight): 26.39
            (By Volume): 16.76
   Pigment to Binder Ratio: 1.63
   Pigment Volume Concentration: 42.46
   Weight Per Gallon: 9.07
   Theoretical VOC (Pounds/Gallon): 3.50
              (Grams/Liter): 419
   Water:Cosolvent Weight Ratio: 82.00:18.00
   pH: 8.0-8.5
   Viscosity #4 Ford Cup (Seconds): 57
   Anti-Sag Resistance: Excellent
Typical Cured Film Properties:
   Cure Schedule:¦ 7 Days Air Dry
   Substrate: Cold Rolled Steel
   Gloss (60/20): 5/0
   Pencil Hardness: Pass 4B
   % Crosshatch Adhesion: 50
   Impact (In. Lbs.):
     Direct: 30
     Reverse: <5

SOURCE: Cargill, Inc.: CARGILL Formulary: Formula P1887-58A

Gloss Black Enamel

Sand Mill (30 minutes):	Parts by Weight
Cymel 303	135.6
Tamol 731 (25%)	1.2
Butyl Cellosolve	21.0
Isopropanol	21.0
Water	84.0
DMAE (20% in water)	1.5
Raven 1035	35.7

Letdown:	
Sand mill grind	150.3
Acrysol WS-68 (38%)	715.5
PTSA* (10%)	6.8
L-5310 (30% in Butyl Cellosolve)	2.2
n-Butanol	5.2

Enamel Constants:
   Total solids, %: 40.7
   Pigment/binder ratio: 5/95
   Acrylic/melamine ratio: 80/20
   % Catalyst (PTSA) on binder solids: 0.2
   % L-5310 on binder solids: 0.2
   Water/organic ratio:
      By volume: 89.8/10.2
      By weight: 91.1/8.9
   pH: 7.6
   Viscosity, sec. (No. 4 Ford Cup): 23
   Volatile Organic Compounds (VOC): 130 gms./l.

   * 10 PTSA/10 DMAE/80 Water

SOURCE: Rohm and Haas Co.: Industrial Coatings: ACRYSOL WS-68:
        Formulation WS-68-3

Gloss Black Enamel

Materials:	Pounds	Gallons
Grind Portion:		
Kelsol 3960	89.2	10.36
Aquacat	1.9	0.23
Magnacat	3.7	0.48
28% Ammonia	2.2	0.26
Water	142.5	17.18
Patcote 519	0.9	0.12
Patcote 577	0.9	0.10
Raven 420	10.0	0.68
Strontium Chromate	10.0	0.32

Steel ball mill to 7 to 7 1/2 N.S.

Kelsol 3960	)		149.0	17.30
28% Ammonia	)	Premix	3.7	0.44
Water	)		176.4	21.25

Add in order with good agitation:

Rhoplex WL-71	185.0	21.52
Water	31.8	3.83
Butyl Carbitol	21.2	2.67
Water	21.2	2.56
14% Ammonia (to pH 7.3 to 7.5)	----	-----

Note: Reduce for spray with 87/13 water/Butyl Cellosolve

Formulation Constants:
   Pounds/gallon: 8.56
   Non-volatiles (Wt.): 33.1%
                  (Vol.): 29.1%
   Pigment/Binder Ratio: 8/92
   PVC: 3.5%
   Rhoplex WL-71/Kelsol 3960: 30/70
   pH: 7.3 to 7.5
   Viscosity: 400-600 cps.
   VOC: 1.87 lbs./gal.

SOURCE: Rohm and Haas Co.: Industrial Coatings: RHOPLEX WL-71:
        Formulation WL-71-9

Gloss Black Enamel
(No Chromate)

Materials:	Pounds	Gallons
Grind Portion:		
Kelsol 3960	89.2	10.36
Aquacat	1.9	0.23
Magnacat	3.7	0.48
28% Ammonia	2.2	0.26
Water	142.5	17.18
Patcote 519	0.9	0.12
Patcote 577	0.9	0.10
Raven 420	13.3	0.91

Steel ball mill to 7 to 7 1/2 N.S.

		Pounds	Gallons
Kelsol 3960	)	149.0	17.30
28% Ammonia	) Premix	3.7	0.44
Water	)	176.4	21.25

Add in order with good agitation:

	Pounds	Gallons
Rhoplex WL-71	185.0	21.52
Water	31.8	3.83
Butyl Carbitol	21.2	2.67
14% Ammonia (to pH 7.3 to 7.5)	-----	-----

Note: Reduce for spray with 87/13 water/Butyl Cellosolve

Formulation Constants:
    Pounds/gallon: 8.5
    Non-volatiles (Wt.): 33.4%
                 (Vol.): 30.1%
    Pigment/Binder: 5/95
    PVC: 3.2%
    Rhoplex WL-71/Kelsol 3960: 30/70
    pH: 7.3 to 7.5
    Viscosity: 200-600 cps.
    Grind: 7 to 7 1/2 N.S.
    VOC: 1.87 lbs./gal.

SOURCE: Rohm and Haas Co.: Industrial Coatings: RHOPLEX WL-71:
        Formulation WL-71-10

Gloss Black Forced-Dry Enamel

Materials:	Pounds	Gallons
Grind Portion:		
Kelsol 3905	89.2	10.36
Aquacat	1.9	0.23
Magnacat	3.7	0.48
28% Ammonia	2.2	0.26
Water	142.5	17.18
Patcote 519	0.9	0.12
Patcote 577	0.9	0.10
Raven 420	13.3	0.91
Steel ball mill to 7 1/2 N.S.		
Kelsol 3905)	149.0	17.30
28% Ammonia) Premix	3.7	0.44
Water)	176.4	21.25

(% Neutralization = 70%, pH 6.0 to 6.3)

Add in order with good agitation:		
Rhoplex WL-71	185.0	21.52
Water	31.8	3.83
Butyl Carbitol	21.2	2.67
14% Ammonia (to pH 7.3 to 7.5)	----	-----
Water	21.2	2.56

Note: Reduce for spray with 87/13 water/Butyl Cellolsolve

Formulation Constants:
    Pounds/gallon: 8.5
    Non-volatiles (Wt.): 32.3%
              (Vol.): 29.1%
    Pigment/Binder Ratio: 5/95
    PVC: 3.2
    Rhoplex WL-71/Kelsol 3905: 30/70
    pH: 7.3 to 7.5
    Viscosity: 200-400 cps.
    VOC: 1.87 lbs./gal.

Source: Rohm and Haas Co.: Industrial Coatings: RHOPLEX WL-71:
      Formulation WL-71-2

Gloss Green Enamel

Materials:	Pounds	Gallons
Grind Portion:		
Chempol 10-0091	88.2	10.02
Aquacat	1.0	0.23
Magnacat	3.7	0.48
28% Ammonia	1.7	0.20
Water	142.3	17.17
Patcote 519	0.9	0.12
Patcote 577	0.9	0.12
Chrome Oxide X-1134	38.3	0.87
Imperial Brazil Yellow X-286C	6.3	0.53
Butyl Cellosolve	5.0	0.67
Butyl Carbitol	5.0	0.63

Steel ball mill to 7 1/2 N.S.

Chempol 10-0091)	147.4	17.14
28% Ammonia) Premix	3.1	0.37
Water)	174.5	21.02

(Neutralization=93%, pH 6.1 to 6.4)
Add in order with good agitation:

Rhoplex WL-71	183.1	21.29
Water	31.5	3.79
Butyl Carbitol	16.0	2.01
14% Ammonia (to pH 7.3 to 7.5)	-----	-----
Water	21.0	2.53

Note: Reduce for spray with 87/13 water/Butyl Cellosolve.

Formulation Constants:
    Pounds/gallon: 8.68
    Non-volatiles (Wt.): 34.5%
                  (Vol.): 28.0%
    Pigment/Binder Ratio: 15/85
    PVC: 4.8%
    Rhoplex WL-71/Chempol 10-0091: 30/70
    Neutralization (final based on Chempol 10-0091 solids): 170%
    pH: 7.3 to 7.5
    Viscosity: 200-400 cps.
    VOC: 2.16 lbs./gal.

SOURCE: Rohm and Haas Co.: Industrial Coatings: RHOPLEX WL-71:
        Formulation WL-71-8

Gloss Maintenance Enamel--Deep Base
A gloss maintenance enamel with good compatibility with Huls
America 896 Color System.

Ingredients:	Pounds	Gallons
Pigment Grind:		
Water	25.0	3.00
Methyl Carbitol Solvent	34.0	4.00
Butyl Carbitol Solvent	23.8	3.00
Dispersant (1)	4.4	0.50
Preservative (2)	1.2	0.13
Nonionic Surfactant (3)	4.7	0.50
Antifoam (4)	2.1	0.25
Urethane Associative Thickener (5)	12.9	1.50
Titanium Dioxide (6)	125.0	3.65
Let Down:		
Water	109.6	13.16
Ucar Thickener SCT-275	21.5	2.50
Ucar Filmer IBT	31.7	4.00
Antifoam (3)	2.1	0.25
Mix:		
Sodium Nitrite	1.0	0.06
Water	4.2	0.50
Ucar Latex 430 (43.0%)	548.1	63.00

Suppliers:
(1) Dispersant - "Tamol" 165 or equivalent
(2) Preservative - "Nuosept" 95 or equivalent
(3) Nonionic Surfactant - "Borchigen" DFN or equivalent
(4) Antifoam - "Foamex" 1488 or equivalent
(5) Urethane Associative Thickener - "Acrysol" RM-1020 or
                                    equivalent
(6) Titanium Dioxide - "Tronox" CR-800 or equivalent

Paint Properties:
   Pigment Volume Concentration (PVC), %: 12.3
   Total Solids, %:
     by Volume: 29.5
     by Weight: 39.5
   Viscosity (equilibrated):
     Stormer, KU: 89
     ICI, Poise: 1.4
   pH, initial: 9.2
   Weight per Gallon, lb: 9.51
   Freeze-Thaw, 3 cycles: Pass
   Heat Stability, 2 wks at 120F (49C): Pass
   Foam Control: Good
   VOC, g/l: 270.7
Film Properties:
   Contrast Ratio 0.003 in: 0.942
   Reflectance, %: 89.3

SOURCE: Ucar Emulsions Systems: UCAR Latex 430: Formulation
     Suggestion M-2205

Gloss White Enamel

Ball Mill Grind:	Parts by Weight
Rutile titanium dioxide (RCL-9)	480.0
Acrysol WS-68 (38%)	126.3
Water	149.4
Butyl Cellosolve	21.4
n-Butanol	20.0
DMAE	2.9

Letdown:	
Ball mill grind	326.4
Acrysol WS-68 (38%)	567.0
Water	48.2
n-Butanol	6.9
L-5310 (30% in Butyl Cellosolve)	1.9
DMAE (10% in water)*	5.1
Cymel 303	58.7
PTSA (10%)**	5.8

Enamel Constants:
   Total solids, %: 48.0
   Pigment/Binder ratio: 40/60
   Acrylic/Melamine ratio: 80/20
   % Catalyst (PTSA) on binder solids: 0.2
   Water/organic ratio, on volume: 91/9
   pH: 7.7
   Viscosity, sec. (No. 4 Ford Cup): 23
   Volatile Organic Compounds (VOC): 124 gms./l.

   * Add slowly with good agitation to achieve desired viscosity
   ** 10 PTSA/10 DMAE/80 Water

SOURCE: Rohm and Haas Co.: Industrial Coatings: ACRYSOL WS-68:
         Formulation WS-68-1

Gloss White Enamel

Sand MIll (30 minutes):	Parts by Weight
Cymel 303	63.0
Tamol 731 (25%)	4.2
DMAE (20% in water)	5.0
Butyl Cellosolve	10.4
Isopropanol	5.6
Water	101.8
Rutile titanium dioxide (RCL-9)	210.0

Letdown:	
Sand mill grind	373.0
Acrysol WS-68 (38%)	618.4
Water	4.0
DMAE (20% in water)	10.8
L-5310 (30% in Butyl Cellosolve)	2.0
PTSA (10%)*	5.8
n-Butanol	6.0

Enamel Constants:
   Total solids, %: 48.0
   Pigment/binder ratio: 40/60
   Acrylic/melamine ratio: 80/20
   % Catalyst (PTSA) on binder solids: 0.2
   % L-5310 on binder solids: 0.2
   Water/organic ratio:
      By volume: 91.1/8.9
      By weight: 92.2/7.8
   pH: 7.6
   Viscosity, sec. (No. 4 Ford Cup): 25
   Volatile Organic Compounds (VOC): 121 gms./l.

   * 10 PTSA/10 DMAE/80 Water

SOURCE: Rohm and Haas Co.: Industrial Coatings: ACRYSOL WS-68:
        Formulation WS-68-2

Gloss White Enamel

Material:		Pounds	Gallons
Part A:			
Epi-Rez 510 Epoxy Resin		125.8	13.04
Heloxy 8 Epoxy Functional Modifier		42.0	5.68
Isopropyl Alcohol		8.4	1.28
Part B:			
CMD WJ60-8537 Curing Agent		160.0	17.78
Defoamer (2)		4.0	0.50
Rutile Titanium Dioxide (3)		250.0	7.31
Disperse to a texture of 9-10 P.C.S., then add at reduced speed:			
CMD WJ60-8537 Curing Agent		79.0	8.78
Defoamer		4.0	0.50
Water		375.9	45.13
Composite Blend:	Viscosity		
Part A	55 KU	176.2	20.00
Part B	78 KU	872.9	80.00
Parts A and B	93 KU		

2) Colloid 640
3) Titanox 2020 or Tronox CR-800

Typical Coating Properties:
    Total Weight Solids, Percent: 53.5
    Total Volume Solids, Percent: 40.7
    PVC, Percent: 17.9

Resin Composition (based on solids):
    Epoxy Resin, Percent: 54.0
    Curing Agent, Percent: 46.0

SOURCE: Shell Chemical Co.: 24-315 Gloss White Enamel

Gloss White Enamel for Electrostatic Disc Spray

Sand Mill (30 minutes):	Parts by Weight
Cymel 303	94.5
Tamol 731 (25%)	6.0
Butyl Cellosolve	12.0
Isopropanol	7.5
Water	130.0
DMAE (10% in water)	15.0
Rutile titanium dioxide (RCL-9)	315.0

Letdown:	
Sand mill grind	250.1
Acrysol WS-68 (38%)	428.9
Water	68.6
Dipropylene glycol	75.0
L-5310 (30% in Butyl Cellosolve)	1.4
DMAE (10% in water)	132.0
PTSA* (10%)	4.0
n-Butanol	10.0

Enamel Constants:
    Total solids, %: 35.0
    Pigment/binder ratio: 40/60
    Acrylic/melamine ratio: 80/20
    % Catalyst (PTSA) on binder solids: 0.2
    % L-5310 on binder solids: 0.2
    Water/organic ratio:
        By volume: 80.4/19.6
        by weight: 81.1/18.9
    pH: 8.4
    Viscosity, sec. (No. 4 Ford Cup): 45
        (Use more DMAE to increase spray viscosity; use less DMAE
        to decrease spray viscosity.)
    Volatile Organic Compounds (VOC): 372 gms./l.

    * 10 PTSA/10 DMAE/80 Water

SOURCE: Rohm and Haas Co.: Industrial Coatings: ACRYSOL WS-68:
        Formulation WS-68-4

Gloss White Enamel for Electrostatic Spray Application

Cowles Grind:	Parts by Weight
Acrysol WS-68 (38%)	12.4
Water	26.5
DMAE (10% in water)	3.3
Tamol 731 (25%)	2.7
Triton X-114	2.0
n-Butanol	4.1

Mix the above ingredients, then add slowly:

Rutile titanium dioxide (RCL-9)	134.5

Mix the grind for 20 minutes, then let down under good agitation
as follows:

Letdown:	
Water) premix	108.1
Butyl Cellosolve)	4.7
n-Butanol	6.3
Acrysol WS-68 (38%)	414.0
L-5310 (30% in Butyl Cellosolve)	1.4
Cymel 303	40.4
Dipropylene glycol	74.7
DMAE (10% in water)	131.8
PTSA* (10%)	4.0

Physical Constants:
Total solids, %: 35.0
Pigment/binder ratio: 40/60
Acrylic/melamine ratio: 80/20
% Catalyst (PTSA) on binder solids: 0.2
Water/organic ratio, by volume: 81.4/18.6
pH: 8.4
Viscosity, sec. (No. 4 Ford Cup): 43
Volatile Organic Compounds (VOC): 362 gms./l.

* 10 PTSA/10 DMAE/80 Water

SOURCE: Rohm and Haas Co.: Industrial Coatings: ACRYSOL WS-68:
Formulation WS-68-5

Gloss Yellow Enamel

Materials:	Pounds	Gallons
Grind Portion:		
Chempol 10-0091	88.2	10.02
Aquacat	1.9	0.23
Magnacat	3.7	0.48
28% Ammonia	1.7	0.20
Water	142.3	17.17
Patcote 519	0.9	0.12
Patcote 577	0.9	0.12
Krolor Yellow KY-7810	58.3	1.77
Yellow Iron Oxide	3.8	0.11
Butyl Cellosolve	5.0	0.67
Butyl Carbitol	5.0	0.63

Steel ball mill to 7 1/2 N.S.

Chempol 10-0091	)	147.4	17.14
28% Ammonia	) Premix	3.1	0.37
Water	)	174.5	21.02

(Neutralization=93%, pH 6.1 to 6.4)
Add in order with good agitation:

	Pounds	Gallons
Rhoplex WL-71	183.1	21.29
Water	31.5	3.79
Butyl Carbitol	16.0	2.01
14% Ammonia (to pH 7.3 to 7.5)	----	----
Water	21.0	2.53

Note: Reduce for spray with 87/13 water/butyl Cellosolve.

Formulation Constants:
    Pounds/gallon: 8.82
    Non-volatiles (Wt.): 35.7%
                 (Vol.): 32.0%
    Pigment/Binder Ratio: 20/80
    PVC: 6.3%
    Rhoplex WL-71/Chempol 10-0091: 30/70
    Neutralization (final based on Chempol 10-0091 solids): 170%
    pH: 7.3 to 7.5
    Viscosity: 200-400 cps.
    VOC: 2.14 lbs./gal.

SOURCE: Rohm and Haas Co.: Industrial Coatings: RHOPLEX WL-71:
        Formulation WL-71-3

Gray Deck Enamel

Ingredient:	Lbs/100 Gal
Propyl Carbitol	27.2
Methyl Propasol	27.2
Diethylamine	8.2
Drew L-475	1.4
Santicizer 160	5.4
DSX 1514	20.4
Water	36.7

Mix well, then add:
Pliolite 7104	91.1

Mix well, then add:
Titanox 2160	197.2
Minusil 5	204.0
Aquasperse Brown Oxide Paste	16.3
Liquakote Yellow Oxide	16.3
Liquakote Black	4.1

HSD to 5 N.S. then drop to fully assembled letdown with mixing:

Let Down Phase:
Pliolite 7104	360.4
Butyl Dipropasol	20.4
Santicizer 160	15.0

Prepare the following premix; then add to letdown:

Water	74.8
A-187	2.7

    VOC: 221 g/l or 1.84 lb/gal
    Density: 11.3 lb/gallon
    Oven Stability: Pass 4 weeks at 120 degrees F
    Freeze Thaw: Pass 5 cycles
    KU Viscosity: 100-110
    NVV: 42%
    PVC: 43%

SOURCE: The Goodyear Tire & Rubber Co.: Formula #2065-106

Gray Low Gloss Enamel

Material:	Lbs.	Gals.
Bentone EW Solution 2%	110.67	13.02
Kelsol 3961-B2G-75	154.09	17.51
Ammonium Hydroxide	13.89	1.86
Butoxy Ethanol	17.20	2.29
n-Butanol	17.21	2.55
Titanox 2101	116.55	3.50
106 Lo Micron Barytes	187.25	5.13
OK 412 Silica	9.81	0.54
Zinc Phosphate J0852	92.58	3.42
Alcophor 827	10.19	0.49
Raven 1255	4.37	0.30
Water	204.92	24.60

Steel ball mill to 6-7 Hegman grind.

Letdown:		
Kelsol 3961-B2G-75	50.60	5.75
Ammonium Hydroxide	6.20	0.83

Premix the following and add to above:		
Cobalt Hydrocure II	3.10	0.41
Manganese Hydrocure II	6.38	0.85
Butoxy Ethanol	8.79	1.17
n-Butanol	6.55	0.97
Water	123.37	14.81

Analysis:
    Pigment Volume Concentration, Percent: 45.4
    Pigment/Binder Ratio: 2.74/1
    Percent Solids, Weight: 49.7
    Percent Solids, Volume: 24.4
    Viscosity @ 25C, #4 Ford, Secs.: 35-45
    Pounds/Gallon: 11.44
    pH: 8.3-8.6
    VOC (excluding water):
        Grams/Liter: 275
        Pounds/Gallon: 2.29

SOURCE: Reichhold Chemicals, Inc.: Waterborne Handbook:
        Formulation 2419-22

Green Chrome Free Gloss Enamel

Material:	Lbs.	Gals.
Arolon 557-B-70	119.73	14.42
Ethylene Glycol Monobutyl Ether	9.21	1.23
Triton X-405	1.84	.20
Byk 301	1.84	.23
Dimethylethanol Amine (DMEA)	9.21	1.24
Water	92.10	11.06
Y-5775 Fast Yellow	27.63	1.92
A-4434 Tacoma Blue	1.84	.15
Titanox 2020	92.10	2.70

Pebble mill or sand mill to 7 Hegman grind.

Letdown:		
Arolon 557-B-70	122.49	14.76
Cymel 350	73.68	7.37
Tributyl Phosphate	9.21	1.02
Ethylene Glycol Monobutyl Ether	27.63	3.68
DMEA	7.37	1.00
Water	352.10	39.02

Analysis:
    Pigment Volume Concentration, Percent: 15.56
    Pigment/Binder Ratio: .5/1
    Percent Solids, Weight: 39.71
    Percent Solids, Volume: 30.61
    Viscosity @ 25C, KU: 80-85
    Pounds/Gallon: 9.21
    pH: 8.2-8.6
    VOC (excluding water):
        Grams/Liter: 328.9
        Pounds/Gallon: 2.74
    Resin/Melamine Ratio: 70/30
    % Reduction to 30 Secs., #2 Zahn, with water: 30

SOURCE: Reichhold Chemicals, Inc.: Waterborne Handbook:
        Formulation 2185-63

Green Enamel

Material:	Lbs.	Gals.
Water	184.11	22.10
2-Butoxy Ethanol	21.05	2.80
n-Butyl Alcohol	21.06	3.12
Ammonium Hydroxide	6.98	0.93
Kelsol 3990-B2G-70	127.15	15.05
Byk P104S	1.27	0.16
Titanox 2101	29.30	0.88
YIO-2288D	29.30	0.87
Sunfast Green 264-8142	12.94	0.74
Dalamar Yellow YT820D	11.86	1.01

Steel ball mill to a 7 Hegman grind.

Letdown:		
Water	313.16	37.59
Ektasolve EP	18.06	2.38
Ammonium Hydroxide	5.64	0.76
Kelsol 3990-B2G-70	84.79	10.03

Premix:		
Cobalt Hydrocure II	7.29	0.95
Activ-8	1.71	0.21
Ektasolve EP	3.00	0.39

Analysis:
    Pigment Volume Concentration, Percent: 17.08
    Pigment/Binder Ratio: 0.55/1
    Percent Solids, Weight: 26.83
    Percent Solids, Volume: 20.53
    Viscosity @ 25C, #4 Ford, Secs.: 35-45
    Pounds/Gallon: 8.78
    pH: 8.2-8.5
    VOC (excluding water):
        Grams/Liter: 401
        Pounds/Gallon: 3.34

SOURCE: Reichhold Chemicals, Inc.: Waterborne Handbook:
          Formulation 2578-50

Green Fast Dry Gloss Enamel

Material:	Lbs.	Gals.
Kelsol 3960-B2G-75	76.71	8.88
Cobalt Hydrocure II	3.02	0.39
Manganese Hydrocure II	3.78	0.50
Butoxy Ethanol	13.69	1.82
Ammonium Hydroxide	3.35	0.45

Add above materials under agitation.

Yellow Iron Oxide 2288D	76.60	2.29
Dalamar Yellow YT820D	14.36	1.22
Palomar Green G5406	15.71	0.67
Water	168.03	20.17

Pebble mill to 7 Hegman grind.

Letdown:		
Kelsol 3960-B2G-75	324.31	37.54
Ammonium Hydroxide	7.14	0.96
Normal Butanol	23.52	3.48
Secondary Butanol	25.19	3.74
Butoxy Ethanol	42.01	5.59
Water	78.97	9.48

Premix above materials and add to mill base under agitation.

Arolon 840-W-39	108.89	12.52
Water	16.79	2.02
Butoxy Ethanol	4.19	0.56

Premix above materials and add to batch under agitation. Adjust pH to 8.0-8.5 with ammonium hydroxide.

Analysis:
    Pigment Volume Concentration, Percent: 10.2
    Pigment/Binder Ratio: .35/1
    Percent Solids, Weight: 40.8
    Percent Solids, Volume: 36.8
    Pounds/Gallon: 8.9
    pH: 8.0-8.5
    VOC (excluding water):
        Grams/Liter: 355
        Pounds/Gallon: 2.96

SOURCE: Reichhold Chemicals, Inc.: Waterborne Handbook:
        Formulation 2175-76C

High Gloss Ford Blue Enamel

Ingredients:	Pounds	Gallons
Chempol 10-0091	297.94	33.86
Butyl Cellosolve	22.35	2.98
28% NH3	8.94	1.20
6% Calcium Drier	4.47	0.55
Coroc A2678M	4.47	0.51
Uvinul N539	4.47	0.51
Byk 035	1.12	0.15
Tiona RCL-628	49.62	1.49
Phthalo Blue	24.81	1.78
Deionized Water	44.70	5.37

Pebble or sandmill to Hegman 7+, then let down with:

Butyl Cellosolve	11.18	1.49
5% Cobalt Hydrocure II	2.46	0.32
12% Zirconium Hydrocem	3.69	0.44
Activ-8	0.45	0.05
Exkin #2	1.12	0.15

Mix well and continue let down with:

Butyl Cellosolve	11.18	1.49
Deionized Water	55.88	6.72

Premix the following and add slowly with rapid mixing:

Acrysol ASE60	1.12	0.13
Deionized Water	55.88	6.72

Adjust pH to 8.0-8.5 and continue let down with:

Deionized Water	279.04	33.47

Properties:
  % Weight Solids: 35.06
  % Volume Solids: 28.38
  P/B Weight Ratio: 0.33
  % PVC: 11.5
  VOC, #/gallon, calculated: 2.78 minus water
  Viscosity, # 4 Ford Cup*: 33-36 seconds
  * Volume reduction 100 parts 10-0091-H with 18-20 parts of
    deionized water.

Film Properties:
  Spray applied over B1000 CRS and air dried two weeks before
testing.
  Dry film thickness, mils: 1.3
  Pencil hardness: 2B
  Crosshatch adhesion: excellent - no tape off
  60/20 gloss: 89/76
  MEK double rubs to metal: 45
  Direct impact, inch-lbs.: >80

SOURCE: Freeman Polymers Division: CHEMPOL 10-0091-H:
        Suggested Starting Formulation

High Gloss Interior Enamel

Material:	Lbs.	Gals.
Water	55.5	6.66
Texanol	10.5	1.33
Colloid 261	9.6	1.00
Igepal CO-630	1.6	0.18
Foamaster R	0.5	0.07
Nuosept 95	0.8	0.08
AMP 95	1.0	0.13
Tronox CR-800	219.0	6.40
High speed disperser, 15 minutes maximum.		
Water	46.8	5.62
Synthemul 40-424	600.0	69.77
Texanol	10.0	1.27
n-Butyl Propionate	50.5	6.93
FC-120	0.3	0.03
RM-825, 10%	4.5	0.53

Analysis:
   Pigment Volume Concentration, Percent: 19.4
   Percent Solids, Weight: 45.4
   Percent Solids, Volume: 33.0
   Initial Viscosity, KU: 96.0
   Pounds/Gallon: 10.11
   pH: 6.8
   VOC (excluding water):
      Grams/Liter: 197
   ICI, Poise: 1.6
   Gloss @ 60/20: 83/65
   Flow and Leveling: Exc.
   Viscosity, 7 Days @ 135F: 91

SOURCE: Reichhold Chemicals, Inc.: Waterborne Handbook:
         Formulation 424-01

High-hiding Eggshell Enamel

Ingredients:	Pounds	Gallons
Pigment Grind:		
Water	226.7	27.21
Cellosize Hydroxyethyl Cellulose ER-4400	2.0	0.18
Preservative (1)	2.3	0.25
Dispersant (2)	9.2	1.00
Triton Nonionic Surfactant N-57	2.3	0.25
2-Amino-2-methyl-1-propanol (3)	1.0	0.13
Propylene Glycol	51.8	6.00
Antifoam (4)	1.9	0.25
Titanium Dioxide (5)	225.0	6.54
Thermo-optic Clay (6)	100.0	5.46
Hydrous Clay (7)	25.0	1.15
Letdown:		
Ucar Filmer IBT	11.9	1.50
Butyl Carbitol Solvent	11.9	1.50
Ucar Thickener SCT-275	17.2	2.00
Ucar Latex 379	418.2	46.21
Antifoam (4)	2.8	0.37

Suppliers:
(1) Preservative - "Nuosept" 145 or equivalent
(2) Dispersant - "Tamol" 731 or equivalent
(3) 2-Amino-2-methyl-1-propanol - AMP-95 or equivalent
(4) Antifoam - "Colloid" 643 or equivalent
(5) Titanium Dioxide - "Tronox" CR-800 or equivalent
(6) Thermo-optic Clay - "Optiwhite" or equivalent
(7) Hydrous Clay - Burgess No. 98 or equivalent

Comparative Properties:
    Pigment Volume Concentration, %: 35.6
    Solids, %:
        by Volume: 37.6
        by Weight: 53.4
    Weight per Gallon, lb: 11.09
Paint Properties:
    Viscosity, initial:
        Stormer, KU: 104
        ICI, poise: 1.4
    pH, initial: 8.2
    Freeze-Thaw Stability, 3 cycles: Pass
    Heat Stability, 2 weeks at 120F: Pass
Film Properties:
    Contrast Ratio, 6-mil bar (1): 0.968
    Reflectance, %: 91.9
    Gloss, 60, 6-mil bar: 8
    Sheen, 85, 6-mil bar: 22
    Flow and Leveling: Brushed: Good
    (1) Opacity increased with toning

SOURCE: Ucar Emulsion Systems: UCAR Latex 379: Formulation
       Suggestion I-1998

High Performance Bake White Enamels

	Parts by weight	
Materials:	PELA	TOFA
Resin WB-3823, 75% NVM	257.0	290.4
Dimethylethanolamine	10.0	11.2
Amino Crosslinker A (1)	48.0	----
Amino Crosslinker B (2)	----	39.4
Titanium dioxide (3)	245.0	258.4
Water	440.0	400.6

Properties:		
Pigment/binder ratio	0.98/1.0	0.98/1.0
Melamine/alkyd ratio	20/80	15/85
Non-volatile material (NVM), wt %	50	52.5
Viscosity, #4 Ford cup, sec	53	55
pH	7.5	7.65

(1) Uformite MM 83
(2) Cymel 301
(3) Ti-Pure R-900

Compounding procedure for the enamels:
1. Solubilize resin in a mixing vessel containing a solution of dimethylethanolamine and deionized water.
2. Charge resin solution, pigment and amino crosslinker a pebble mill. Grind overnight to a Hegman Grind of 7+.
3. After milling, adjust to correct pH and viscosity.

Mechanical Properties of WB-3823 enamels:

	PELA(1)	
Properties	B-37(3)	CRS(4)
Pencil hardness	H	F
Crosshatch adhesion	5B	5B
1/8-inch conical bend, % pass	100	100
Impact resistance, in-lb:		
direct	140	100
reverse	80	60

	TOFA(2)	
	B-37(3)	CRS(4)
Pencil hardness	2H	3H
Crosshatch adhesion	5B	5B
1/8-inch conical bend, % pass	100	100
Impact resistance, in-lb:		
direct	120	140
reverse	80	60

(1) 1.0 to 1.2 mil film thickness, cured 20 minutes at 121C
(2) 1.0 to 1.2 mil film thickness, cured 30 minutes at 177C
(3) Substrate: Bonderite 37 treated cold rolled steel
(4) Substrate: Cold rolled steel

SOURCE: Amoco Chemical Co.: Water-Borne Bake Alkyds Based on AMOCO TMA and IPA: Formula Table 8 and Table 9

High Solids Polyester/Melamine Enamel

Grind:	Weight	Solids
Resin WS-3-2C (85% NV)	258.0	219.3
Titanium Dioxide Pigment	256.3	256.3
Hexamethoxymethyl Melamine	94.0	94.0
Dimethylethanolamine (DMEA)	18.1	-----
Flow Control Additive	3.1	0.3

Let Down:		
Demineralized Water	354.0	-----
n-Butyl Alcohol	16.5	-----

Physical Constants of Enamel:
   Resin/Crosslinker Ratio: 70/30
   Binder/Pigment, wt %: 55/45
   Nonvolatiles, wt %: 57
      Water/Organic, Wt %: 82/18
      pH: 8.0-8.5
   Viscosity, No. 4 Ford Cup, sec: 55-60

Physical Properties of Enamel Coating (b,c):
   Cure Temperature/Time, C/min: 177/20
   Initial Gloss, Gardner, %:
      60: 98
      20: 83
   Impact Resistance, in.-lb (cm-kg):
      Direct: 140 (161)
      Reverse: 90 (104)
   Pencil Hardness: 4H
   Solvent Resistance, MEK double rubs: 200+
   Scribed Adhesion, % pass: 100
   Conical Mandrel Flexibility, ASTM D-522, 1/8 in., % pass: 100
   Stain Resistance: 5-10
   Detergent Resistance (d), 1 1/2% standard detergent solution,
      240 hr, 165F (74C):
      60 gloss retention, %: 100
      blisters, frequency/size: None
   Salt Spray, 500 hr:
      60 gloss retention, %: 100
      Scribe creepage, in. (mm): <1/32
      blisters: None
   Storage Stability:
      Initial: pH/Viscosity, cP: 8.3/320
      After 500 hours at 120F; pH/Viscosity, cP: 7.0/400
b) Unprimed, zinc phosphatized, cold-rolled steel panels coated
   with a 1-1 1/2 mil film were used in these evaluations
c) Test methods are those commonly used in the polymers and
   coatings industry.
d) Coatings were applied over an epoxy primer for detergent
   resistance tests.

SOURCE: Eastman Chemical Products, Inc.: Enamel Prepared from
      Polyester Resin WS-3-2C

Light Heliogen Green Enamel

	Pounds	Gallons
Cargill Water Reducible Silicone Modified Alkyd 74-7435	155.2	17.25
Aqueous Ammonia 28%	6.7	0.89
Ethylene Glycol Monobutyl Ether	13.2	1.76
Byk-301	3.3	0.41
Heliogen Green L8690	6.7	0.36
Titanium Dioxide Ti-Pure R-900	194.0	5.54
Deionized Water	118.7	14.27

Sand Mill to a 7 Hegman

Cargill Water Reducible Silicone Modified Alkyd 74-7435	110.9	12.32
Aqueous Ammonia 28%	6.5	0.87

Premix the following, then add:

Ethylene Glycol Monobutyl Ether	15.9	2.11
Intercar 6% Cobalt	4.0	0.52
Intercar 6% Zirconium	5.0	0.69
Activ-8	1.3	0.17

Deionized Water	288.3	34.58

Deionized Water (Hold for Viscosity Adjustment)	68.8	8.26

Paint Properties:
   % Nonvolatile (By Weight): 40.63
              (By Volume): 27.14
   Pigment to Binder Ratio: 0.98
   Pigment Volume Concentration: 22.00
   Weight per Gallon: 9.98
   Theoretical VOC (Pounds/Gallon): 2.58
              (Grams/Liter): 309
   Water:Cosolvent Weight Ratio: 82.80:18.20
   pH: 8.0-8.5
   Viscosity Krebs-Stormer (KU): 70+-5

Typical Cured Film Properties:
   Cure Schedule: 7 Days Air Dry
   Substrate: Cold Rolled Steel
   Gloss (60/20): 86/60
   Pencil Hardness: Pass F
   % Crosshatch Adhesion: 100
   Impact (In. Lbs.):
     Direct: 30
     Reverse: <5
   Dry Time Linear Drying Recorder (Hours): 2

SOURCE: Cargill, Inc.: CARGILL Formulary: Formula P1843-227

Non-Leafing Aluminum Enamel

Material:	Lbs.	Gals.
Add, mix in order listed:		
Water	463.40	55.63
Ammonium Hydroxide	11.20	1.49
Kelsol 3906-B2G-75	241.00	27.54
Premix, add to above:		
Letdown:		
Cobalt Hydrocure II	5.55	0.72
Activ-8	2.21	0.28
Butoxy Ethanol	11.27	1.50
Manganese Hydrocure	3.04	0.40
Final Letdown:		
Butoxy Ethanol	26.59	3.54
Hydro-Paste 82255	111.25	8.90

Analysis:
   Pigment Volume Concentration, Percent: 15.0
   Pigment/Binder Ratio: .37/1
   Percent Solids, Weight: 30.6
   Percent Solids, Volume: 24.6
   Pounds/Gallon: 8.76
   pH: 8.2-8.5
   VOC (excluding water):
      Grams/Liter: 389
      Pounds/Gallon: 3.24

SOURCE: Reichhold Chemicals, Inc.: Waterborne Handbook:
      Formulation 1652-40

Quality Gloss White Enamel
A quality high-hiding gloss enamel

Ingredients:	Pounds	Gallons
Pigment Grind:		
Water	50.10	6.02
Propylene Glycol	43.30	5.00
Dispersant (1)	9.20	1.00
Nonionic Surfactant (2)	2.20	0.25
Preservative (3)	1.20	0.13
Antifoam (4)	0.25	0.03
Titanium Dioxide (5)	250.00	7.31
Let Down:		
Anionic Surfactant (6)	3.10	0.37
Butyl Carbitol Solvent	19.80	2.50
Ucar Filmer IBT	39.60	5.00
Antifoam (7)	5.60	0.75
Ucar Latex 430 (43.0%)	604.70	69.51
Ucar Thickener SCT-275	6.50	0.75
Premix:		
Anionic Thickener (8)	3.30	0.37
Water	8.30	1.00
Ammonium Hydroxide, 28% Aqueous Solution	0.10	0.01

Suppliers:
(1) Dispersant - "Tamol" 731 or equivalent
(2) Nonionic Surfactant - "Triton" CF-10 or equivalent
(3) Preservative - "Nuosept" 95 or equivalent
(4) Antifoam - "Nalco" 2300 or equivalent
(5) Titanium Dioxide - "Tronox" CR-800 or equivalent
(6) Anionic Surfactant - "Triton" GR-7M or equivalent
(7) Antifoam - "Byk" 035 or equivalent
(8) Anionic Thickener - "Acrysol" RM-5 or equivalent

Paint Properties:
    Pigment Volume Concentration (PVC), %: 20.3
    Total Solids, %:
        by Volume: 35.9
        by Weight: 49.8
    Viscosity (equilibrated):
        Stormer, KU: 86
        ICI, Poise: 1.2
    pH, initial: 9.4
    Weight per Gallon, lb: 10.47
    Freeze-Thaw, 3 cycles: Pass
    Heat Stability, 2 wks at 120F (49C): Pass
    Flow and Leveling, brushed: Good
    Sag Resistance, brushed: Excellent
    Foam Control: Good
    VOC, g/l: 248

SOURCE: Ucar Emulsion Systems: UCAR Latex 430: Formulation
       Suggestion E-2201

Quality Gloss White Enamel

A quality gloss enamel with an outstanding balance of properties

Ingredients:	Pounds	Gallons
Pigment Grind:		
Water	82.90	9.95
Propylene Glycol	34.60	4.00
Cellosize Hydroxyethyl Cellulose ER-15M	0.75	0.07
Dispersant (1)	9.60	1.00
2-Amino-2-methyl-1-propanol (2)	3.00	0.37
Antifoam (3)	2.10	0.25
Preservative (4)	1.20	0.13
Nonionic Surfactant (5)	2.10	0.25
Nonionic Surfactant (6)	7.10	0.75
Titanium Dioxide (7)	225.00	6.58
Let Down:		
Butyl Carbitol Solvent	15.80	2.00
Mix:		
Butyl Carbitol Solvent	5.90	0.75
Ucar Thickener SCT-275	6.50	0.75
Ucar Latex 430 (43%)	590.70	67.90
Ucar Filmer IBT	39.60	5.00
Antifoam (3)	2.10	0.25

Suppliers:
(1) Dispersant - "Colloid" 261 or equivalent
(2) 2-Amino-2-methyl-1-propanol - AMP-95 or equivalent
(3) Antifoam - "Foamex" 1488 or equivalent
(4) Preservative - "Nuosept" 95 or equivalent
(5) Nonionic Surfactant - "Surfynol" 104E or equivalent
(6) Nonionic Surfactant - "Borchigen" DFN or equivalent
(7) Titanium Oxide - "Tronox" CR-800 or equivalent

Paint Properties:
  Pigment Volume Concentration (PVC), %: 19.0
  Total Solids, %:
    by Volume: 34.5
    by Weight: 48.3
  Viscosity (equilibrated):
    Stormer, KU: 97
    ICI, Poise: 1.8
  pH, initial: 9.4
  Weight per Gallon, lb: 10.29
  Freeze-Thaw, 3 cycles: Pass
  Heat Stability, 2 wks at 120F (49C): Pass
  Flow and Leveling, brushed: Good
  Sag Resistance, brushed: Excellent
  Foam Control: Excellent
  VOC, g/l: 245.6
Film Properties:
  Contrast Ratio 0.003 in: 0.973
  Reflectance %: 92.0

SOURCE: Ucar Emulsion Systems: UCAR Latex 430: Formulation
        Suggestion E-2202

Quality Gloss White Enamel
A quality enamel with exceptional application properties.

Ingredients:	Pounds	Gallons
Pigment Grind:		
Water	42.00	5.04
Butyl Carbitol Solvent	47.50	6.00
Propylene Glycol	43.20	5.00
Cellosize Hydroxyethyl Cellulose ER-15M	0.75	0.07
Dispersant (1)	6.10	0.62
Nonionic Surfactant (2)	2.20	0.25
Nonionic Surfactant (3)	2.10	0.25
Antifoam (4)	0.70	0.09
Preservative (5)	1.20	0.13
Titanium Dioxide (6)	225.00	6.58
Let Down:		
Mix:		
Ucar Thickener SCT-270	13.00	1.50
Butyl Carbitol Solvent	11.90	1.50
Ucar Latex 430 (43.0%)	600.00	68.97
Plasticizer (7)	23.60	3.00
Antifoam (8)	7.50	1.00

Suppliers:
(1) Dispersant - "Tamol" 1124 or equivalent
(2) Nonionic Surfactant - "Triton" N-57 or equivalent
(3) Nonionic Surfactant - "Surfynol" 104E or equivalent
(4) Antifoam - "Colloid" X-0137-MC-33 or equivalent
(5) Preservative - "Nuosept" 95 or equivalent
(6) Titanium Dioxide - "Tronox" CR-800 or equivalent
(7) Plasticizer - "Kodaflex" TXIB or equivalent
(8) Antifoam - "Colloid" 640 or equivalent

Paint Properties:
   Pigment Volume Concentration (PVC), %: 18.8
   Total Solids, %:
      by Volume: 34.9
      by Weight: 48.2
   Viscosity (equilibrated):
      Stormer, KU: 86
      ICI, Poise: 1.2
   pH, initial: 8.7
   Weight per Gallon, lb: 10.27
   Freeze-Thaw, 3 cycles: Pass
   Heat Stability, 2 wks at 120F (49C): Pass
   Flow and Leveling, brushed: Good
   Sag Resistance, brushed: Excellent
   Foam Control: Excellent
   VOC, g/l-plasticizer: 226.4

SOURCE: Ucar Emulsion Systems: UCAR Latex 430: Formulation
      Suggestion E-2203

Red Acrylic Emulsion Enamel

Ingredients:	Pounds	Gallons
Coroc A-2678-M	45.81	5.21
2-butoxyethanol	7.71	1.03
Surfynol 104BC	8.05	1.07
Surfynol DF-75	1.32	0.16
Water	30.27	3.63
Mapico Red #347 Iron Oxide	20.55	0.48
13-3060 Novoperm A20 Red	30.83	2.73

Pre-mix first four ingredients, then add water with agitation. Load mixture into ball mill followed by pigments.

Porcelain ball grind to a #7 Hegman. Adjust grind and/or wash mill with water from below if necessary.

Let down with:		
Chempol 20-4301	572.76	65.09
2-butoxyethanol	31.96	4.26
Butyl Carbitol	20.44	2.57
Dibutyl Phthalate	7.06	0.81
Nuocure CK-10	2.01	0.25
Michem 39235	30.18	3.62
Water (hold to adjust viscosity)	75.72	9.09

Pre-mix co-solvents, plasticizers and drier. Add this mixture to the vortex of the agitating emulsion <u>VERY SLOWLY.</u> If the mixture becomes heavy, add a small amount of water from the "adjust viscosity." Check mixture for kick-out or seeding. Add the wax dispersion to the "let down" with agitation. After mixing well, add the grind to the "let down" with good agitation. Mix well and allow the paint to set overnight.

Check the pH and adjust with ammonia if necessary. Avoid pH values above 7.6, as this will cause the viscosity of the paint to increase significantly. Use the "adjust viscosity" water to bring the viscosity of the coating to spec at room temperature.

Properties:
    Total solids, % by weight: 34.42%
    Total solids, % by volume: 29.42%
    Weight/gallon: 8.85 lbs./gal.
    PVC: 10.90
    Pigment/binder ratio: 0.20
    VOC: 2.04 lbs./gal.
    Viscosity: 38" #2 Zahn
    pH: 7.6
    Substrate: B-1000 CRS
    Cure: 1 week air dry
    Set to touch: 17'
    Dust free: 20'
    Tack free: 75'-90'
    Dry through: 75'-90'

SOURCE: Freeman Polymers Division: CHEMPOL 20-4301-D Formulation

Red Enamel

Material:	Lbs.	Gals.
Water	302.47	36.31
2-Butoxy Ethanol	5.33	0.71
n-Butyl Alcohol	8.33	1.23
Ektasolve EP	8.33	1.10
Ammonium Hydroxide	8.36	1.12
Kelsol 3990-B2G-70	140.04	16.57
Byk P104S	1.37	0.17
Novaperm Red F3RK-70	40.76	3.79
Copperas Red R-7098	8.43	0.20

Steel ball mill to a 7 Hegman grind.

Letdown:		
Water	212.52	25.51
Ammonium Hydroxide	3.49	0.47
Kelsol 3990-B2G-70	93.38	11.05

Premix:		
Cobalt Hydrocure II	8.97	1.17
Activ-8	1.51	0.19
2-Butoxy Ethanol	3.00	0.40

Analysis:
   Pigment Volume Concentration, Percent: 17.51
   Pigment/Binder Ratio: 0.29/1
   Percent Solids, Weight: 25.67
   Percent Solids, Volume: 22.78
   Viscosity @ 25C, #4 Ford, Secs.: 35-45
   pH: 8.2-8.5
   Pounds/Gallon: 8.46
   VOC (excluding water):
      Grams/Liter: 335
      Pounds/Gallon: 2.79

SOURCE: Reichhold Chemicals, Inc.: Waterborne Handbook:
        Formulation 2578-49

Red Gloss Acrylic Enamel

Material:	Lbs.	Gals.
Arolon 559-G4-70	73.42	8.41
Butoxy Ethanol	18.38	2.45
Dimethylethanol Amine (DMEA)	5.48	.74
Water	73.39	8.81
Disperbyk	1.44	.16
X-1843 Madras Orange	4.10	3.98
13-3690 Novoperm Bordeaux HF3R	22.00	1.90

Pebble or sand mill to 7+ Hegman Grind.

Letdown:		
Arolon 559-G4-70	109.93	21.87
Resimene 740	100.16	10.22
Butoxy Ethanol	14.25	1.90
n-Butanol	17.62	2.61
DMEA	9.92	1.34
Water	296.63	35.61

Analysis:
    Pigment Volume Concentration, Percent: 17.3
    Pigment/Binder Ratio: .24/1
    Percent Solids, Weight: 39.10
    Percent Solids, Volume: 34.02
    Viscosity @ 25C, KU: 70-75
    Pounds/Gallon: 8.68
    % Reduction to 30 Secs., #2 Zahn, with water: 30
    pH: 8.2-8.6
    VOC (excluding water):
        Grams/Liter: 338
        Pounds/Gallon: 2.82

SOURCE: Reichhold Chemicals, Inc.: Waterborne Handbook:
        Formulation 2185-83

Red Gloss Enamel

Material:	Lbs.	Gals.
Water	157.82	18.95
Ammonium Hydroxide	10.52	1.40
Kelsol 3910-B2G-75	134.67	15.37
Butoxy Ethanol	21.04	2.81
Cobalt Hydrocure II	6.24	0.83
Activ-8	1.04	0.13
Novaperm F5RK	14.03	1.24
Novaperm F3RK70	15.43	1.44
Oswego Orange X-2065	56.81	4.77
TiPure R-902	47.70	1.43

Ball mill to 6-7 Hegman grind.

Letdown:		
Kelsol 3910-B2G-75	142.38	16.25
n-Butanol	22.44	3.32
Ammonium Hydroxide	11.22	1.49
Water	56.81	6.82
Versaflow 102	2.10	0.25
Butoxy Ethanol	34.37	4.57
Water	163.43	19.62

Analysis:
   Pigment Volume Concentration, Percent: 28.9
   Pigment/Binder Ratio: .65/1.0
   Percent Solids, Weight: 38.3
   Percent Solids, Volume: 30.7
   Viscosity @ 25C, #4 Ford, Secs.: 75-80
   Pounds/Gallon: 8.93
   pH: 8.2-8.5
   VOC (excluding water):
      Grams/Liter: 334
      Pounds/Gallon: 2.79

SOURCE: Reichhold Chemicals, Inc.: Waterborne Handbook:
         Formulation 2116-71A

Red High Gloss Enamel

Material:	Lbs.	Gals.
Water	142.35	17.10
Ammonium Hydroxide	6.57	0.88
Kelsol 3906-B2G-75	82.13	9.50
Premix driers and solvent:		
Cobalt Hydrocure	2.30	0.30
Activ-8	1.10	0.14
Butoxy Ethanol	5.48	0.73
Rex Orange X-2806	84.32	1.82
Garnet Toner X-2433	26.28	2.05
Sand mill to 6 Hegman grind.		
Letdown:		
Water	328.50	39.42
Ammonium Hydroxide	8.76	8.17
Kelsol 3906-B2G-75	219.00	25.36
10% DC-29 in Butyl Cell	5.04	0.61

Analysis:
   Pigment Volume Concentration, Percent: 13.6
   Pigment/Binder Ratio: .5/1
   Percent Solids, Weight: 36.9
   Percent Solids, Volume: 28.4
   Pounds/Gallon: 9.19
   pH: 8.2-8.6
   VOC (excluding water):
      Grams/Liter: 265
      Pounds/Gallon: 2.2

SOURCE: Reichhold Chemicals, Inc.: Waterborne Handbook:
         Formulation 174-1

Red Rose Air Dry Enamel

	Pounds	Gallons
Cargill Water Reducible Chain Stopped Alkyd 74-7495	134.6	15.65
Aqueous Ammonia 28%	4.4	0.58
Byk-301	1.0	0.13
Byk-020	1.9	0.26
sec-Butanol	4.5	0.67
Ethylene Glycol Monobutyl Ether	6.4	0.85
Quindo Magenta RV-6803	20.7	1.72
Synthetic Red Iron Oxide RO-3097	7.4	0.18
Titanium Dioxide Ti-Pure R-900	9.4	0.27
Deionized Water	70.6	8.47

Sandmill to a 7.5 Hegman

	Pounds	Gallons
Cargill Water Reducible Chain Stopped Alkyd 74-7495	137.8	16.02
Aqueous Ammonia 28%	7.0	0.94

Premix the following, then add:

	Pounds	Gallons
sec-Butanol	5.6	0.83
Ethylene Glycol Monobutyl Ether	4.0	0.53
Intercar 6% Cobalt	3.6	0.46
Intercar 6% Zirconium	2.6	0.37
Activ-8	0.9	0.12
Deionized Water	432.7	51.95

Paint Properties:
    % Nonvolatile (By Weight): 27.09
                (By Volume): 22.80
    Pigment to Binder Ratio: 0.08
    Pigment Volume Concentration: 2.19
    Weight Per Gallon: 8.55
    Theoretical VOC (Pounds/Gallon): 2.90
                (Grams/Liter): 348
    Water:Cosolvent Weight Ratio: 82.04:17.96
    pH: 8.7
    Viscosity Krebs-Stormer (KU): 65-75

Typical Cured Film Properties:
    Cure Schedule: 7 Days Air Dry
    Substrate: Cold Rolled Steel
    Gloss (60/20): 95/85
    Dry Time Linear Drying Recorder (Minutes): 64

SOURCE: Cargill, Inc.: CARGILL Formulary: Formula P1843-257B

Satin White Air Dry Enamel

	Pounds	Gallons
Cargill Water Reducible Acrylic Modified Alkyd 74-7432	174.0	20.47
Aqueous Ammonia 28%	10.1	1.34
Triethylamine	1.0	0.17
Titanium Dioxide Ti-Pure R-900	174.0	4.97
Imsil A-8	98.6	4.47
Byk-301	1.5	0.18
Deionized Water	149.0	17.89

Sand Mill to a 7 Hegman

	Pounds	Gallons
Cargill Water Reducible Acrylic Modified Alkyd 74-7432	74.6	8.77
Aqueous Ammonia 28%	4.3	0.58
Triethylamine	0.4	0.07

Premix the following, then add:

	Pounds	Gallons
Intercar 6% Cobalt	2.9	0.38
Intercar 6% Zirconium	2.9	0.40
Activ-8	0.8	0.10
Deionized Water	335.0	40.21

Paint Properties:
    % Nonvolatile (By Weight): 43.71
                  (By Volume): 28.62
    Pigment to Binder Ratio: 1.54
    Pigment Volume Concentration: 33.14
    Weight Per Gallon: 10.29
    Theoretical VOC (Pounds/Gallon): 2.09
                    (Grams/Liter): 250
    Water:Cosolvent Weight Ratio: 85.34:14.66
    pH: 8.80
    Viscosity #4 Ford Cup (Seconds): 45

Typical Cured Film Properties:
    Cure Schedule: 7 Days Air Dry
    Substrate: Cold Rolled Steel
    Gloss (60): 29
    Pencil Hardness: Pass B
    % Crosshatch Adhesion: 100
    Impact (in. Lbs.):
        Direct: 20
        Reverse: <5

SOURCE: Cargill, Inc.: CARGILL Formulary: Formula P1887-32-1

Semigloss Enamel

Ingredients:	Pounds	Gallons
Pigment Grind:		
Water	268.2	32.20
Cellosize Hydoxyethyl Cellulose ER-15M	2.0	0.18
Preservative (1)	2.4	0.25
Dispersant (2)	9.2	1.00
Triton Nonionic Surfactant N-57	2.1	0.25
2-Amino-2-methyl-1-propanol (3)	1.0	0.13
Propylene Glycol	43.2	5.00
Antifoam (4)	1.0	0.13
Titanium Dioxide (5)	250.0	7.27
Letdown:		
UCAR Thickener SCT-275, premix (a)	29.9	3.48
Ucar Latex 379	428.4	47.28
Ucar Filmer IBT	20.0	2.52
Antifoam (4)	1.9	0.25
Triton Anionic Surfactant GR-7M	0.5	0.06

Suppliers:
(1) Preservative - "Nuosept" 145 or equivalent
(2) Dispersant - "Tamol" 731 or equivalent
(3) 2-Amino-2-methyl-1-propanol - AMP-95 or equivalent
(4) Antifoam - "Byk" 035 or equivalent
(5) Titanium Dioxide - "Tronox" CR-800 or equivalent

(a) Premix consists of 1:1:1 blend of propylene glycol:UCAR
    Thickener SCT-275:water.

Comparative Properties:
    Pigment Volume Concentration, %: 23.0
    Solids, %:
        by Volume: 31.8
        by Weight: 46.0
    Weight per Gallon, lb: 10.60

Paint Properties:
    Viscosity, initial:
        Stormer, KU: 82
        ICI, poise: 0.4
    pH, initial: 8.5
    Freeze-thaw Stability, 3 cycles: Pass
    Heat Stability, 2 weeks at 120F: Pass

Film Properties:
    Contrast Ratio, 6-mil bar: 0.967
    Reflectance, %: 89.5
    Gloss, 60: Drawdown, 6-mil bar: 59
    Flow and Leveling: Brushed: Very Good
                       Drawdown, NYPC Blade: 8

SOURCE: Ucar Emulsion Systems: UCAR Latex 379: Formulation
        Suggestion I-1989

Semigloss Maintenance Enamel-White

A semigloss maintenance enamel with anti-corrosive pigment.

Ingredients:	Pounds	Gallons
Pigment Grind:		
Water	79.9	9.59
Dispersant (1)	3.4	0.37
2-Amino-2-methyl-1-propanol (2)	3.0	0.37
Antifoam (3)	2.1	0.25
Nonionic Surfactant (4)	2.1	0.25
Nonionic Surfactant (5)	7.1	0.75
Titanium Dioxide (6)	200.0	5.85
Anti-corrosive Pigment (7)	25.0	1.00
Let Down:		
Antifoam (3)	2.1	0.25
Propylene Glycol	34.6	4.00
Butyl Carbitol Solvent	20.8	2.62
Mix:		
Ucar Thickener SCT-275	5.4	0.62
Butyl Carbitol Solvent	5.0	0.62
Ucar Filmer IBT	39.6	5.00
Mix:		
Sodium Nitrite	1.0	0.06
Water	4.2	0.50
Ucar Latex 430 (43.0%)	590.7	67.90

Suppliers:
(1) Dispersant - "Borchigen" ND or equivalent
(2) 2-Amino-2-methyl-1-propanol - AMP-95 or equivalent
(3) Antifoam - "Foamex" 1488 or equivalent
(4) Nonionic Surfactant - "Surfynol" 10XE or equivalent
(5) Nonionic Surfactant - "Borchigen" DFN or equivalent
(6) Titanium Dioxide - "Tronox" CR-800 or equivalent
(7) Anti-corrosive Pigment - "Moly-white" MZAP or equivalent

Paint Properties:
   Pigment Volume Concentration (PVC), %: 19.7
   Total Solids, %:
      by Volume: 34.8
      by Weight: 48.3
   Viscosity (equilibrated):
      Stormer, KU: 88
      ICI, Poise: 0.5
   pH, initial: 9.3
   Weight per Gallon, lb: 10.26
   VOC, g/l: 249.2

Film Properties:
   Control Ratio, 0.003 in: 0.966
   Reflectance, %: 91.5
   Gloss, 60: 38.1

SOURCE: Ucar Emulsion Systems: UCAR Latex 430: Formulation
      Suggestion M-2204

Semigloss White Enamel

Ingredients:	Pounds	Gallons
Pigment Grind		
Water	268.2	32.20
Cellosize Hydroxyethyl Cellulose ER-15M	2.0	0.18
Preservative (1)	2.3	0.25
Dispersant (2)	9.2	1.00
Triton Nonionic Surfactant N-57	2.1	0.25
2-Amino-2-methyl-1-propanol (3)	1.0	0.13
Propylene Glycol	43.2	5.00
Antifoam (4)	0.9	0.13
Titanium Dioxide (5)	250.0	7.27
Letdown:		
Butyl Carbitol Solvent	15.8	2.00
Thickener, premix (a)	21.5	2.50
Ucar Latex 367	428.4	47.28
Ucar Filmer IBT	11.9	1.50
Antifoam (4)	1.8	0.25
Triton Anionic Surfactant GR-7M	0.6	0.06

Suppliers:
(1) Preservative - "Nuosept" 95 or equivalent
(2) Dispersant - "Tamol" 731 or equivalent
(3) 2-Amino-2-methyl-1-propanol - AMP-95 or equivalent
(4) Antifoam - "BYK" 035 or equivalent
(5) Titanium Dioxide - "Tronox" CR-800 or equivalent

(a) Premix consists of 1:1:1 blend of propylene glycol:Rohm and
    Haas' "Acrysol SCT-275" Thickener:water, respectively.

Paint Properties:
   Pigment Volume Concentration (PVC), %: 23.0
   Solids, %:
      by Volume: 31.6
      by Weight: 46.8
   Viscosity (initial):
      Stormer, KU: 85
      ICI, poise: 1.0
   pH (initial): 8.1
   Weight per Gallon, lb.: 10.59
   Freeze-Thaw Stability, 3 cycles: Pass (+2 KU)
   Heat Stability, 2 weeks at 120F: Pass (+5 KU)

Film Properties:
   Contrast Ratio, 3 mil film: 0.968
   Reflectance, %: 92.6
   Gloss, 60:
      Drawdown: 61.0
      Brushout: 67.0
   Flow and Leveling, Brushed: Very Good

SOURCE: Ucar Emulsion Systems: UCAR Latex 367/UCAR Latex 376:
        Formulation Suggestion I-2219

Semigloss White Trim Enamel

Ingredients:	Pounds	Gallons
Pigment Grind:		
Water	246.0	29.51
Phenyl Mercuric Acetate, 100%	1.0	0.05
Cellosize Hydroxyethyl Cellulose QP-4400	3.5	0.32
Dispersant (1)	9.7	1.00
Nonionic Surfactant (2)	2.0	0.25
Propylene Glycol	51.8	6.00
Antifoam (3)	0.9	0.13
Titanium Dioxide (4)	250.0	7.26
Let Down:		
Ucar Acrylic 525	471.9	52.43
Ucar Filmer IBT	17.8	2.25
Antifoam (3)	1.9	0.25
Anionic Surfactant (5)	1.1	0.13
Ammonium Hydroxide, 28% Aqueous Solution	3.0	0.42

Suppliers:
(1) Dispersant - "Polywet" ND-2 or equivalent
(2) Nonionic Surfactant - "Igepal" CO-630 or equivalent
(3) Antifoam - "Colloid" 681-F or equivalent
(4) Titanium Dioxide - "Zopaque" RCL-9 or equivalent
(5) Anionic Surfactant - "Triton" GR-7M or equivalent

Paint Properties:
   Pigment Volume Concentration, %: 22.0
   Total Solids, %:
      by Volume: 33.0
      by Weight: 48.5
   Viscosity, initial:
      Stormer, KU: 90-95
      ICI, poise: 1.1
   pH, initial: 9.4
   Weight per Gallon, lb: 10.61
   Freeze-Thaw, 3 cycles: Pass
   Heat Stability, 2 wk at 120F: Pass

Film Property:
   Gloss, 60: 45-50

SOURCE: Ucar Emulsion Systems: UCAR Acrylic 525: Formulation
        Suggestion E-2077

Steel Blue Gloss Enamel

Material:	Lbs.	Gals.
Kelsol 3907-G2B-75	45.71	5.34
Ammonium Hydroxide	2.32	0.31
n-Butanol	3.85	0.57
Butoxy Ethanol	3.83	0.51
Monarch Blue X3367	4.50	0.34
Raven 1020	3.21	0.22
Yellow Iron Oxide 2288D	2.35	0.07
Ti-Pure R-902	40.01	1.17
Water	76.22	9.15
Steel ball mill to 7 Hegman grind.		

Letdown:		
Kelsol 3907-G2B-75	254.30	29.71
n-Butanol	13.97	2.07
Ammonium Hydroxide	7.62	1.02
Water	403.34	48.42

Premix:		
Activ-8	1.28	0.16
Cobalt Hydrocure II	6.35	0.84
Butoxy Ethanol	0.75	0.10

Analysis:
   Pigment Volume Concentration, Percent: 6.62
   Pigment/Binder Ratio: .22/1
   Percent Solids, Weight: 32.91
   Percent Solids, Volume: 27.17
   Viscosity @ 25C, #4 Ford, Secs.: 35-40
   Pounds/Gallon: 8.7
   pH: 8.2-8.5
   VOC (excluding water):
       Grams/Liter: 294
       Pounds/Gallon: 2.45

SOURCE: Reichhold Chemicals, Inc.: Waterborne Handbook:
        Formulation 2103-71

Tintable Eggshell Enamel

	lbs/100 gal	g/l
Premix on Kadyzolver at 2,000 rpm for 10 min:		
Propylene glycol	51.8	62.1
Water	50.0	59.9
Thickener	1.5	1.8
Tamol 731 (25%)	7.0	8.4
Nopco NDW	2.0	2.4
Add and grind at 2,000 rpm for 20 min:		
Ti-Pure R-902	250.0	299.5
Imsil A-25	75.0	89.8
Let down (low speed on an air stirrer):		
Water	83.0	99.4
Nopco NDW	4.0	4.8
100% acrylic latex, 46.5%	450.0	539.1
Texanol)	12.6	15.1
Propylene glycol) premix	17.3	20.7
Water)	24.9	29.8
Dowicil 75) premix	1.3	1.6
Triton N-57	4.0	4.8
Triton X-114	4.0	4.8
Triton X-207	2.0	2.4
Thickener (2.5% solution)	60.1	72.0

Tint with Cal/Ink 8800 series colorants at 4 oz/gal

Physical Properties:
    PVC, %: 33.2
    Volume solids, %: 32.3
    Weight solids, %: 48.50
    Density, lbs/gal (kg/l): 11.00 (1.32)
    Initial pH: 8.2-8.5
    Stormer viscosity (overnight), KU: 92-96

SOURCE: Aqualon Co.: NATROSOL B: Formula Table IV

Velvet Latex Enamel
Optiwhite/Burgess No. 98 Approach

	Lbs./100 Gal.
Water	398
KTPP	1.00
Natrosol 250 MBR	6.50
Nalco #2300	1.00
Tamol 1254	3.28
Triton X-100	2.00
AMP-95	3.00
Protel GXL	1.00
R-900	200
Optiwhite	75
Burgess No. 98	75
Ethylene Glycol	28
Ucar 367	205
Ucar 522	109
Texanol	15.84
Foamaster G	7.60

Evaluation Results:
Hiding 3 Mil Bird Blade on Opacity Panel:
   Reflectance: .908
   Contrast Ratio: .974
Tint Strength .5 Blue/200 grams:
   Reflectance: .702
Sheen:
   85: 19.4
   60: 3.6
Viscosity KU: 95

SOURCE: Burgess Pigment Co.: Formula 71 OPT/98

Waterborne Polyester Enamel

Ingredients:	Weight
WS-17-1T Polyester, 75% solids	33.42
Cymel 303 melamine resin	8.33
N,N-Dimethylaminoethanol (DMEA)	2.33
p-Toluenesulfonic acid, 40% in isopropanol	0.25
Fluorad FC-430, 20% in water	0.25
Demineralized water	55.42

Enamel Properties:
   pH: 8.42
   Viscosity, cP: 810
   Theoretical solids, %: 31.3

Coatings Properties-Initial:
   Substrate: 20 gauge CRS with Bonderite 37
   Cure: 30 minutes at 177C (350F)

   Pencil hardness, cut/scratch: 3-4H/H
   Direct/reverse impact resistance, in.-lb: 160/160
   MEK double rubs: 200+
   5 Minute I2 stain test: No stain
   500 Hours salt spray--mm creepage from scribe: 1
   Humidity resistance--96 hours at 60C (140F): No blisters

Coatings Properties-After Aging Solution 4 wks at 49C (120F):
   Substrate: 20 gauge CRS with Bonderite 37
   Cure: 30 minutes at 177F (350F)

   Pencil hardness, cut/scratch: 4H/H
   Direct/reverse impact resistance, in.-lb: 160/160
   MEK double rubs: 200+
   5 Minute I2 stain test: No stain
   500 Hours salt spray--mm creepage from scribe: 1
   Humidity resistance--96 hours at 60C (140F): No blisters

SOURCE: Eastman Chemical Products, Inc.: Enamel Prepared
          from Polyester Resin WS-17-1T

Waterborne Polyester/Melamine Enamel
Clear Enamel Composition

Ingredients:	Weight
Predispersion:	
WB-17-1NS Polyester Resin (neat)	100.0
Isopropyl Alcohol	20.0
Distilled Water	20.0
Cymel 303 or Resimene 745 Melamine Resin	42.9
Distilled Water	136.0

Enamel Preparation:
   Enamels can be conveniently prepared by weighing the predispersion into a suitable stirred vessel and heating to a temperature slightly below reflux (approximately 70-80C). Begin by adding hot distilled water slowly to reduce the percent solids to 50% if it is not already at that level. Reduce the heat to approximately 60C and add the melamine resin. Agitate the mixture well and add the remainder of the warm distilled water. These nonalkaline dispersions can have shelf stability in excess of one year.

Enamel Properties:
   Polyester/Melamine: 70/30
   Density, lb/gal (g/L): 8.81 (1057)
   Calculated Nonvolatiles, wt %: 44.2
   Calculated VOC, wt VOC/vol coating: 1.6 (192)
      (minus H2O), lb/gal (g/L)
   Viscosity by No. 4 Ford Cup, sec: 11

Coating Properties:
   Film Thickness, mils (um): 0.8-1.0 (20-25)
   Pencil Hardness, mar/cut: F/H
   Impact Resistance:
      Direct, in./lb: 160
      Reverse, in./lb: 150
   MEK Double-rub Solvent Resistance: >250
   1/8-in. Conical Mandrel Flexibility, % pass: 100
   Cleveland Humidity, 100 h at 60C (140F):
      % Gloss Retention at 20: 88
      Blistering: None
   Stain Resistance (water/acetone rinse, 5=no stain):
      5 min: 4
      30 min: 3
   Gloss, %:
      at 20: 80
      at 60: 99

a) On 20-gauge, cold-rolled, Bonderite 37 pretreated steel
   baked 30 minutes at 150C (302F).

SOURCE: Eastman Chemical Co.: Waterborne Polyester Resin
         WB-17-INS

Water Reducible Acrylic Enamel-IIB
K-FLEX UD-320W Modified

This formulation demonstrates several advantages when modify-
ing a water reducible acrylic system with K-FLEX UD-320W. The
advantages include lower VOC, harder films and improved salt
spray, humidity and QUV resistance. Properties generally
improved as the level of UD-320W was increased.

Pebble Mill Grind:

Titanox 2020	480.00
Acrysol WS-68	63.12
K-Flex UD-320W	26.64
Water	144.44
Butyl Cellosolve	43.76
n-Butanol	40.64
DMEA	1.40

Letdown:
Grind:

Acrysol WS-68	1073.60
Water	282.24
n-Butanol	16.88
Dow Corning #57 (30% in Butyl Cellosolve)	4.64
DMEA (10% in Water)	9.36
Cymel 303	216.00
K-Flex UD-320W	80.00

Enamel Characteristics:
  Acrylic/K-Flex/Melamine: 60/10/30
  Total Solids, %: 48.3
  Total Resin Solids, %: 29.0
  Pigment/Binder: 40/60
  VOC, calc. (lbs/gal): 0.50
  pH: 7.27
  Viscosity, 25C, cps: 89

Film Properties:
  Substrate: Bonderite 1000 (Iron Phosphated CRS)
  Cure: 30 min/250F (121C)
  Catalyst: 1.6% Nacure X49-110 (on TRS)
  Film Thickness: 0.9 mil DFT
Acrylic/K-Flex/Melamine: 60/10/30
  Pencil Hardness: F-H
  Knoop Hardness: 6.4
  MEK Rubs: 80
  Impact, Rev., in/lb: 5-10
        , Dir., in/lb: 40-50
  Crosshatch Adhesion, %: 100
  Initial Gloss, 60: 89.6
             , 20: 72.1

SOURCE: King Industries: Formulation UDW-4-II-B

Water Reducible Acrylic Enamel-II-C-II

Pebble Mill Grind:

Titanox 2020	480.00
Acrysol WS-68	63.12
K-Flex UD-320W	26.64
Water	144.44
Butyl Cellosolve	43.76
n-Butanol	40.64
DMEA	1.40

Letdown:
Grind:

Acrysol WS-68	789.44
Water	452.72
n-Butanol	16.88
Dow Corning #57 (30% in Butyl Cellosolve)	4.64
DMEA (10% in Water)	7.04
Cymel 303	252.00
K-Flex UD-320W	160.00

Enamel Characteristics:
    Acrylic/K-Flex/Melamine: 45/20/35
    Total Solids, %: 48.3
    Total Resin Solids, %: 29.0
    Pigment/Binder: 40/60
    VOC, calc. (lbs/gal): 0.60
    pH: 7.28
    Viscosity, 25C., cps: 35

Film Properties:
    Substrate: Bonderite 1000 (Iron Phosphated CRS)
    Cure: 30 min/250F (121C)
    Catalyst: 1.6% Nacure X49-110 (on TRS)
    Film Thickness: 0.9 mil DFT
Acrylic/K-Flex/Melamine: 45/20/35
    Pencil Hardness: H-2H
    Knoop Hardness: 7.0
    MEK Rubs: 100
    Impact, Rev., in/lb: 5-10
         , Dir., in/lb: 30-40
    Crosshatch Adhesion, %: 100
    Initial Gloss, 60: 87.0
                 , 20: 78.9
    Salt Spray, 126 hrs mm Creepage/Blister: 7-9/NA
    Humidity, 60C, 250 hrs: Gloss, 60: 11.4
                                  , 20:   2.4
                          Blister: Med-Den/Mic
    QUV**, 250 hrs: Gloss, 60: 78.9
                         , 20: 42.7
    ** QUV schedule of 8 hours UV at 50C., 4 hours humidity
       at 40C.

SOURCE: King Industries: Formulation UDW-4-II-C-II

Water Reducible Air Dry Gloss White Enamel

Ingredients:	Pounds	Gallons
Chempol 10-0173	106.81	12.33
Byk P-104S	2.94	0.37
Surfynol 104BC	2.11	0.28
2-Butoxyethanol	8.46	1.12
Ammonium hydroxide (28%)	8.44	1.13
Water	84.56	10.15
Ti-Pure R-900	146.71	4.27

Weigh in first four ingredients and mix well. Add ammonia and water separately with agitation.
Add pigment slowly with continued agitation.
Sandmill to #7 Hegman keeping grind temperature below 120F with water jacket.
Let down with:

Chempol 10-0173	149.17	17.22
Ammonium hydroxide (28%)	12.67	1.70
2-Butoxyethanol	11.84	1.58
Nuocure CK-10	7.17	0.88
Byk 301	0.42	0.05
Water	333.52	40.04
Water (Hold to adjust viscosity)	73.97	8.88

In the "let down", pre-blend resin and ammonium hydroxide with agitation. Premix the 2-butoxyethanol, Nuocure CK-10, and Byk 301 and add this premix to the resin/ammonia mixture slowly with good agitation. Slowly add the first amount of water and mix well. Now add the "let down" portion to the grind with good agitation. Mix well and allow the paint to set overnight.
Check the pH and adjust with ammonium hydroxide if necessary. Use the "adjust viscosity" water to bring the viscosity of the coating to spec at room temperature.

Properties:
    Total solids, % by weight: 35.06
    Total solids, % by volume: 23.77
    Weight/gallon, lbs.: 9.49
    PVC: 17.97
    Pigment/binder weight ratio: 0.79/1.00
    VOC, less water: 2.79 lbs/gal.
    Viscosity #2 Zahn: 30"-50"
    pH: 8.0-8.5
    DFT: 0.9-1.0 mils
    Cure: 1 week air dry
    Gloss, 60: 91
          20: 76
    Pencil hardness: B
    Crosshatch adhesion: 100% retention, taped

SOURCE: Freeman Polymers Division: CHEMPOL 10-0173-A Formula

Water Reducible, Low Gloss Black Enamel

10-0091-E is a water reducible, air dry enamel, based on CHEMPOL 10-0091 alkyd resin. It was especially formulated for dip, flow or spray application over high profile, poorly treated steel substrate. 10-0091-E will provide maximum corrosion resistance by providing a high-build film designed to bridge over irregular surfaces. Films of 10-0091-E dry quickly (15-30 minutes) and "tie-down" for maximum protection of metal substrate. Films exposed to salt spray show no loss of adhesion after 1 hour recovery.

Ingredients:	Pounds	Gallons
Chempol 10-0091	133.67	15.19
Ammonium hydroxide (28% NH3 min)	6.84	0.94
DMAMP-80	6.56	0.83
Ektasolve EP	50.55	6.66
Anti-Sag WR 300	3.15	0.42
Carbon Black-Sterling R	15.18	1.04
Strontium Chromate	9.61	0.31
Barmite XF	137.83	3.85
Duramite	67.95	3.02
Atomite	67.95	3.02
Surfynol PC	2.48	0.31
Bentone 27) Premix	10.37	0.73
Methyl alcohol)	6.90	1.04
MPA 1075) Premix	30.15	4.37
Butyl Cellosolve)	10.14	1.35
Propasol P	33.71	4.58
Butyl Cellosolve	15.62	2.08
Bykumen	3.07	0.42
50% Gilsonite in Xylene	42.83	5.20

Combine CHEMPOL 10-0091, ammonium hydroxide, DMAMP-80, and Ektasolve EP. Mix in Anti-Sag WR 300 and then add pigments. Add the remaining ingredients and mix well. Grind in a steel ball mill to a 6 Hegman.

Let Down With:		
Chempol 10-0091	117.22	13.32
5% Cobalt Hydrocure II	1.62	0.21
5% Calcium Hydrochem	3.39	0.42
5% Manganese Hydrochem	1.58	0.21
Exkin #2	3.98	0.52
Water	159.52	19.15
Texanol	20.57	2.60
Ammonium hydroxide (28% NH3 min)	2.26	0.31
Syloid 74x3500	10.35	0.62
Neocryl A-640	62.61	7.28

Mix "Let Down" ingredients in order listed. Add the Syloid 74x3500 under agitation, then add Neocryl A-640 slowly while continuing agitation. Adjust to a final pH of 8.0-8.5 with ammonium hydroxide.

SOURCE: Freeman Polymers Division: CHEMPOL 10-0091-E Formulation

Water-Reducible Polyester Enamel A

Grind (attritor)	Weight %
Pigment (R-900 TiO2)	20.57
Resin WS-3-1C (76.5% N.V.)	23.00
Dimethylethanol amine	1.53
Hexamethoxymethyl melamine	7.54
Cyzac 4040 catalyst	0.12
Flow contol agent (L-5310, 10% in n-butanol)	0.24

Let Down:	
Distilled Water	47.00

Physical Constants:
Viscosity, #4 Ford Cup, sec: 27
pH: 8.2
Nonvolatile, wt %: 45.7
Polyester/Crosslinker, wt %: 70/30
Pigment/Binder, wt %: 45/55
Water/Organic, wt %: 87/13
Cure cycle: 350F (177C)

Physical Properties:
Film thickness: 1.0-1.5 mil
Gloss, 60/20: 93/79
Pencil hardness: 4H
Conical mandrel flexibility, 1/8 in.,
    ASTM D-522, % Pass: 100
Impact resistance, Direct, in.-lb (cm-kg): >160 (>185)
                  Reverse, in.-lb (cm-kg): 140 (162)
Solvent resistance, Double MEK Rubs: 200+
Crosshatch adhesion, % Pass: 100
Stain resistance, 24 hr at room temp (b):
   Tomato juice: 10
   Black shoe polish: 10
   Mustard: 9
   Iodine (5 minutes): 7
Detergent resistance (c):
1-1/2% AATCC Solution at 165+-2F:
Blistering: 240 hr: Very Few #8
          500 hr: ----
% Gloss Retention, 60/20:
            240 hr: 100/87
            500 hr: ------

(b) Visually rated using the following scale: 10=No stain,
   8=Slight, 6=Moderate, 4=Considerable.
NOTE: Substrate used was cold-rolled steel panel treated with
    zinc phosphate.
(c) Epoxy primed, zinc phosphate treated, cold-rolled steel
    panels were used in this evaluation.

SOURCE: Eastman Chemical Products, Inc.: Enamels Prepared
    from Polyester Resin WS-3-1C

Water-Reducible Polyester Enamel B

Grind (attritor):	Weight %
Pigment (R-900 TiO2)	19.17
Resin WS-3-1C (76.5% N.V.)	21.43
Dimethylethanol amine	1.43
Benzoguanamine (85% N.V.)	8.27
Cyzac 4040 catalyst	0.12
Flow control agent (L-5310, 10% in n-butanol)	0.24

Let Down:	
Distilled Water	49.34

Physical Constants:
    Viscosity, #4 Ford Cup, sec: 30
    pH: 8.4
    Nonvolatile, wt %: 42.6
    Polyester/Crosslinker, wt %: 70/30
    Pigment/Binder, wt %: 45/55
    Water/Organic, wt %: 86/14
    Cure cycle: 350F (177C) for 30 minutes

Physical Properties:
    Film thickness: 1.0-1.5 mil
    Gloss, 60/20: 92/78
    Pencil hardness: 4H
    Conical mandrel flexibility, 1/8 in.,
        ASTM D-552, % Pass: 100
    Impact resistance, Direct, in.-lb (cm-kg): 140 (162)
                       Reverse, in.-lb (cm-kg): 80 (92)
    Solvent resistance, Double MEK Rubs: 200+
    Crosshatch adhesion, % Pass: 100
    Stain resistance, 24 hr at room temp (b):
        Tomato juice: 10
        Black shoe polish: 10
        Mustard: 9
        Iodine (5 minutes): 7
    Detergent resistance (c):
    1-1/2% AATCC Solution at 165+-2F:
    Blistering: 240 hr: None
                500 hr: Very Few #8
    % Gloss Retention, 60/20:
                       240 hr: 100/89
                       500 hr: 92/72

b) Visually rated using the following scale: 10= No stain,
    8=Slight, 6=Moderate, 4=Considerable.
NOTE: Substrate used was cold-rolled steel panels treated with
    zinc phosphate.
c) Epoxy primed, zinc phosphate treated, cold-rolled steel panels
    were used in this evaluation.

SOURCE: Eastman Chemical Products, Inc.: Enamel Prepared from
        Polyester Resin WS-3-1C

Water Reducible White Enamel

Formulation 10-1706-A is a water reducible, thermosetting, acrylic enamel which provides good adhesion to both metallic and nonmetallic substrates.

Ingredients:	Pounds	Gallons
Chempol 10-1706	49.28	5.60
Dimethylethanolamine	5.10	0.69
Nalco 2309	4.31	0.59
Bykumen	4.33	0.59
Ektasolve EP	23.16	2.95
Water	53.16	6.39
Titanium Dioxide (R-900)	183.48	5.51

Grind Pebble Mill 7 1/2 - 8 Hegman gauge.
Add for stability:

Chempol 10-1706	49.28	5.60
Dimethylethanolamine	5.10	0.69

Let down with:

Chempol 10-1706	98.65	11.21
Cymel 385	45.53	4.42
Dimethylethanolamine	9.46	1.28
Exkin No. 2	1.53	0.20
Ektasolve EP	42.47	5.41
Water	406.60	48.87

Properties:
   Total solids, % by weight: 38
   Total solids, % by volume: 25
   Viscosity, #2 Zahn cup, seconds: 35
   Pigment/binder weight ratio: 1.0/1.0
   PVC, %: 21.6
   Weight/gallon, lbs.: 9.8
   VOC, lbs./gal., less water: 3.2

Performance Notes:
   Drawndown 3 mils wet on Bonderite 1000 cured at 15'/325F:
      Dry film thickness, mils: 0.8
      60 Gloss: 83
      Pencil hardness: >4H
      Crosshatch adhesion: 100
      MEK resistance: 100
      Forward impact: 20
      Reverse impact: <20

SOURCE: Freeman Polymers Division: CHEMPOL 10-1706-A Formulation

White Acrylic Baking Enamel

Ingredients:	Pounds	Gallons
Cymel 303	43.90	4.39
Coroc A-2678-M	43.90	4.99
Butyl Cellosolve	32.93	4.38
Nalco 2309	1.10	0.15
Deionized water	131.71	15.83
Surfynol 104BC	0.88	0.12
TiPure R700	219.51	6.80

Pebble grind to Hegman 7-1/2+, then let down with the follow-ing pre-mixed:

Chempol 20-4301	426.95	48.52
Deionized Water	75.73	9.10

Mix thoroughly and continue let down with:

Cycat 4040	2.20	0.27
Dimethylethanolamine	0.88	0.12
Deionized water	44.34	5.33

Adjust to desired viscosity using DMEA.

Properties:
    Total solids, % by weight: 43.00
    Total solids, % by volume: 29.40
    P/B weight ratio: 1.11
    % PVC: 23.1
    VOC; lbs/gallon less H2O (theory): 1.28
    Initial Viscosity #4 Ford Cup, secs: 16-17
    pH: 7.6
    Adjusted viscosity, #4 Ford Cup, secs: 45 (pH=8.2)

Film Properties: Spray applied over B1000 treated CRS:
    Suggested bake schedule: 20 mins/325F
    Dry film thickness, mils: 1.0-1.2
    60/20 gloss: 80/40
    Pencil hardness: 3H
    MEK double rubs, 100: no effect
    Crosshatch adhesion: Exc. - 100% no tape off
    Reverse impact, inch-lbs.: <5
    Direct impact, inch-lbs.: 20, sl. cracking but no tape off
    1/8" conical mandrel: 12 mm crack - no adhesion loss

SOURCE: Freeman Polymers Division: CHEMPOL 20-4301-G Formulation

White Acrylic Emulsion Enamel

Ingredients:	Pounds	Gallons
Coroc A-2678-M	21.70	2.47
2-butoxyethanol	7.30	0.97
Surfynol 104BC	7.62	1.01
Surfynol DF-75	1.25	0.15
Water	28.67	3.44
Ti-Pure R-900	152.49	4.44

Pre-mix first four ingredients, then add water with agitation. Add pigment slowly with agitation.

Cowles to a #7 Hegman with a high speed dispersator. Adjust grind viscosity with water from below if necessary.

Let down with:		
Chempol 20-4301	572.63	65.07
2-butoxyethanol	30.28	4.03
Butyl Carbitol	19.36	2.44
Dibutyl Phthalate	6.69	0.77
Nuocure CK-10	1.90	0.23
Michem 39235	28.59	3.43
Water (hold to adjust viscosity)	96.19	11.55

Premix co-solvents, plasticizers and drier. Add this mixture to the vortex of the agitating emulsion VERY SLOWLY. If the mixture becomes heavy, add a small amount of water from the "adjust viscosity." Check mixture for kick-out or seeding. Add the wax dispersion to the "let down" with agitation. After mixing well, add the grind to the "let down" with good agitation. Mix well and allow the paint to set overnight.

Check the pH and adjust with ammonia if necessary. Avoid pH values above 7.6, as this will cause the viscosity of the paint to increase significantly. Use the "adjust viscosity" water to bring the viscosity of the coating to spec at room temperature.

Properties:
    Total solids, % by weight: 40.25%
    Total solids, % by volume: 29.26%
    Weight/gallon: 9.75 lbs./gal.
    PVC: 15.18
    Pigment/binder ratio: 0.64
    VOC: 1.88 lbs./gal.
    Viscosity: 30" #2 Zahn
    pH: 7.5
    Substrate: B-1000 CRS
    Cure: 1 week air dry
    Set to touch: 12'
    Dust free: 14'
    Tack free: 30'
    Dry through: 30'
    DFT: 1.2-1.5 mil

SOURCE: Freeman Polymers Division: CHEMPOL 20-4301-B Formulation

White Acrylic Emulsion Enamel

Ingredients:	Pounds	Gallons
Acrysol I-62	44.20	5.08
Ammonia, 28%	4.83	0.65
2-butoxyethanol	8.03	1.07
Surfynol 104BC	8.03	1.07
Surfynol DF-75	1.32	0.16
Water	28.71	3.45
DuPont R-960 VHG	150.76	4.64

Add ammonia to Acrysol I-62 slowly and with good agitation until the solution turns clear and viscous. Add remaining ingredients one at a time with continued agitation.

Cowles to a #7 Hegman with a high speed dispersator. Adjust grind viscosity with water from below if necessary.

Let down with:

	Pounds	Gallons
Chempol 20-4301	603.57	68.59
2-butoxyethanol	31.91	4.25
Butyl Carbitol	20.40	2.57
Dibutyl Phthalate	7.05	0.81
Nuocure CK-10	2.00	0.25
Michem 39235	30.13	3.62
Water (hold to adjust viscosity)	31.57	3.79

Premix co-solvents, plasticizers and drier. Add this mixture to the vortex of the agitating emulsion <u>VERY SLOWLY.</u> If the mixture becomes heavy, add a small amount of water from the "adjust viscosity." Check mixture for kick-out or seeding. Add the wax dispersion to the "let down" with agitation. After mixing well, add the grind to the "let down" with good agitation. Mix well and allow the paint to set overnight.

Check the pH and adjust with ammonia if necessary. Avoid pH values above 7.6, as this will cause the viscosity of the paint to increase significantly. Use the "adjust viscosity" water to bring the viscosity of the coating to spec at room temperature.

Properties:
   Total solids, % by weight: 42.58%
   Total solids, % by volume: 32.06%
   Weight/gallon: 9.73 lbs./gal.
   PVC: 14.47
   Pigment/binder ratio: 0.57
   VOC: 1.76 lbs./gal.
   Viscosity: 31" #2 Zahn
   pH: 7.6
   Substrate: B-1000 CRS
   Cure: 1 week air dry
   Set to touch: 10'-12'
   Dust free: 13'-15'
   Tack free: 25'-30'
   Dry through: 25'-30'

SOURCE: Freeman Polymers Division: CHEMPOL 20-4301-C Formulation

White Air Dry Enamel

	Pounds	Gallons
Cargill Water Reducible Silicone Modified Alkyd 74-7435	197.9	21.94
L-5310*	2.5	0.30
Ethylene Glycol Monobutyl Ether	11.9	1.59
Aqueous Ammonia 28%	8.3	0.96
Deionized Water	118.7	14.25
Titanium Dioxide Ti-Pure R-900	237.6	6.97

High Speed to a 7 Hegman

	Pounds	Gallons
Cargill Water Reducible Silicone Modified Alkyd 74-7435	118.7	13.16
Aqueous Ammonia 28%	5.3	0.61

Premix the following, then add:

	Pounds	Gallons
Ethylene Glycol Monobutyl Ether	14.3	1.90
Intercar 6% Cobalt	3.6	0.45
Intercar 6% Zirconium	4.5	0.62
Activ-8	1.2	0.16

	Pounds	Gallons
Deionized Water	248.4	29.64

	Pounds	Gallons
Deionized Water (Hold for Viscosity Adjustment)	62.1	7.45

*50% in Ethylene Glycol Monobutyl Ether

Paint Properties:
   % Nonvolatile (By Weight): 46.4
             (By Volume): 30.7
   Pigment to Binder Ratio: 0.98
   Weight Per Gallon: 10.35
   Theoretical VOC (Pounds/Gallon): 2.39
             (Grams/Liter): 286
   Water:Cosolvent Weight Ratio: 75.5:24.5
   pH: 8.5
   Viscosity Krebs-Stormer (KU): 78

Typical Cured Film Properties:
   Cure Schedule: 7 Days Air Dry
   Substrate: Cold Rolled Steel
   Dry Film Thickness (Mils): 1.5
   Gloss (60/20): 94/78
   Humidity (Hours): 96
      Gloss (60/20): 89/75
      Blisters: Medium Dense #4
      Observations: None
   QUV (Hours): 500
      Gloss (60/20): 35/4

SOURCE: Cargill, Inc.: CARGILL Formulary: Formula P1686-189

White Air Dry Enamel

	Pounds	Gallons
Cargill Water Reducible Styrenated Copolymer Alkyd 74-7425	213.1	24.63
Ethylene Glycol Monobutyl Ether	11.2	1.49
Byk-301	2.2	0.28
Aqueous Ammonia 28%	10.1	1.35
Titanium Dioxide Ti-Pure R-900	201.8	5.77
Deionized Water	72.9	8.75

High Speed Disperse to a 7.5 Hegman

Cargill Water Reducible Styrenated Copolymer Alkyd 74-7425	78.5	9.07

Premix the following, then add:

Ethylene Glycol Monobutyl Ether	11.2	1.49
sec-Butanol	5.6	0.83
Intercar 6% Cobalt	1.1	0.14
Intercar 6% Manganese	1.3	0.17
Activ-8	0.6	0.07
Aqueous Ammonia 28%	3.4	0.45
Deionized Water	319.6	38.37
Exkin No. 2	1.1	0.15
Aqueous Ammonia 28% (Adjust to pH 8.0-8.5)	2.2	0.26
Deionized Water (Hold for Viscosity Adjustment)	56.1	6.73

Paint Properties
   % Nonvolatile (By Weight): 41.2
                (By Volume): 28.2
   Pigment to Binder Ratio: 0.98
   Weight Per Gallon: 9.92
   Theoretical VOC (Pounds/Gallon): 2.70
                (Grams/Liter): 324
   pH: 8.0-8.5
   Viscosity Krebs-Stormer (KU): 70-75

Typical Cured Film Properties:
   Cure Schedule: 7 Days Air Dry
   Substrate: Cold Rolled Steel
   Gloss (60): 87
   Pencil Hardness (1 Day): Pass 5B
                (7 Days): Pass 3B
   Impact (In. Lbs.):
      Direct: 20
   Dry Time (Hours): 1/3
      500g Zapon: 60-65

SOURCE: Cargill, Inc.: CARGILL Formulary: Formula P1863-D1

White Air Dry Enamel

	Pounds	Gallons
Cargill Water Reducible Chain Stopped		
Alkyd 74-7470	90.6	10.42
Aqueous Ammonia 28%	6.0	0.80
Titanium Dioxide Ti-Pure R-902	186.7	5.46
Deionized Water	146.9	17.64

Sand Mill to a 7 Hegman

Cargill Water Reducible Chain Stopped		
Alkyd 74-7470	159.5	18.34
Aqueous Ammonia 28%	7.9	1.06

Premix the following, then add:

Ethylene Glycol Monobutyl Ether	28.1	3.74
5% Cobalt Hydro Cure II	1.8	0.24
Manganese Hydor Cure II	2.2	0.28
Deionized Water	350.2	42.02

Paint Properties:
    % Nonvolatile (By Weight): 38.41
                (By Volume): 25.57
    Pigment to Binder Ratio: 0.99
    Pigment Volume Concentration: 21.47
    Weight Per Gallon: 9.80
    Theoretical VOC (Pounds/Gallon): 2.46
                (Grams/Liter): 295
    Water:Cosolvent Weight Ratio: 84.03:15.97
    pH: 8.13
    Viscosity Krebs-Stormer (KU): 76

Typical Cured Film Properties:
    Cure Schedule: 7 Days Air Dry
    Substrate: Cold Rolled Steel
    Dry Film Thickness (Mils): 1.5
    Gloss (60/20): 91/78
    Humidity (Hours): 96
        Gloss (60/20): 92/70
        Blisters: Medium #4
        Observations: None
    South Florida Exposure (Months): 6
        Gloss (60/20):
        Washed: 76/39
        Unwashed: 71/38

SOURCE: Cargill, Inc.: CARGILL Formulary: Formula P1843-30E

White Air Dry Enamel

	Pounds	Gallons
Cargill Water Reducible Fatty Acid Alkyd		
74-7472	96.9	11.27
Aqueous Ammonia 28%	7.7	1.03
Titanium Dioxide Ti-Pure R-902	208.2	6.09
Deionized Water	125.9	15.11

Sand Mill to a 7 Hegman

	Pounds	Gallons
Cargill Water Reducible Fatty Acid Alkyd		
74-7472	180.6	21.00
Aqueous Ammonia 28%	7.3	0.98

Premix the following, then add:

	Pounds	Gallons
Propylene Glycol Monopropyl Ether	21.8	2.96
5% Cobalt Hydro Cure II	2.1	0.27
Manganese Hydro Cure II	3.1	0.40
Deionized Water	340.6	40.89

Paint Properties:
    % Nonvolatile (By Weight): 42.14
                 (By Volume): 28.83
    Pigment to Binder Ratio: 0.99
    Pigment Volume Concentration: 21.25
    Weight Per Gallon: 9.94
    Theoretical VOC (Pounds/Gallon): 2.29
                    (Grams/Liter): 275
    Water:Cosolvent Weight Ratio: 82.98:17.02
    pH: 8.52
    Viscosity #4 Ford Cup (Seconds): 33

Typical Cured Film Properties:
    Cure Schedule: 7 Days Air Dry
    Substrate: Cold Rolled Steel
    Dry Film Thickness (Mils): 1.5
    Gloss (60/20): 87/60
    QUV (Hours): 500
       Gloss (60/20): 18/1
    South Florida Exposure (Months): 6
       Gloss (60/20)
       Washed: 64/17
       Unwashed: 58/16

SOURCE: Cargill, Inc.: CARGILL Formulary: Formula P1843-30H

White Air Dry Enamel

	Pounds	Gallons
Cargill Water Reducible Fatty Acid Alkyd 74-7476	129.2	15.02
Aqueous Ammonia 28%	8.2	1.10
Titanium Dioxide Ti-Pure R-902	297.4	8.70
Deionized Water	134.4	16.14

Sand Mill to a 7 Hegman

	Pounds	Gallons
Cargill Water Reducible Fatty Acid Alkyd 74-7476	167.6	19.49
Aqueous Ammonia 28%	8.5	1.14

Premix the following, then add:

	Pounds	Gallons
Ethylene Glycol Monobutyl Ether	23.2	3.08
5% Cobalt Hydro Cure II	2.1	0.27
Manganese Hydro Cure II	3.2	0.41
Deionized Water	288.7	34.65

Paint Properties:
    % Nonvolatile (By Weight): 49.20
                  (By Volume): 32.86
    Pigment to Binder Ratio: 1.32
    Pigment Volume Concentration: 26.58
    Weight Per Gallon: 10.63
    Theoretical VOC (Pounds/Gallon): 2.19
                    (Grams/Liter): 262
    Water:Cosolvent Weight Ratio: 80.62:19.38
    pH: 8.07
    Viscosity #4 Ford Cup (Seconds): 18

Typical Cured Film Properties:
    Cure Schedule: 7 Days Air Dry
    Substrate: Cold Rolled Steel
    Dry Film Thickness (Mils): 1.5
    Gloss (60/20): 87/70
    Humidity (Hours): 96
        Gloss (60/20): 90/80
        Blisters: Dense #6
        Observations: None
    QUV (Hours): 500
        Gloss (60/20): 38/3
    South Florida Exposure (Months): 6
        Gloss (60/20):
        Washed: 75/26
        Unwashed: 66/24

SOURCE: Cargill, Inc.: CARGILL Formulary: Formula P1843-30L

White Air Dry Enamel

	Pounds	Gallons
Cargill Water Reducible Phenolic Modified		
Alkyd 74-7478	150.3	17.48
Aqueous Ammonia 28%	7.0	0.94
Propylene Glycol Monopropyl Ether	7.6	1.03
sec-Butanol	3.8	0.57
Titanium Dioxide Ti-Pure R-902	207.0	6.05
Deionized Water	182.5	21.91

Sand Mill to a 7 Hegman

Cargill Water Reducible Phenolic Modified		
Alkyd 74-7478	126.3	14.68
Aqueous Ammonia 28%	6.7	0.89

Premix the following, then add:

Propylene Glycol Monopropyl Ether	7.6	1.03
sec-Butanol	3.8	0.57
Nuocure CK 10%	4.2	0.51
Exkin No. 2	0.7	0.09
Deionized Water	285.4	34.25

Paint Properties:
    % Nonvolatile (By Weight): 41.98
                 (By Volume): 28.89
    Pigment to Binder Ratio: 0.99
    Pigment Volume Concentration: 21.13
    Weight Per Gallon: 9.93
    Theoretical VOC (Pounds/Gallon): 2.30
                 (Grams/Liter): 275
    Water:Cosolvent Weight Ratio: 83.01:16.99
    pH: 8.16
    Viscosity Krebs-Stormer (KU): 87

Typical Cured Film Properties:
    Cure Schedule: 7 Days Air Dry
    Substrate: Cold Rolled Steel
    Dry Film Thickness (Mils): 1.5
    Gloss (60/20): 89/90
    Humidity (Hours): 96
        Gloss (60/20): 81/68
        Blisters: Medium Dense #8
    QUV (Hours): 500
        Gloss (60/20): 31/2
    South Florida Exposure (Months): 6
        Gloss (60/20):
        Washed: 88/57
        Unwashed: 75/46

SOURCE: Cargill, Inc.: CARGILL Formulary: Formula P1843-30N

White Air Dry Enamel

	Pounds	Gallons
Cargill Water Reducible Chain Stopped Alkyd 74-7487	76.3	8.62
Aqueous Ammonia 28%	2.6	0.35
Titanium Dioxide Ti-Pure R-900	161.1	4.60
Supercoat	49.9	2.21
Surfynol 104 E	1.5	0.18
Ethylene Glycol Monobutyl Ether	8.0	1.06
Byk-020	1.5	0.20
Deionized Water	70.5	8.46

High Speed to a 7 Hegman

	Pounds	Gallons
Cargill Water Reducible Chain Stopped Alkyd 74-7487	175.0	19.75
Aqueous Ammonia 28%	6.1	0.81

Premix the following, then add:

	Pounds	Gallons
Ethylene Glycol Monobutyl Ether	4.2	0.56
5% Cobalt Hydro Cure II	1.3	0.16
Manganese Hydro Cure II	1.3	0.15
Activ-8	0.7	0.09
Exkin No. 2	1.5	0.19
Raybo 62-HydroFlo	4.0	0.48
Ethylene Glycol Monobutyl Ether	8.5	1.13
Deionized Water	255.7	30.71
Aqueous Ammonia 28% (Adjust to pH of 8.3)	4.0	0.53
Deionized Water (Hold for Viscosity Adjustment)	164.6	19.76

Paint Properties:
    % Nonvolatile (By Weight): 41.56
                 (By Volume): 28.55
    Pigment to Binder Ratio: 1.04
    Pigment Volume Concentration: 23.94
    Weight Per Gallon: 9.98
    Theoretical VOC (Pounds/Gallon): 2.09
                    (Grams/Liter): 250
    Water:Cosolvent Weight Ratio: 85.70:14.30
    pH: 8.3
    Viscosity #4 Ford Cup (Seconds): 36

Typical Cured Film Properties:
    Cure Schedule: 7 Days Air Dry
    Substrate: Cold Rolled Steel
    Gloss (60/20): 91/77
    Pencil Hardness: Pass B
    % Crosshatch Adhesion: 100

SOURCE: Cargill, Inc.: CARGILL Formulary: Formula P1887-51B1

White Air Dry Enamel

	Pounds	Gallons
Cargill Water Reducible Chain Stopped		
Alkyd 74-7495	134.2	15.60
Aqueous Ammonia 28%	4.3	0.58
sec-Butanol	4.4	0.66
Ethylene Glycol Monobutyl Ether	6.4	0.85
Byk-301	0.9	0.12
Titanium Dioxide Ti-Pure R-900	182.8	5.22
Deionized Water	56.5	6.78

Sandmill to a 7.5 Hegman

Cargill Water Reducible Chain Stopped		
Alkyd 74-7495	137.3	15.97
Aqueous Ammonia 28%	7.9	1.05

Premix the following, then add:

sec-Butanol	5.6	0.83
Ethylene Glycol Monobutyl Ether	4.0	0.53
Intercar 6% Cobalt	3.5	0.45
Intercar 6% Zirconium	2.6	0.36
Activ-8	0.9	0.12

Deionized Water	423.8	50.88

Paint Properties:
  % Nonvolatile (By Weight): 38.56
              (By Volume): 25.75
  Pigment to Binder Ratio: 0.95
  Pigment Volume Concentration: 20.47
  Weight Per Gallon: 9.75
  Theoretical VOC (Pounds/Gallon): 2.66
              (Grams/Liter): 319
  Water:Cosolvent Weight Ratio: 81.64:18.36
  pH: 8.7
  Viscosity Krebs-Stormer (KU): 65-75

Typical Cured Film Properties:
  Cure Schedule: 7 Days Air Dry
  Substrate: Cold Rolled Steel
  Gloss (60/20): 86/74
  Pencil Hardness: Pass B
  % Crosshatch Adhesion: 100
  Impact (In. Lbs.):
    Direct: 30
    Reverse: <5
  Dry Time Linear Drying Recorder (Hours): 3/4
  Humidity (Hours): 100
    Observations: None

SOURCE: Cargill, Inc.: CARGILL Formulary: Formula P1843-257A

White Enamel

Material:	Lbs.	Gals.
Water	63.14	7.58
Tamol 731	3.40	0.37
Ti-Pure R902	221.11	6.66
Foamaster VL	0.23	0.03

Pebble mill to a 7 Hegman grind.

Letdown:		
Arolon 845-W-45	602.48	69.25
Butoxy Ethanol	60.44	8.07
Water	48.98	5.88
10% Sodium Nitrite	7.09	0.70
Ammonium Hydroxide	11.65	1.56

Analysis:
    Pigment Volume Concentration, Percent: 17.61
    Pigment/Binder Ratio: 0.8/1.00
    Percent Solids, Weight: 48.96
    Percent Solids, Volume: 38.45
    Viscosity @ 25C, #2 Zahn, Secs.: 35-45
    Pounds/Gallon: 10.17
    pH: 8.2-8.5
    VOC (excluding water):
        Grams/Liter: 160.80
        Pounds/Gallon: 1.34

SOURCE: Reichhold Chemicals, Inc.: Waterborne Handbook:
        Formulation 2402-24A

White Enamel

Material:	Pounds	Gallons
Part A:		
Epi-Rez 510 Epoxy Resin	162.1	16.84
Heloxy 8 Epoxy Functional Modifier	53.5	7.21
Part B:		
Epi-Cure W50-8535 Curing Agent	216.0	24.80
Rutile Titanium Dioxide (2)	250.0	7.60
Defoamer (3)	4.0	0.50
High speed disperse to texture 7-8 NS. Reduce speed and add:		
Glacial Acetic Acid	2.0	0.24
Defoamer (3)	4.0	0.50
Water	352.4	42.31

    (2) Titanox 2020
    (3) Colloid 640

Composite Blend:	Viscosity		
Part A	68 KU	215.6	24.05
Part B	60 KU	828.4	75.95
Parts A and B	100 KU		

Typical Coating Properties:
    Total Weight Solids, percent: 44.2
    Total Volume Solids, percent: 45.0
    Volatile Organic Compounds (VOC):
        Pounds/Gallon: 0.14
        Grams/Liter: 16.8

SOURCE: Shell Chemical Co.: 24-212 White Enamel Using EPI-CURE
     W50-8535 Curing Agent

White Enamel

Material:	Pounds	Gallons
Part A:		
Rutile Titanium Dioxide	250.0	7.34
Epi-Rez 510 Epoxy Resin	130.0	13.50
High speed disperse to texture of 10 P.C.S. Add:		
Epi-Rez 510 Epoxy Resin	32.1	3.34
Heloxy 8 Epoxy Functional Modifier	53.5	5.82
Part B:		
Epi-Cure W50-8535 Curing Agent	216.0	24.80
Glacial Acetic Acid	2.0	0.24
Water	374.0	44.96
Composite Blend:		
Part A	465.6	30.00
Part B	592.0	70.00

Typical Coating Properties
    Viscosity:
        Fresh: 75-85 KU
        1 Hour: Off Scale
    Pounds/Gallon: 10.58
    Application: Brush or Spray
    Cure Schedule: Room Temperature
    P.V.C.: 17.5%
    Vehicle Non-Volatile: 45.8%
    Total Solids: 53.2%

Typical Film Properties:
    Gloss (60 Angle): 98 Units
    Pencil Hardness: "B" 72 Hours Cure
    Resin Composition:
        Epoxy Resin: 66.5%
        Curing Agent: 33.5%

SOURCE: Shell Chemical Co.: 24-146 White Enamel

White Enamel Formulation

Materials:	Parts by weight
Resin WB-300 (80% NVM)	288.2
Dimethylethanolamine	16.3
Deionized water	262.6
Melamine (1)	40.7
Flow agent (2)	0.5
Titanium dioxide (3)	244.1
Water for spray viscosity (approx)	121.1
Solvent (4) for spray viscosity (approx)	26.5

Coating Properties:
  Pigment/binder ratio: 0.9/1.0
  Resin/melamine ratio: 15/85
  Water/cosolvent ratio: 82/18
  Non-volatile material (NVM), wt%: 53-54
  Viscosity, #4 Ford, sec: 30-35
  pH: 7.5-8.0

Compounding Procedure:
1. Mix polyester resin and dimethylethanolamine until homo-geneous. Slowly add water until mixed thoroughly. Stir in melamine and adjust pH to 7.5-8.0.
2. Stir in flow agent and pigment.
3. Charge mixture to pebble mill. Roll overnight to Hegman Grind of 7+.
4. Discharge mill and adjust to desired application viscosity.

(1) Cymel 303
(2) FC-430
(3) Ti-Pure R-900
(4) Butyl Carbitol diethylene glycol n-butyl ether

Performance Properties (1):
  Properties: CRS (4):
  Film thickness, mils: 1.3
  60 gloss, %: 92
  Yellowness index (7): -3.83
  Sward hardness: 89
  Pencil hardness: 2H
  Impact resistance, in-lb:
    direct: 80
    reverse: 50
  Conical bend, 1/8-inch, % pass: 100
  Crosshatch adhesion: 4B
  (1) Cure cycle: 20 minutes at 177C
  (2) Commercial solvent-applied, oil-free, flexible polyester
  (4) Cold rolled steel
  (7) Yellowness index:[(A-B)/G]×100

SOURCE: Amoco Chemical Co.: Water-Borne Bake Polyesters Based on AMOCO IPA and TMA: Formula Table 11 and 12

White Fast Dry Enamel

Material:	Lbs.	Gals.
Arolon 970-G4-70	148.93	16.93
Premix:		
Cobalt Hydrocure	4.27	.56
Activ-8	.72	.09
Ti-Pure R-902	100.93	3.03
Roller mill to 7 Hegman grind.		
Letdown:		
Arolon 970-G4-70	55.82	6.34
Ammonium Hydroxide	7.82	1.04
Propasol P	7.24	.98
Butoxy Ethanol	17.16	2.29
sec-Butanol	16.29	2.43
n-Butanol	17.59	2.61
Lodyne S-107	1.55	.16
Water	528.81	63.48

Analysis:
   Pigment Volume Concentration, Percent: 16.5
   Pigment/Binder Ratio: .7/1
   Percent Solids, Weight: 26.9
   Percent Solids, Volume: 18.4
   Viscosity @ 25C, #4 Ford, Secs.: 45-50
   Pounds/Gallon: 9.08
   pH: 8.2-8.5
   VOC (excluding water):
      Grams/Liter: 409
      Pounds/Gallon: 3.41

SOURCE: Reichhold Chemicals, Inc.: Waterborne Handbook:
      Formulation 2053-35

White Gloss Enamel

Material:	Lbs.	Gals.
Arolon 860-W-45	285.12	32.44
Surfynol 104H	2.59	0.32
Colloid 679	2.81	0.39
Water	60.48	7.26
Ammonium Hydroxide	2.16	0.28
Titanox 2101	226.80	6.80
Cowles grind to a 6 Hegman.		

Letdown:		
Arolon 860-W-45	341.28	38.84
Butoxy Ethanol	28.08	3.74
Diethylene Glycol n-Butyl Ether	28.08	3.54
Byk 301	1.08	0.13
10% Sodium Nitrite	4.10	0.50
Water	43.20	5.18
Ammonium Hydroxide	4.32	0.58

Analysis:
    Pigment Volume Concentration, Percent: 18.2
    Pigment/Binder Ratio: .8/1
    Percent Solids, Weight: 49.4
    Percent Solids, Volume: 37.5
    Viscosity @ 25C, #2 Zahn, Secs.: 35-45
    Pounds/Gallon: 10.30
    pH: 8.2-8.5
    VOC (excluding water):
        Grams/Liter: 155
        Pounds/Gallon: 1.29

SOURCE: Reichhold Chemicals, Inc.: Waterborne Handbook:
        Formulation 2402-17A

White Gloss Enamel

Material:	Lbs.	Gals.
Kelsol 3960-B2G-75	183.30	21.08

Premix:		
Activ-8	0.92	0.12
Cobalt Hydrocure	5.68	0.75

Add then add:		
Butoxy Ethanol	9.17	1.22
Titanox 2101	188.80	5.67

Roller mill to 7 Hegman grind.

Letdown:		
Kelsol 3960-B2G-75	69.65	8.01
n-Butanol	13.44	2.00
sec-Butanol	12.83	1.92
Ammonium Hydroxide	13.44	1.80
Butoxy Ethanol	3.67	0.50
Water	474.14	56.93

Analysis:
    Pigment Volume Concentration, Percent: 22.0
    Pigment/Binder Ratio: 1/1
    Percent Solids, Weight: 32.2
    Percent Solids, Volume: 25.7
    Viscosity @ 25C, #4 Ford, Secs.: 53-56
    Pounds/Gallon: 11.74
    pH: 8.2-8.5
    VOC (excluding water):
        Grams/Liter: 302
        Pounds/Gallon: 2.52

SOURCE: Reichhold Chemicals, Inc.: Waterborne Handbook:
        Formulation 2079-7

White Gloss Enamel

Material:	Lbs.	Gals.
Kelsol 3990-B2G-70	124.98	14.79
Ammonium Hydroxide	6.72	0.90
Butoxy Ethanol	16.67	2.22
Diacetone Alcohol	8.37	1.07
Propyl Dipropasol	8.31	1.07

Premix:		
Cobalt Hydrocure II	5.07	0.67
Activ-8	0.80	0.10
Butoxy Ethanol	2.48	0.33
Titanox 2101	171.50	5.15
Water	208.80	24.98

Pebble mill to a 7 Hegman grind.

Letdown:		
Kelsol 3990-B2G-70	119.91	14.19
Ammonium Hydroxide	4.11	0.55
Water	283.05	33.98

Analysis:
    Pigment Volume Concentration, Percent: 21.2
    Pigment/Binder Ratio: 1/1
    Percent Solids, Weight: 36.07
    Percent Solids, Volume: 24.66
    Viscosity @ 25C, #4 Ford, Secs.: 35-45
    pH: 8.2-8.5
    Pounds/Gallon: 9.60
    VOC (excluding water):
        Grams/Liter: 334
        Pounds/Gallon: 2.79

SOURCE: Reichhold Chemicals, Inc.: Waterborne Handbook:
        Formulation 2470-59

White, High Gloss, Air Dry or Force Dry Water Reducible Enamel

10-1300-A is formulated to demonstrate outstanding performance of CHEMPOL 10-1300 in high gloss, air dry and force dry
application for use over treated or untreated metal surfaces.
This enamel develops early water resistance within 2-4 hours
after application.

10-1300-A is recommended for either interior or exterior
applications. Salt spray tests have demonstrated the superior
corrosion resistance properties of 10-1300-A.

This enamel can be prepared without first having to solubilize CHEMPOL 10-1300. It is recommended that the formulation
procedure be closely followed. Changes in pigment wetting agents
or driers may cause variation in gloss and performance.

10-1300-A contains 20% organic solvent and amine, 80% water by
volume, which meets Rule 66 requirements for water reducible
coatings containing organic solvents.

Ingredients:	Pounds	Gallons
Premix in order listed:		
Chempol 10-1300	66.71	8.07
Triethylamine	4.86	0.82
Polycompound W-953	2.84	0.37
Nalco 43J-36	1.49	0.15
Water	91.06	10.91
Bykumen	2.84	0.37
Titanium Dioxide (Ti-Pure R-900)	186.76	5.38

Grind Pebble Mill to 7 1/2 Hegman.
Add to mill the following and redisperse one hour:

Chempol 10-1300	66.71	8.07
Triethylamine	4.86	0.82

Let down with:
Premix in order listed and add while under agitation:

Chempol 10-1300	133.41	16.41
Triethylamine	11.35	1.87
Exkin No. 2	1.49	0.15
Paint Additive #14	1.49	0.15
Hydrocure	5.60	0.75
Water	382.99	45.98

Properties:
   Total solids, % by weight: 38.72
   Total solids, % by volume: 26.00
   Viscosity #4 Ford cup, seconds: 30-45
   Reduction for spray application: Use as is
   Pigment/binder weight ratio: 1.0/1.0
   PVC, %: 21
   pH: 8.5-9.5
   Gloss, 60 Gardner: 100+
   Weight/gallon, lbs.: 9.64

SOURCE: Freeman Polymers Division: CHEMPOL 10-1300-A Formulation

Suggested Starting Formulation

10-1300-F is a reformulation of 10-1300-A which uses recent developments in water-reducible technology. Higher solids and enhanced paint stability are the improved properties achieved with 10-1300-F.

Ingredients:	Pounds	Gallons
Chempol 10-1300	69.81	8.46
Triethylamine	5.02	0.79
Coroc A2678M	2.93	0.33
Surfynol 104BC	0.42	0.06
Deionized Water	115.12	13.84
Acrysol ASE60	1.05	0.12
TiPure R700	195.50	6.05

Grind pebble mill to Hegman 7 or better.
Let down with:

Chempol 10-1300	69.81	8.46
NH4OH (28% NH3)	3.66	0.49
Deionized Water	26.16	3.14
Coroc A2678M	4.19	0.48

Premix the following and continue let down under agitation:

Chempol 10-1300	139.61	16.92
Triethylamine	6.07	0.96
NH4OH (28% NH3)	4.19	0.56
Exkin #2	1.57	0.20
Byk 301	0.52	0.06
Cobalt Hydrocure II	7.33	0.96
Manganese Hydrocure II	3.14	0.37
Deionized Water	313.97	37.74

Properties:
   Total solids, % by weight: 41.30
   Total solids, % by volume: 28.06
   Viscosity #4 Ford Cup, secs.: 50.00
   Reduction for spray application: Use as is
   P/B weight ratio: 1.00
   PVC, %: 21.57
   pH: 8.5-9.0
   VOC, #/gallon minus water: 2.35

Film Properties:
   3.0 mil wet drawdown over glass; 78F; 75% relative humidity:
      Set to touch: 25-30 minutes
      Zapon (100 grams): 60-70 minutes
      Dry hard: 4-5 hours
      60 gloss: 90

SOURCE: Freeman Polymers Division: CHEMPOL 10-1300-F Formulation

White High Gloss Brushing Enamel

Material:	Lbs.	Gals.
Water	85.0	10.20
Triethylamine	7.5	1.24
Propasol P	12.0	1.63
Kelsol 3922-G-80	90.0	10.43
Patcote 531	1.0	0.14
Ti-Pure R-902	275.0	8.04

Adjust pH to 8.2-8.6.
Grind in high speed disperser for 15-20 minutes (6+ Hegman).

Letdown:		
Kelsol 3922-G-80	196.0	22.71
Aroplaz 1272	12.0	1.44
Ammonium Hydroxide	13.0	1.73
Water	368.0	44.18

Pre-blend:		
Water	7.0	0.84
Acrysol ASE-60	7.0	0.80

Adjust pH to 8.2-8.6 before proceeding.

Premix:		
Propasol P*	14.0	1.90
Activ-8	0.5	0.06
Cobalt Hydrocure II	3.5	0.46
Calcium Hydrocem	3.5	0.44
Zirconium Hydrocem	3.0	0.36

* Methyl Carbitol may be substituted.

Analysis:
   Pigment Volume Concentration, Percent: 22.7
   Pigment/Binder Ratio: 1.14/1
   Percent Solids, Weight: 47.0
   Percent Solids, Volume: 33.0
   Viscosity @ 25C, #4 Ford, Secs.: 92-96
   Pounds/Gallon: 10.30
   pH: 8.2-8.6
   VOC (excluding water):
      Grams/Liter: 235
      Pounds/Gallon: 1.96

SOURCE: Reichhold Chemicals, Inc.: Waterborne Handbook:
        Formulation 2093-44-2AA

White Semi-Gloss Trade Sales Enamel

Material:	Lbs.	Gals.
Arolon 580-W-2	135.0	15.17
Butoxy Ethanol	40.0	5.33
Ethylene Glycol	15.0	1.60
Potassium Tripolyphosphate	0.5	0.01
Tamol 731	5.0	0.63
6% Cobalt Cyclodex	1.0	0.12
Igepal CO-630	2.0	0.25
Triton GR7	2.0	0.25
Titanox 2101	275.0	7.88
Disperse to 7+ Hegman grind. Then add:		
Letdown:		
Arolon 580-W-42	160.0	18.00
Thickener LN	38.0	4.00
Water	38.0	4.00
Nopco NDW	4.0	0.50
Rhoplex AC 707	190.0	21.11
Water	175.0	21.00

Analysis:
    Pigment Volume Concentration, Percent: 21.5
    Pigment/Binder Ratio: 1.1/1
    Percent Solids, Weight: 48.4
    Percent Solids, Volume: 34.0
    Viscosity @ 25C, KU: 75-85
    Pounds/Gallon: 10.83
    pH: 7.0-7.5
    VOC (excluding water):
        Grams/Liter: 149
        Pounds/Gallon: 1.24

SOURCE: Reichhold Chemicals, Inc.: Waterborne Handbook:
        Formulation 31111-42-25

WR Polyester/Melamine Enamel
K-FLEX Modified

Cowles Grind:	50/15/35
Cargill 7203	156.4
Dimethylethanol amine	10.8
Cymel 303	54.8
Byk 301	1.3
K-Flex 188	23.5
DI Water	30.0
Titanox 2020	313.1

    7+ on Hegman

Letdown:	
Cargill 7203	52.1
Dimethylethanol amine	3.6
Cymel 303	54.7
K-Flex 188	23.4
DI Water	285.2
DI Water (Hold for Adjustment)	242.9

Enamel Characteristics:
    % Solids (by wt): 50
    % TRS: 25
    % Neutralization: 100
    Enamel pH: 8.0

Film Properties:
    Substrate: Bonderite 1000 Iron Phosphated CRS
    Catalyst: None
    Cure: 15' @ 325F.
    Film Thickness: 0.75-0.80 ml
    Pencil Hardness: F-H
    Double MEK Rubs: 100+
    Impact Rev. in. lb.: 120-130
        Dir. in. lb.: 130-140
    Crosshatch Adhesion % Loss: 0
    Gloss 60: 90
    Water Soak 50C. 72 hr.: 65/NA
      60 Gloss/Blister 240 hr.: 60/NA
                480 hr.: 45/NA
    Salt Spray 240 hr. mm Creep/Blister: 3-5/V. Lt. 9

SOURCE: King Industries: Water Reducible Formulation #9

Yellow Gloss Enamel

Material:	Lbs.	Gals.
Water	206.40	24.77
Ammonium Hydroxide	6.19	0.83
Kelsol 3904-BG4-75	103.20	11.94
Add in order listed:		
Aquacat	2.06	0.24
12% Manganese Cem-All	2.06	0.26
Butoxy Ethanol	10.32	1.37
Titanox 2101	15.48	0.45
Mapico Yellow 1050	46.44	1.37
Chrome Yellow X3356	144.48	3.11
Molybdate Orange YE-637-D	6.19	0.13
Sand mill to 6 Hegman grind.		
Letdown:		
Water	275.54	33.12
Ammonium Hydroxide	6.19	0.83
Butoxy Ethanol	5.16	0.69
Kelsol 3904-BG4-75	180.60	20.89

Analysis:
Pigment Volume Concentration, Percent: 18.0
Pigment/Binder Ratio: 1/1
Percent Solids, Weight: 42.7
Percent Solids, Volume: 28.0
Pounds/Gallon: 10.1
pH: 8.2-8.5
VOC (excluding water):
Grams/Liter: 259
Pounds/Gallon: 2.16

SOURCE: Reichhold Chemicals, Inc.: Waterborne Handbook:
Formulation 1619-22

Yellow Iron Oxide High Gloss Enamel

Material:	Lbs.	Gals.
Kelsol 3960-B2G-75	55.98	6.48
Ammonium Hydroxide	4.03	0.54
Butoxy Ethanol	15.99	2.13
Titanox 2101	4.99	0.15
Yellow Iron Oxide 2288D	96.14	2.87
Water	103.96	12.48

Pebble mill to a 7+ Hegman grind.

Letdown:		
Kelsol 3960-B2G-75	233.02	26.97
Ammonium Hydroxide	10.98	1.47
Butoxy Ethanol	14.94	1.99
n-Butanol	10.97	1.63
Cobalt Hydrocure II	6.50	0.86
Activ-8	1.12	0.14
Water	352.27	42.29

Analysis:
    Pigment Volume Concentration, Percent: 11.49
    Pigment/Binder Ratio: 0.47/1.0
    Percent Solids, Weight: 35.76
    Percent Solids, Volume: 26.87
    Viscosity @ 25C, #4 Ford, Secs.: 35-45
    Gloss (60/20): 93/87
    Pounds/Gallon: 9.11
    VOC (excluding water):
        Grams/Liter: 322.80
        Pounds/Gallon: 2.69

SOURCE: Reichhold Chemicals, Inc.: Waterborne Handbook:
        Formulation 2403-47

Section IV
Enamels—Baking

Black Anti-Corrosive Dip Enamel

	Pounds	Gallons
Cargill Water Reducible Chain Stopped		
Alkyd 74-7474	68.1	7.74
Ethylene Glycol Monobutyl Ether	7.5	1.00
Aerosil R972	2.6	0.14
Special Black 4	11.3	0.76
Strontium Chromate 30-AC-3008	28.3	1.00
Byk-020	1.6	0.22
Byk-301	1.3	0.16
Aqueous Ammonia 28%	1.9	0.25
Triethylamine	1.5	0.25
Deionized Water	121.5	14.58

Sand Mill to a 7 Hegman

Cargill Water Reducible Chain Stopped		
Alkyd 74-7474	181.5	20.63
Ethylene Glycol Monopropyl Ether	39.0	5.15

Premix the following, then add:

Butyl Carbitol	5.0	0.63
Manganese Hydro Cure II	1.4	0.17
5% Cobalt Hydro Cure II	2.3	0.30
Activ-8	1.6	0.20

Surfynol 104A	1.9	0.27
Exkin No. 2	2.4	0.31
Aqueous Ammonia 28%	8.0	1.06
Deionized Water	376.2	45.18

Paint Properties:
    % Nonvolatile (By Weight): 27.06
                (By Volume): 22.03
    Pigment to Binder Ratio: 0.22
    Pigment Volume Concentration: 8.80
    Weight Per Gallon: 8.65
    Theoretical VOC (Pounds/Gallon): 3.20
                (Grams/Liter): 384
    Water:Cosolvent Weight Ratio: 80.00:20.00
    Viscosity #4 Ford Cup (Seconds): 51

SOURCE: Cargill, Inc.: CARGILL Formulary: Formula P1848-162B

Black Baking Dip Enamel

	Pounds	Gallons
Cargill Water Reducible Epoxy Ester		
73-7331	106.8	12.64
Ethylene Glycol Monopropyl Ether	3.9	0.51
Dimethylethanolamine	6.9	0.93
Byk-301	1.5	0.19
Special Black 4	19.5	1.32
Aerosil R972	6.7	0.37
Patcote 577	2.4	0.34
Deionized Water	181.0	21.73

Sand Mill to a 7 Hegman

	Pounds	Gallons
Cargill Water Reducible Epoxy Ester		
73-7331	147.5	17.45
Patcote 577	2.4	0.34
Ethylene Glycol Monopropyl Ether	6.5	0.85
Isobutanol	8.6	1.28
Diethylene Glycol Monobutyl Ether	5.7	0.72
Dimethylethanolamine	10.1	1.37
Cargill High Solids Monomeric Methylated		
Melamine 23-2347	24.5	2.42
Nacure 155	1.9	0.23
Deionized Water	310.7	37.31

Paint Properties:
    % Nonvolatile (By Weight): 27.38
                (By Volume): 24.43
    Pigment to Binder Ratio: 0.13
    Pigment Volume Concentration: 6.89
    Weight Per Gallon: 8.46
    Theoretical VOC (Pounds/Gallon): 3.00
                (Grams/Liter): 360
    Water:Cosolvent Weight Ratio: 80.00:20.00
    pH: 8.37
    Viscosity #4 Ford Cup (Sesonds): 25

SOURCE: Cargill, Inc.: CARGILL Formulary: Formula P1848-180

Black Gloss Baking Enamel

Material:	Lbs.	Gals.
Arolon 970-G4-70	79.06	9.07
Ethylene Glycol Monobutyl Ether	15.47	2.06
Dimethylethanol Amine (DMEA)	4.30	0.58
Surfynol 104E	3.44	0.41
Water	137.50	16.51
Raven 1250	21.48	1.47

Steel ball mill to a 7 Hegman grind.

Letdown:		
Arolon 970-G4-70	230.32	26.41
Cymel 303	68.75	6.88
n-Butanol	21.48	3.18
DMEA	6.88	0.93
Water	270.71	32.50

Analysis:
   Pigment Volume Concentration, Percent: 4.65
   Pigment/Binder Ratio: .07/1
   Percent Solids, Weight: 35.90
   Percent Solids, Volume: 31.68
   Viscosity @ 25C, KU: 95-100
   pH: 8.2-8.6
   Pounds/Gallon: 8.59
   Resin/Melamine Ratio: 76/24
   % Reduction to 30 Secs., #2 Zahn: 80
   VOC (excluding water):
      Grams/Liter: 335.7
      Pounds/Gallon: 2.8

SOURCE: Reichhold Chemicals, Inc.: Waterborne Handbook:
        Formulation 2185-67

Black Low Gloss Baking Enamel

Material:	Lbs.	Gals.
Kelsol 3910-B2G-75	71.0	8.20
Triethylamine	4.3	0.71
Water	248.5	29.87
Butoxy Ethanol	10.65	1.42
Strontium Chromate J-1365	17.75	0.57
106 Lo Micron White Barytes	35.5	0.96
Talc 399	35.5	1.55
Novacite L207A	35.5	1.61
Raven 1020	7.1	0.49

Steel ball mill to a 7 Hegman grind.

Letdown:		
Kelsol 3910-B2G-75	124.30	14.35
Triethylamine	5.68	0.94
Cymel 327	39.76	4.06
Water	275.48	33.10
Butoxy Ethanol	16.33	2.17

Analysis:
    Pigment Volume Concentration, Percent: 21.33
    Pigment/Binder Ratio: .72/1
    Percent Solids, Weight: 33.77
    Percent Solids, Volume: 24.42
    Viscosity @ 25C, #4 Ford, Secs.: 30-40
    Pounds/Gallon: 9.28
    pH: 8.2-8.5
    VOC (excluding water):
        Grams/Liter: 235
        Pounds/Gallon: 1.96

SOURCE: Reichhold Chemicals, Inc.: Waterborne Handbook:
        Formulation 2106-30

Bone White Semi-Gloss Baking Enamel

	Pounds	Gallons
Cargill Water Reducible Short Tofa Alkyd		
74-7451	181.9	20.67
Triethylamine	12.3	2.02
Ethylene Glycol Monobutyl Ether	16.5	2.19
Byk-020	2.3	0.31
Titanium Dioxide Ti-Pure R-900	153.2	4.38
Synthetic Iron Oxide Yellow 1103	6.9	0.20
Flatting Agent TS-100	15.3	0.84
Aerosil R972	2.5	0.14
Raybo 62-HydroFlo	1.3	0.16

Sand Mill to a 6-6 1/2 Hegman

	Pounds	Gallons
Cargill Water Reducible Short Tofa Alkyd		
74-7451	116.9	13.28
Ethylene Glycol Monobutyl Ether	16.2	2.15
Triethylamine	6.9	1.14
Cargill High Solids Monomeric Methylated		
Melamine 23-2347	33.9	3.35
Butyl Carbitol	4.6	0.58
Byk-020	1.5	0.21
Byk-Catalyst-451	6.6	0.87
Sec-Butanol	16.1	2.39
Deionized Water	216.4	25.97

	Pounds	Gallons
Deionized Water (Hold for Viscosity		
Adjustment)	159.5	19.15

Paint Properties:
   % Nonvolatile (By Weight): 43.58
                (By Volume): 31.08
   Pigment to Binder Ratio: 0.73
   Pigment Volume Concentration: 17.88
   Weight Per Gallon: 9.71
   Theoretical VOC (Pounds/Gallon): 3.13
                   (Grams/Liter): 375
   Water:Cosolvent Weight Ratio: 68.65:31.35
   pH: 8.0-8.5
   Viscosity #4 Ford Cup: 78

Typical Cured Film Properties:
   Cure Schedule: 10 Minutes at 350F
   Substrate: Cold Rolled Steel
   Gloss (60/20): 75/25
   Pencil Hardness: Pass 2H
   % Crosshatch Adhesion: 100
   Impact (In. Lbs.):
      Direct: 150
      Reverse: 10

SOURCE: Cargill, Inc.: CARGILL Formulary: Formulation P1887-102C

Brown High Gloss Baking Enamel

Material:	Lbs.	Gals.
Cymel 303	67.20	6.75
n-Butanol	13.10	1.93
2-Butoxy Ethanol	13.05	1.74
Hoover Brown 7148	86.40	2.68
Byk 104S	2.00	0.25

Sand or pebble mill to 7+ Hegman grind.

Letdown:		
Arolon 465-G4-80	335.05	34.05
Triethylamine	30.04	4.51
Water	399.17	47.92
Byk 301	1.54	0.19

Analysis:
   Pigment Volume Concentration, Percent: 7.6
   Pigment/Binder Ratio: 0.26/1
   Percent Solids, Weight: 44.49
   Percent Solids, Volume: 35.04
   Viscosity @ 25C, KU: 62-65
   pH: 8.2-8.6
   VOC (excluding water):
      Grams/Liter: 288
      Pounds/Gallon: 2.4
   Resin/Melamine Ratio: 80/20
   Percent Reduction to 30", #2 Zahn Cup with water: 25

SOURCE: Reichhold Chemicals, Inc.: Waterborne Handbook:
        Formulation 2552-1A

Clear Low VOC Baking Enamel

Material:	Lbs.	Gals.
Kelsol 4069-WG4-55	486.6	54.25
Cymel 303	90.0	9.00
Byk 301	2.1	0.25
Water	304.1	36.50

Add the above materials in order under good agitation.
Cure Schedule: 30 minutes at 325F.

Analysis:
  Percent Solids, Weight: 40.4
  Percent Solids, Volume: 36.0
  Viscosity @ 25C, #4 Ford, Secs.: 110
  Pounds/Gallon: 8.8
  VOC (excluding water):
    Grams/Liter: 168
    Pounds/Gallon: 1.4

  Formulation 2316-73

White High Gloss Baking Enamel

Material:	Lbs.	Gals.
Kelsol 4069-WG4-55	119.0	13.27
Water	149.4	17.94
Ti-Pure R-960	261.9	7.86

Premix the above materials, then sandmill or pebble mill grind to
  7+ Hegman.

Letdown:

	Lbs.	Gals.
Kelsol 4069-WG4-55	327.4	36.50
Cymel 303	81.9	8.19
Byk 301	2.7	0.33
Water	132.5	15.91

Cure Schedule: 30 minutes at 325F.
Note: Reduce to spray viscosity 20-30 Secs., #2 Zahn with water.

Analysis:
  Pigment Volume Concentration, Percent: 19.3
  Pigment/Binder Ratio: 0.8/1.0
  Percent Solids, Weight: 54.8
  Percent Solids, Volume: 40.7
  Viscosity @ 25C, KU: 110
  Pounds/Gallon: 10.7
  VOC (excluding water):
    Grams/Liter: 146
    Pounds/Gallon: 1.2

SOURCE: Reichhold Chemicals, Inc.: Waterborne Handbook:
      Formulation 2316-68B

275 Grams/Liter Gloss Black Baking Enamel

	Pounds	Gallons
Cargill Water Reducible Chain Stopped Alkyd 74-7487	116.4	13.14
Ethylene Glycol Monobutyl Ether	4.2	0.56
Triethylamine	7.5	1.23
Byk-301	1.6	0.20
Special Black 4	12.8	0.87
Deionized Water	155.3	18.64

Sand Mill to a 7 Hegman

	Pounds	Gallons
Cargill Water Reducible Chain Stopped Alkyd 74-7487	160.7	18.14
Ethylene Glycol Monobutyl Ether	9.1	1.21
Triethylamine	11.0	1.81
Cargill High Solids Polymeric Methylated Melamine 23-2347	36.1	3.76
Deionized Water	336.9	40.44

Paint Properties:
    % Nonvolatile (By Weight): 31.20
                (By Volume): 27.48
    Pigment to Binder Ratio: 0.05
    Pigment Volume Concentration: 3.16
    Weight Per Gallon: 8.52
    Theoretical VOC (Pounds/Gallon): 2.29
                (Grams/Liter): 275
    Water:Cosolvent Weight Ratio: 84.00:16.00
    Viscosity #4 Ford Cup (Seconds): 61

Typical Cured Film Properties:
    Cure Schedule: 20 Minutes at 275F
    Substrate: Cold Rolled Steel
    Gloss (60/20): 97/89
    Pencil Hardness: Pass B
    % Crosshatch Adhesion: 100

SOURCE: Cargill, Inc.: CARGILL Formulary: Formulation P1848-L3

Gloss Black Baking Enamel

	Pounds	Gallons
Cargill Water Reducible Chain Stopped Alkyd 74-7487	116.4	13.14
Ethylene Glycol Monobutyl Ether	4.2	0.56
Triethylamine	7.5	1.23
Byk-301	1.6	0.20
Special Black 4	12.8	0.87
Deionized Water	155.3	18.64

Sand Mill to a 7 Hegman

	Pounds	Gallons
Cargill Water Reducible Chain Stopped Alkyd 74-7487	160.7	18.14
Ethylene Glycol Monobutyl Ether	9.1	1.21
Triethylamine	11.0	1.81
Cargill High Solids Polymeric Methylated Melamine 23-2347	36.1	3.76
Deionized Water	336.9	40.44

Paint Properties:
    % Nonvolatile (By Weight): 31.20
                   (By Volume): 27.48
    Pigment to Binder Ratio: 0.05
    Pigment Volume Concentration: 3.16
    Weight Per Gallon: 8.52
    Theoretical VOC (Pounds/Gallon): 2.29
                    (Grams/Liter): 275
    Water:Cosolvent Weight Ratio: 84.00:16.00
    Viscosity #4 Ford Cup (Seconds): 61

Typical Cured Film Properties:
    Cure Schedule: 20 Minutes at 275F
    Substrate: Cold Rolled Steel
    Gloss (60/20): 97/89
    Pencil Hardness: Pass B
    % Crosshatch Adhesion: 100

SOURCE: Cargill, Inc.: CARGILL Formulary: Formula P1848-L3

Gloss White Baking Enamel

	Pounds	Gallons
Cargill Water Reducible Oil Free Polyester 72-7289	141.1	15.34
Dimethylethanolamine	8.7	1.18
Cargill High Solids Monomeric Methylated Melamine 23-2347	49.2	4.87
Dow Corning 14 Additive	3.0	0.45
Deionized Water	28.1	3.37
Titanium Dioxide Ti-Pure R-902	246.0	7.19

High Speed to a 7 Hegman

	Pounds	Gallons
Cargill Water Reducible Oil Free Polyester 72-7289	121.0	13.15
Dimethylethanolamine	12.9	1.75
Deionized Water	439.0	52.70

Paint Properties:
    % Nonvolatile (By Weight): 46.91
                (By Volume): 31.65
    Pigment to Binder Ratio: 1.00
    Pigment Volume Concentration: 22.72
    Weight Per Gallon: 10.49
    Theoretical VOC (Pounds/Gallon): 2.04
                (Grams/Liter): 245
    Water:Cosolvent Weight Ratio: 83.87:16.13
    pH: 8.0-8.5
    Viscosity #2 Zahn Cup (Seconds): 50-60

Typical Cured Film Properties:
    Cure Schedule: 10 Minutes at 350F
    Substrate: Cold Rolled Steel
    Gloss (60/20): 88/66
    Pencil Hardness: Pass 3H
    % Crosshatch Adhesion: 100
    Impact (In. Lbs.):
        Direct: 150
        Reverse: 20
    Humidity (Hours): 166
        Gloss (60/20): 86/62
        Pencil Hardness: Pass 2H
        % Crosshatch Adhesion: 100
    QUV (Hours): 150
        Gloss (60/20): 65/21
    Salt Spray (Hours): 166
        Scribe Creep (mm): 2
        Scribe Blisters: Medium Dense #4
        Scribe Corrosion (mm): 1
        Surface Blisters: None
        Tape Pull Removal (mm): 4

SOURCE: Cargill, Inc.: CARGILL Formulary: Formulation P1877-49D

Gloss White Baking Enamel

	Pounds	Gallons
Cargill Water Reducible Short Soya Alkyd		
74-7450	150.2	17.78
Cargill High Solids Monomeric Methylated		
Melamine 23-2347	35.8	3.54
Ethylene Glycol Monobutyl Ether	10.0	1.33
Byk-301	3.0	0.37
Titanium Dioxide Ti-Pure R-900	282.4	8.07

High Speed to a 7 Hegman

Cargill Water Reducible Short Soya Alkyd		
74-7450	217.3	25.72
Triethylamine	22.0	3.64
Ethylene Glycol Monobutyl Ether	18.0	2.40
Deionized Water	309.5	37.15

Paint Properties:
    % Nonvolatile (By Weight): 55.04
                  (By Volume): 38.97
    Pigment to Binder Ratio: 0.96
    Pigment Volume Concentration: 20.71
    Weight Per Gallon: 10.48
    Theoretical VOC (Pounds/Gallon): 2.58
                    (Grams/Liter): 309
    Water:Cosolvent Weight Ratio: 65.66:34.34
    Viscosity Krebs-Stormer (KU): 65

Typical Cured Film Properties:
    Cure Schedule: 10 Minutes at 350F
    Substrate: Cold Rolled Steel
    Gloss (60/20): 95/82
    Pencil Hardness: Pass 2H
    % Crosshatch Adhesion: 100
    Impact (In. Lbs.):
       Direct: 120
       Reverse: 20
    Salt Spray (Hours): 96
       Scribe Creep (mm): 8

SOURCE: Cargill, Inc.: CARGILL Formulary: Formula P1883A-7450

Gloss White Baking Enamel

	Pounds	Gallons
Cargill Water Reducible Chain Stopped		
Alkyd 74-7474	104.4	11.87
Dimethylethanolamine	6.4	0.86
Ethylene Glycol Monobutyl Ether	14.4	1.92
Titanium Dioxide Ti-Pure R-900	215.2	6.15
Byk-020	1.7	0.23
Byk-301	1.1	0.14
Deionized Water	196.0	23.54

High Speed to a 7 Hegman

	Pounds	Gallons
Cargill Water Reducible Chain Stopped		
Alkyd 74-7474	148.5	16.87
Dimethylethanolamine	9.0	1.22
Ethylene Glycol Monobutyl Ether	13.7	1.82
Cargill High Solids Monomeric Methylated		
Melamine 23-2347	25.9	2.57
Surfynol 104 E	1.7	0.21
Deionized Water	271.5	32.60

Paint Properties:
    % Nonvolatile (By Weight): 42.83
             (By Volume): 28.74
    Pigment to Binder Ratio: 0.99
    Pigment Volume Concentration: 21.40
    Weight Per Gallon: 10.10
    Theoretical VOC (Pounds/Gallon): 2.50
             (Grams/Liter): 300
    Water:Cosolvent Weight Ratio: 81.00:19.00
    Viscosity #4 Ford Cup (Seconds): 47

Typical Cured Film Properties:
    Cure Schedule: 20 Minutes at 275F
    Substrate: Cold Rolled Steel
    Gloss (60/20): 91/75
    Pencil Hardness: Pass HB
    Impact (In. Lbs.):
      Direct: 150
      Reverse: 35

SOURCE: Cargill, Inc.: CARGILL Formulary: Formula P1848-157

Green Gloss Force Cure Enamel

Material:	Lbs.	Gals.
Water	111.62	13.40
Triethylamine	6.74	1.11
Kelsol 3902-BG4-75	167.56	19.15
Add in order listed:		
Propasol P	5.60	0.76
Cobalt Hydrocure	2.24	0.29
12% Manganese Cem-All	2.56	0.29
Medium Chrome Yellow X-3356	94.82	2.04
Heucophthal Blue RF BT-627-D	39.09	2.97
Lactimon WS	4.14	0.49
Foamkill 639Q	4.47	0.63
Sand mill to 7 Hegman grind.		
Letdown:		
Water	334.95	40.21
Triethylamine	6.74	1.11
Kelsol 3902-BG4-75	139.65	15.96
sec-Butanol	5.58	0.83
Propasol P	5.59	0.76

Analysis:
   Pigment Volume Concentration, Percent: 11.5
   Pigment/Binder Ratio: .48/1
   Percent Solids, Weight: 39.9
   Percent Solids, Volume: 28.0
   Pounds/Gallon: 9.31
   pH: 8.2-8.5
   VOC (excluding water):
      Grams/Liter: 293
      Pounds/Gallon: 2.46

SOURCE: Reichhold Chemicals, Inc.: Waterborne Handbook:
          Formulation 1701-3

High Gloss Black Baking Enamel

	Pounds	Gallons
Cargill Water Reducible Epoxy Ester 73-7331	113.4	13.42
Ethylene Glycol Monobutyl Ether	4.1	0.55
Dimethylethanolamine	7.3	0.99
Byk-301	1.6	0.19
Special Black 4	12.5	0.85
Deionized Water	166.9	20.03

Sand Mill to a 7 Hegman

	Pounds	Gallons
Cargill Water Reducible Epoxy Ester 73-7331	156.6	18.53
Ethylene Glycol Monobutyl Ether	17.8	2.37
Dimethylethanolamine	10.7	1.45
Cargill High Solids Monomeric Methylated Melamine 23-2347	26.0	2.57
Nacure 155	2.0	0.25
Deionized Water	323.2	38.80

Paint Properties:
  % Nonvolatile (By Weight): 27.24
            (By Volume): 24.80
  Pigment to Binder Ratio: 0.06
  Pigment Volume Concentration: 3.40
  Weight Per Gallon: 8.42
  Theoretical VOC (Pounds/Gallon): 2.98
            (Grams/Liter): 357
  Water:Cosolvent Weight Ratio: 80.00:20.00
  pH: 8.46
  Viscosity #4 Ford Cup (Seconds): 41

Typical Cured Film Properties:
  Cure Schedule: 10 Minutes at 300F
  Substrate: Cold Rolled Steel
  Gloss (60/20): 95/75
  Pencil Hardness: Pass HB
  Impact (In. Lbs.):
     Direct: 160
     Reverse: 140

SOURCE: Cargill, Inc.: CARGILL Formulary: Formula P1848-170

High Gloss Gray Baking Enamel

	Pounds	Gallons
Cargill Water Reducible Short Tofa Alkyd		
74-7461	78.4	9.23
Dimethylethanolamine	4.9	0.66
Titanium Dioxide Ti-Pure R-900	113.0	3.23
Special Black 4	5.6	0.38
Patcote 577	2.4	0.34
Byk-301	2.1	0.27
Ethylene Glycol Monobutyl Ether	1.3	0.18
Deionized Water	97.0	11.64

Sand Mill to a 7-8 Hegman

	Pounds	Gallons
Cargill Water Reducible Short Tofa Alkyd		
74-7461	202.4	23.81
Cargill High Solids Monomeric Methylated		
Melamine 23-2347	27.3	2.71
Dimethylethanolamine	15.0	2.04
Nacure 155	2.1	0.26
Ethylene Glycol Monobutyl Ether	7.2	0.96
Deionized Water	368.9	44.29

Paint Properties:
    % Nonvolatile (By Weight): 37.23
                (By Volume): 27.39
    Pigment to Binder Ratio: 0.52
    Pigment Volume Concentration: 13.18
    Weight Per Gallon: 9.28
    Theoretical VOC (Pounds/Gallon): 2.64
                (Grams/Liter): 317
    Water:Cosolvent Weight Ratio: 80.00:20.00
    pH: 8.7
    Viscosity #4 Ford Cup (Seconds): 44

Typical Cured Film Properties:
    Cure Schedule: 10 Minutes at 325F
    Substrate: Cold Rolled Steel
    Gloss (60/20): 93/76
    Pencil Hardness: Pass HB
    Impact (In. Lbs.):
        Direct: 160
        Reverse: 60

SOURCE: Cargill, Inc.: CARGILL Formulary: Formula P1848-169A

High Gloss Light Blue Baking Enamel

	Pounds	Gallons
Cargill Water Reducible Short Soya Alkyd 74-7450	124.6	14.74
Dimethylethanolamine	9.9	1.34
Heliogen Blue L 6875 F	0.9	0.06
Patcote 577	1.3	0.18
Titanium Dioxide Ti-Pure R-900	154.4	4.41
Deionized Water	151.8	18.22

Sand Mill to a 7 Hegman

	Pounds	Gallons
Cargill Water Reducible Short Soya Alkyd 74-7450	117.6	13.92
Cargill High Solids Monomeric Methylated Melamine 23-2347	23.3	2.31
Dimethylethanolamine	7.0	0.96
Nacure 155	1.8	0.22
Ethylene Glycol Monobutyl Ether	29.4	3.91
Deionized Water	330.9	39.73

Paint Properties:
   % Nonvolatile (By Weight): 36.68
                (By Volume): 24.84
   Pigment to Binder Ratio: 0.80
   Pigment Volume Concentration: 18.02
   Weight Per Gallon: 9.53
   Theoretical VOC (Pounds/Gallon): 2.87
                   (Grams/Liter): 344
   Water:Cosolvent Weight Ratio: 80.00:20.00
   pH: 8.82
   Viscosity #4 Ford Cup (Seconds): 48

Typical Cured Film Properties:
   Cure Schedule: 10 Minutes at 325F
   Substrate: Cold Rolled Steel
   Dry Film Thickness (Mils): <1.5
   Gloss (60/20): 90/70
   Pencil Hardness: Pass 2H
   % Crosshatch Adhesion: 100
   Impact (In.Lbs.):
      Direct: 130
      Reverse: 5
   MEK Double Rubs: Pass 100

SOURCE: Cargill, Inc.: CARGILL Formulary: Formula P1900-4B

High Gloss White Baking Enamel

	Pounds	Gallons
Cargill Water Reducible Coconut Oil Alkyd 74-7455	84.7	9.41
Dimethylethanolamine	5.4	0.73
Cargill High Solids Monomeric Methylated Melamine 23-2347	24.6	2.44
Titanium Dioxide Ti-Pure R-900	174.3	4.98
Byk P 104 S	1.5	0.19
Deionized Water	126.1	15.13

High Speed to a 7 Hegman

Cargill Water Reducible Coconut Oil Alkyd 74-7455	189.7	21.08
Dimethylethanolamine	10.8	1.47
Byk-020	3.5	0.48
Ethylene Glycol Monopropyl Ether	16.5	2.18
Deionized Water	349.2	41.91

Paint Properties:
    % Nonvolatile (By Weight): 39.76
                (By Volume): 27.06
    Pigment to Binder Ratio: 0.80
    Pigment Volume Concentration: 18.41
    Weight Per Gallon: 9.86
    Theoretical VOC (Pounds/Gallon): 2.77
                (Grams/Liter): 332
    Water:Cosolvent Weight Ratio: 80.00:20.00
    pH: 8.35
    Viscosity #4 Ford Cup (Seconds): 52

Typical Cured Film Properties:
    Cure Schedule: 10 Minutes at 325F
    Substrate: Cold Rolled Steel
    Gloss (60/20): 91/70
    Pencil Hardness: Pass HB
    % Crosshatch Adhesion: 100
    Impact (In. Lbs.):
        Direct: 40
        Reverse: <5

SOURCE: Cargill, Inc.: CARGILL Formulary: Formula P1848-182B

High Gloss White Baking Enamel

	Pounds	Gallons
Cargill Water Reducible Acrylic 17-7240	43.88	5.07
Dimethylethanolamine	4.43	0.60
Deionized Water	109.70	13.17
CT-136	9.62	1.14
Titanium Dioxide Ti-Pure R-902	134.72	3.94

Sand Mill to a 7 Hegman

	Pounds	Gallons
Cargill Water Reducible Acrylic 17-7240	148.57	17.17
Dimethylethanolamine	14.82	2.01
Cargill High Solids Monomeric Methylated Melamine 23-2347	33.68	3.34
Deionized Water	429.16	51.52
Ethylene Glycol Monobutyl Ether	13.47	1.79
Byk 306	1.93	0.25

Paint Properties:
    % Nonvolatile (By Weight): 32.64
                  (By Volume): 22.26
    Pigment to Binder Ratio: 0.78
    Pigment Volume Concentration: 17.70
    Weight Per Gallon: 9.44
    Theoretical VOC (Pounds/Gallon): 2.75
                    (Grams/Liter): 329.06
    Water:Cosolvent Weight Ratio: 88.60:11.40
    pH: 8.40
    Viscosity #4 Ford Cup (Seconds): 36

Typical Cured Film Properties:
    Cure Schedule: 10 Minutes @ 350F
    Substrate: B1000
    Dry Film Thickness: 1.0
    Gloss (60/20): 91/71
    Pencil Hardness: 2H
    % Crosshatch Adhesion: 100
    Impact (In. Lbs.):
        Direct: 90
        Reverse: 5
    Humidity (Hours): 250
        Blisters: none
        % Crosshatch Adhesion: 100
    Salt Spray (Hours): 250
        Scribe Creep (mm): 2.5

SOURCE: Cargill, Inc.: CARGILL Formulary: Formula 1855-91

High-Gloss White Baking Finish
Roller Mill Grind

```
Roller mill grind: (3 passes):
Acryloid WR-97 (as supplied)                           300.0
Dimethylaminoethanol                                     7.8
Rutile titanium dioxide                                300.0

Letdown:
Roller mill grind above                                607.8
Acryloid WR-97 (as supplied)                           127.5
Uformite MM-83 (as supplied)                            93.6
Dimethylaminoethanol                                     3.3
Water                                                  782.4
```

Formulation Constants:
   Pigment/binder ratio: 44.5/55.5
   Acrylic/melamine ratio: 80/20
   Solids content: 41.7%
   Water/solvent ratio, by volume: 80/20
   Spray viscosity, No. 4 Ford cup: 30 seconds
   VOC, lbs/gal (calculated): 2.2

Experimental Formulation CTS-197-1

High-Gloss White Baking Finish
Cowles Dissolver Grind

```
Charge to the tank and mix at low speed:
Acryloid WR-97 (50% solids reduced with water)         150.0
Water                                                  110.0
Add gradually while increasing speed:
Rutile titanium dioxide                                340.0
Grind at high speed for 12 to 15 minutes then reduce speed and
   letdown with the following:
Water                                                  734.1
Dimethylaminoethanol (DMAE)                              9.8
Acryloid WR-97 (as supplied)                           377.6
Uformite MM-83 (as supplied)                           106.1
```

Formulation Constants:
   Pigment/binder ratio: 44.5/55.5
   Acrylic/melamine ratio: 80/20
   Solids content: 41.7%
   Water/solvent ratio, by volume: 80/20
   Spray viscosity, No. 4 Ford cup: 30 seconds
   VOC, lbs/gal (calculated): 2.2

SOURCE: Rohm and Haas Co.: Industrial Coatings: ACRYLOID WR-97:
   Experimental Formulation CTS-197-2

High-Gloss White Baking Finish
Sand Mill Grind

Charge to sand mill and grind (15 minutes):
Acryloid WR-97 (50% solids reduced with water)	20
Rutile titanium dioxide	50
Water	30
Sand	100

Filter then letdown as follows:
Filtered grind above	89.00
Acryloid WR-97 (as supplied)	50.71
Uformite MM-83 (as supplied)	13.88
Dimethylaminoethanol (DMAE)	1.32
Water	84.62

Formulation Constants:
   Pigment/binder ratio: 44.5/55.5
   Acrylic/melamine ratio: 80/20
   Solids content: 41.7%
   Water/solvent ratio, by volume: 80/20
   Viscosity, No. 4 Ford cup: 30 seconds
   VOC, lbs/gal (calculated): 2.2

High-Gloss White Baking Finish
Pebble Mill Grind

Pebble mill grind: (16 hours)
Acryloid WR-97 (50% solids reduced with water)	20.0
Rutile titanium dioxide	50.0
Water	30.0

Letdown:
Pebble mill grind above	100.0
Acryloid WR-97 (as supplied)	57.0
Uformite MM-83 (as supplied)	15.6
Dimethylaminoethanol	1.5
Water	95.0

Formulation constants:
   Pigment/binder ratio: 44.5/55.5
   Acrylic/melamine ratio: 80/20
   Solids content: 41.7%
   Water/solvent ratio, by volume: 80/20
   Spray viscosity, No. 4 Ford cup: 30 seconds
   VOC, lbs/gal (calculated): 2.2

SOURCE: Rohm and Haas Co.: Industrial Coatings: ACRYLOID WR-97:
        Experimental Formulation CTS-197-4

Low Temperature Cure Black Baking Enamel

	Pounds	Gallons
Cargill Water Reducible Short Tofa Alkyd 74-7451	62.4	7.09
Triethylamine	4.1	0.68
Byk-301	1.0	0.12
Special Black 4	12.1	0.82
Deionized Water	91.4	10.97

Sand Mill to a 7 Hegman

	Pounds	Gallons
Cargill Water Reducible Short Tofa Alkyd 74-7451	180.1	20.47
Triethylamine	13.2	2.17
Cargill High Solids Polymeric Methylated Melamine 23-2317	86.6	9.02
Ethylene Glycol Monobutyl Ether	8.8	1.17
Byk-Catalyst-451	8.4	1.11
Deionized Water	386.3	46.38

Paint Properties:
    % Nonvolatile (By Weight): 30.11
               (By Volume): 25.84
    Pigment to Binder Ratio: 0.05
    Pigment Volume Concentration: 3.17
    Weight Per Gallon: 8.54
    Theoretical VOC (Pounds/Gallon): 2.80
                 (Grams/Liter): 336
    Water:Cosolvent Weight Ratio: 80.00:20.00
    pH: 8.30
    Viscosity #4 Ford Cup (Seconds): 31

Typical Cured Film Properties:
    Cure Schedule: 20 Minutes at 250F
    Substrate: Cold Rolled Steel
    Gloss (60/20): 97/72
    Pencil Hardness: Pass H
    % Crosshatch Adhesion: 65
    Impact (In. Lbs.):
       Direct: 80
       Reverse: <5

SOURCE: Cargill, Inc.: CARGILL Formulary: Formula P1887-30B

Red Baking Enamel

	Pounds	Gallons
Cargill Water Reducible Chain Stopped		
Alkyd 74-7487	115.5	13.04
Triethylamine	6.0	0.99
Cargill High Solids Monomeric Methylated		
Melamine 23-2347	27.8	2.75
Ethylene Glycol Monobutyl Ether	22.2	2.96
Byk-301	2.0	0.25
13-3061 Novo Perm Red F5RK	40.2	3.55
X-2806 Red Orange	30.2	0.65
Deionized Water	277.8	33.34

Pebble Mill to a 7.5 Hegman

	Pounds	Gallons
Cargill Water Reducible Chain Stopped		
Alkyd 74-7487	145.5	16.43
Triethylamine	6.0	0.99
Ethylene Glycol Monobutyl Ether	7.8	1.04
Deionized Water	200.0	24.01

Paint Properties:
    % Nonvolatile (By Weight): 34.97
               (By Volume): 29.32
    Pigment to Binder Ratio: 0.30
    Pigment Volume Concentration: 14.34
    Weight Per Gallon: 8.81
    Theoretical VOC (Pounds/Gallon): 2.23
               (Grams/Liter): 268
    Water:Cosolvent Weight Ratio: 83.38:16.62
    Resin/Crosslinker Ratio: 88/12
    pH: 8.1-8.5
    Viscosity Krebs-Stormer (KU): 60-65

Typical Cured Film Properties:
    Cure Schedule: 12 Minutes at 350F
    Substrate: Cold Rolled Steel
    Gloss (60/20): 86/61
    Pencil Hardness: Pass F
    Impact (In. Lbs.):
       Direct: 120
    Humidity (Hours): 120
       Blisters: None
       % Crosshatch Adhesion: 99
    Salt Spray (Hours): 96
       Scribe Creep (mm): 8
       Surface Blisters: None
       Surface Corrosion: None

SOURCE: Cargill, Inc.: CARGILL Formulary: Formulation P1883-B5

Semi-Gloss Gray Baking Enamel

	Pounds	Gallons
Cargill Water Reducible Polyester 72-7203	156.9	17.63
Triethylamine	7.7	1.27
Cargill High Solids Monomeric Methylated		
Melamine 23-2347	46.3	4.59
Deionized Water	55.2	6.62
Dow Corning 14	2.7	0.41
Titanium Dioxide Ti-Pure R-902	145.9	4.27
Imsil A-10	97.3	4.41
Special Black 4	2.9	0.20

Sand Mill to a 7 Hegman

Cargill Water Reducible Polyester 72-7203	91.8	10.32
OK 500	9.7	0.62
Triethylamine	13.0	2.15
Deionized Water	336.1	40.34

Deionized Water (Hold for Viscosity Adjustment)	59.7	7.17

Paint Properties:
   % Nonvolatile (By Weight): 47.69
               (By Volume): 33.12
   Pigment to Binder Ratio: 1.10
   Pigment Volume Concentration: 28.65
   Weight Per Gallon: 10.25
   Theortical VOC (Pounds/Gallon): 1.86
              (Grams/Liter): 223
   Water:Cosolvent Weight Ratio: 84.09:15.91
   pH: 8.0-8.5
   Viscosity #2 Zahn Cup (Seconds): 80

Typical Cured Film Properties:
   Cure Schedule: 15 Minutes at 350F
   Substrate: Cold Rolled Steel
   Gloss (60/20): 58/9
   Pencil Hardness: Pass 2H
   % Crosshatch Adhesion: 100
   Impact (In. Lbs.):
      Direct: 160
      Reverse: 30
   Humidity (Gloss 60/20):
      Initial: 58/9
      170 Hours: 65/15
      Observations: Swelling
   Salt Spray (Hours): 150
      Scribe Creep (mm): 6
      Scribe Blisters: Dense #4
      Scribe Corrosion (mm): 1

SOURCE: Cargill, Inc.: CARGILL Formulary: Formula P1877-56D

Silver Metallic Bake Enamel

Material:	Lbs.	Gals.
Hydro-Paste 8726	9.87	0.77
Butoxy Ethanol	16.17	2.16

Reduce paste with solvent slowly with agitation.

Arolon 557 (50 NV)	319.52	38.22
See neutralization below		

Cymel 373	81.06	7.72
Tri-n-butyl Phosphate	9.70	.82
10% DC-29 in BuCell	5.40	.54
Water	413.99	49.77

Resin Neutralization:	Lbs.
Arolon 557-B-70	714.2
Dimethylethanolamine	50.0
Water	235.8

Percent Nonvolatile: 50.0
pH: 8.5
Pounds Per Gallon: 8.3

Analysis:
Pigment Volume Concentration, Percent: 2.0
Pigment/Binder Ratio: .03/1
Percent Solids, Weight: 27.5
Percent Solids, Volume: 23.8
Viscosity @ 25C, #4 Ford, Secs.: 65-75
Pounds/Gallon: 8.55
pH: 8.2-8.5
VOC (excluding water):
Grams/Liter: 349.0
Pounds/Gallon: 2.91

SOURCE: Reichhold Chemicals, Inc.: Waterborne Handbook:
Formulation 1961-65

Tan Semi-Gloss Baking Enamel

Material:	Lbs.	Gals.
Kelsol 3961-B2G-75	103.32	12.15
Butoxy Ethanol	10.33	1.38
Triethylamine	8.27	1.36
Water	103.32	12.40
Triton X-405	2.07	0.26
Raven 1020	1.03	0.07
R.I.O. R-4098	1.03	0.02
Y.I.O. Y-2288	10.33	0.31
Titanox 2020	123.98	3.64
Atomite	151.87	6.75
Cab-O-Sil M-5	1.03	0.06

Pebble mill or sand mill to 6+ Hegman grind.

Letdown:		
Kelsol 3961-B2G-75	103.32	12.15
Cymel 370	70.25	7.17
Butoxy Ethanol	10.33	1.38
n-Butanol	20.66	3.06
Triethylamine	8.27	1.36
Water	303.75	36.48

Analysis:
  Pigment Volume Concentration, Percent: 33.06
  Pigment/Binder Ratio: 1.4/1
  Percent Solids, Weight: 47.98
  Percent Solids, Volume: 32.80
  Viscosity @ 25C, KU: 65-70
  Pounds/Gallon: 10.33
  pH: 8.2-8.6
  VOC (excluding water):
     Grams/Liter: 284
     Pounds/Gallon: 2.37
  Resin/Melamine Ratio: 70/30
  % Reduction to 30", #2 Zahn cup, with water: 10

SOURCE: Reichhold Chemicals, Inc.: Waterborne Handbook:
        Formulation 2185-45

Very Low VOC Gloss Gray Baking Enamel

	Pounds	Gallons
Cargill Waterborne Polyester Dispersion 108-1775	175.1	19.38
Cargill High Solids Monomeric Methylated Melamine 23-2347	56.8	5.62
Titanium Dioxide Ti-Pure R-902	164.0	4.79
Special Black 4A	3.3	0.22
Aerosil R972	2.1	0.12

Sand Mill to a 7 Hegman

Cargill Waterborne Polyester Dispersion 108-1775	595.6	67.68
Dow Corning 14	3.0	0.45
Deionized Water	8.0	0.96
Dimethylethanolamine (Note #1 and #2)	2.0	0.27

Formulating Notes:
1. The amount of dimethylethanolamine necessary to obtain the final paint pH of 8.0-8.3 will vary more than typical water reducible systems.
2. Adjust the paint to its final pH immediately after processing to prevent settling.

Paint Properties:
    % Nonvolatile (By Weight): 45.33
                (By Volume): 33.92
    Pigment to Binder Ratio: 0.59
    Pigment Volume Concentration: 15.13
    Weight Per Gallon: 10.10
    Theoretical VOC (Pounds/Gallon): 0.70
                (Grams/Liter): 84
    pH: 8.0-8.3
    Viscosity #2 Zahn (Seconds): 50-60
Typical Cured Film Properties:
    Cure Schedule: 10 Minutes at 350F
    Substrate: Cold Rolled Steel
    Gloss (60/20): 87/71
    Pencil Hardness: 3H
    % Crosshatch Adhesion: 100
    Impact (In. Lbs.):
        Direct: 160
        Reverse: 120
    QCT Humidity (672 hours):
        Blisters: None
        Percent Gloss Retention: 93/82
    Salt Spray (168 hours):
        Scribe Creep (mm): 3-6
        Scribe Blisters: #8 Medium
        Surface Blisters: None
        Surface Corrosion: None

SOURCE: Cargill, Inc.: CARGILL Formulary: Formula P1877-68G

Very Low VOC Gloss Rose Baking Enamel

	Pounds	Gallons
Cargill Waterborne Polyester Dispersion 108-1775	200.6	22.80
Cargill High Solids Monomeric Methylated Melamine 23-2347	56.1	5.55
Titanium Dioxide Ti-Pure R-902	34.9	1.02
Monastral Red Y RT-759-D	35.1	2.91
Aerosil R972	2.0	0.11

Sand Mill to a 7 Hegman

Cargill Waterborne Polyester Dispersion 108-1775	560.6	63.70
Dow Corning 14	3.3	0.50
Dimethylethanolamine	2.2	0.30
Deionized Water (Hold for viscosity adjustment)	25.9	3.11

Paint Properties:
    % Nonvolatile (By Weight): 38.76
                (By Volume): 32.48
    Pigment to Binder Ratio: 0.25
    Pigment Volume Concentration: 12.45
    Weight Per Gallon: 9.21
    Theoretical VOC (Pounds/Gallon): 0.73
                (Grams/Liter): 88
    pH: 8.0-8.3
    Viscosity #2 Zahn (Seconds): 50-60

Typical Cured Film Properties:
    Cure Schedule: 10 Minutes at 350F
    Substrate: Cold Rolled Steel
    Gloss (60/20): 87/59
    Pencil Hardness: 2H
    % Crosshatch Adhesion: 100
    Impact Resistance (In Lbs.):
        Direct: 160
        Reverse: 60
    Humidity (188 hours):
        Blisters: None
        % Crosshatch Adhesion: 99
    Salt Spray (188 hours):
        Scribe Creep (mm): 1-9
        Scribe Blisters: #2-#4 Medium Dense
        Surface Blisters: None
        Surface Corrosion: None

SOURCE: Cargill, Inc.: CARGILL Formulary: Formula P1977-74J

Very Low VOC Gray Baking Enamel

	Pounds	Gallons
Cargill Waterborne Polyester Dispersion		
72-7231	172.3	19.80
72-7231 must be at a minimum pH of 6.7 for dispersing		
Cargill High Solids Monomeric Methylated		
Melamine 23-2347	55.9	5.53
Titanium Dioxide Ti-Pure R-902	161.4	4.72
Special Black 4A	3.2	0.22
Aerosil R972	2.1	0.11

Sand Mill to a 7 Hegman

Cargill Waterborne Polyester Dispersion		
72-7231	586.1	67.37
Dow Corning 14 Additive	2.9	0.44

Premix the following, then add:

Additol XW 395	4.9	0.57
Deionized Water	4.9	0.59
Triethylamine (Note #2 and #3)	3.9	0.65

Formulating Notes:
1. 72-7231 must be at a minimum pH of 6.7 for dispersing.
2. The amount of triethylamine necessary to obtain the final pH of 8.0 to 8.3 will vary.
3. Adjust the paint to its final pH immediately after processing to prevent settling.

Paint Properties:
  % Nonvolatile (By Weight): 45.39
             (By Volume): 34.41
  Pigment to Binder Ratio: 0.58
  Pigment Volume Concentration: 14.68
  Weight Per Gallon: 9.98
  Theoretical VOC (Pounds/Gallon): 0.76
               (Grams/Liter): 91
  pH: 8.0-8.3
  Viscosity #2 Zahn (Seconds): 35-45

Typical Cured Film Properties:
  Cure Schedule: 10 minutes at 350F
  Substrate: Cold Rolled Steel
  Gloss (60/20): 86/70
  Pencil Hardness: Pass H
  % Crosshatch Adhesion: 100
  Impact (In. Lbs.):
     Direct: 160
     Reverse: 120

SOURCE: Cargill, Inc.: CARGILL Formulary: Formula P1877-68B

Very Low VOC Gray Baking Enamel

	Pounds	Gallons
Cargill Waterborne Polyester Dispersion		
72-7232	135.0	16.71
72-7232 must be at a minimum pH of 6.7 for dispersing		
Deionized Water	22.5	2.71
Cargill High Solids Monomeric Methylated		
Melamine 23-2347	51.1	5.06
Titanium Dioxide Ti-Pure R-902	147.6	4.32
Special Black 4A	3.0	0.20
Aerosil R972	1.9	0.10

Sand Mill to a 7 Hegman

Cargill Waterborne Polyester Dispersion		
72-7232	459.5	56.86
Deionized Water	76.6	9.19
Byk 301	2.7	0.33

Premix the following, then add:

Additol XW 395	4.5	0.52
Deionized Water	4.5	0.54
Triethylamine (Note #2 and #3)	3.6	0.59
Deionized Water (Hold for Viscosity		
Adjustment)	23.9	2.87

Formulating Notes:
1. 72-7232 must be at a minimum pH of 6.7 for dispersing.
2. The amount of triethylamine necessary to obtain the final
   pH of 8.0 to 8.3 will vary.
3. Adjust the paint to its final pH immediately after processing
   to prevent settling.

Paint Properties:
    % Nonvolatile (By Weight): 44.34
              (By Volume): 32.72
    Pigment to Binder Ratio: 0.58
    Pigment Volume Concentration: 14.12
    Weight Per Gallon: 9.36
    Theoretical VOC (Pounds/Gallon): 0.63
                (Grams/Liter): 76
    pH: 8.0-8.3
    Viscosity #2 Zahn (Seconds): 50-60

Typical Cured Film Properties:
    Cure Schedule: 10 Minutes at 350F
    Substrate: Cold Rolled Steel
    Gloss (60/20): 91/74
    Pencil Hardness: Pass H

SOURCE: Cargill, Inc.: CARGILL Formulary: Formula P1877-68J

Very Low VOC Semi-Gloss Blue Baking Enamel

	Pounds	Gallons
Cargill Waterborne Polyester Dispersion		
72-7230	142.6	16.39
Cargill High Solids Monomeric Methylated		
Melamine 23-2347	53.0	5.25
Heliogen Blue L 7071F	10.0	0.75
Titanium Dioxide Ti-Pure R-902	79.9	2.34
Imsil A-10	93.9	4.26
Aerosil R972	1.9	0.10

Sand Mill to a 7 Hegman

	Pounds	Gallons
Cargill Waterborne Polyester Dispersion		
72-7230	567.3	65.20
OK 500	9.4	0.59
Dow Corning 14 Additive	3.0	0.45
Additol XW 395	7.2	0.83
Triethylamine	6.3	1.04

	Pounds	Gallons
Deionized Water (Hold for Viscosity		
Adjustment)	23.3	2.80

Paint Properties:
    % Nonvolatile (By Weight): 46.60
                (By Volume): 35.73
    Pigment to Binder Ratio: 0.72
    Pigment Volume Concentration: 22.50
    Weight Per Gallon: 9.98
    Theoretical VOC (Pounds/Gallon): 0.77
                (Grams/Liter): 92
    pH: 8.0-8.3
    Viscosity #2 Zahn Cup (Seconds): 35-45

Typical Cured Film Properties:
    Cure Schedule: 15 Minutes at 350F
    Substrate: Sheet Molding Compound
    Dry Film Thickness (Mils): 1.0
    Gloss (60/20): 50/6
    Crosshatch Adhesion (% Remaining):
        24 Hours: 100
        Water Soak (4 Days, 24 Hour Recovery): 100
        Humidity (14 Days, 24 Hour Recovery): 100

SOURCE: Cargill, Inc.: CARGILL Formulary: Formula P1877-21E

Very Low VOC Violet Baking Enamel

	Pounds	Gallons
Cargill Waterborne Polyester Dispersion		
72-7232	124.3	15.38
72-7232 must be at a minimum of 6.7 for dispersing		
Deionized Water	24.8	2.97
Cargill High Solids Monomeric Methylated		
Melamine 23-2347	53.5	5.30
Titanium Dioxide Ti-Pure R-902	99.0	2.90
Monastral Violet R RT-201-D	9.9	0.79

Sand Mill to a 7 Hegman

Cargill Waterborne Polyester Dispersion		
72-7232	492.5	60.96
Byk 301	3.1	0.39
Diethylene Glycol Monobutyl Ether	0.9	0.11
Deionized Water	79.2	9.50

Premix the following, then add:

Additol XW 395	5.0	0.58
Deionized Water	5.0	0.60
Triethylamine (Note #2 and #3)	3.1	0.52

Formulating Notes:
1. 72-7232 must be at a minimum pH of 6.7 for dispersing.
2. The amount of triethylamine necessary to obtain the final pH of 8.0 to 8.3 will vary.
3. Adjust the paint to its final pH immediately after processing to prevent settling.

Paint Properties:
   % Nonvolatile (By Weight): 42.46
              (By Volume): 32.93
   Pigment to Binder Ratio: 0.40
   Pigment Volume Concentration: 11.19
   Weight Per Gallon: 9.00
   Theoretical VOC (Pounds/Gallon): 0.66
              (Grams/Liter): 79
   pH: 8.0-8.3
   Viscosity #2 Zahn (Seconds): 50-60

Typical Cured Film Properties:
   Cure Schedule: 10 Minutes at 350F
   Substrate: Sheet Molding Compound
   Dry Film Thickness (Mils): 1.0
   Gloss (60/20): 86/70
   Crosshatch Adhesion (% Remaining):
      24 Hours: 100
      Water Soak (120 Hours, 24 Hour Recovery): 99

SOURCE: Cargill, Inc.: CARGILL Formulary: Formula P1917-8

Very Low VOC White Baking Enamel

	Pounds	Gallons
Cargill Waterborne Polyester Dispersion 72-7230	160.0	18.39
72-7230 must be at a minimum pH of 6.7 for dispersing		
Cargill High Solids Monomeric Methylated Melamine 23-2347	51.9	5.14
Titanium Dioxide Ti-Pure R-902	220.1	6.44

Sand Mill to a 7 Hegman

	Pounds	Gallons
Cargill Waterborne Polyester Dispersion 72-7230	544.6	62.59
Dow Corning 14 Additive	3.0	0.45
Triethylamine (Note #2 and #3)	3.6	0.60
Deionized Water (Hold for Viscosity Adjustment)	53.3	6.39

Formulating Notes:
1. 72-7230 must be at a minimum pH of 6.7 for dispersing.
2. The amount of triethylamine necessary to obtain the final paint pH of 8.0 to 8.3 will vary.
3. Adjust the paint to its final pH immediately after processing to prevent settling.

Paint Properties:
    % Nonvolatile (By Weight): 46.67
                  (By Volume): 33.49
    Pigment to Binder Ratio: 0.84
    Pigment Volume Concentration: 19.22
    Weight Per Gallon: 10.36
    Theoretical VOC (Pounds/Gallon): 0.70
                    (Grams/Liter): 84
    pH: 8.0-8.3
    Viscosity #2 Zahn Cup (Seconds): 35-45

Typical Cured Film Properties:
    Cure Schedule: 15 Minutes at 350F
    Substrate: Cold Rolled Steel
    Gloss (60/20): 93/73
    Pencil Hardness: Pass 2H
    % Crosshatch Adhesion: 100
    Impact (In. Lbs.):
        Direct: 160
        Reverse: 140
    Humidity (Hours): 150
        Blisters: None
        % Crosshatch Adhesion: 100
    Salt Spray (Hours): 150
        Scribe Creep (mm): 4
        Scribe Blisters: Few #2

SOURCE: Cargill, Inc.: CARGILL Formulary: Formulation P1877-32C

Very Low VOC White Baking Enamel

	Pounds	Gallons
Cargill Waterborne Polyester Dispersion		
72-7231	203.6	19.15
72-7231 must be at a minimum pH of 6.7 for dispersing		
Cargill High Solids Monomeric Methylated		
Melamine 23-2347	54.0	5.35
Titanium Dioxide Ti-Pure R-902	277.1	8.10

Sand Mill to a 7 Hegman

Cargill Waterborne Polyester Dispersion		
72-7231	530.2	65.20
Dow Corning 14 Additive	3.2	0.47

Premix the following, then add:

Additol XW 395	5.1	0.59
Deionized Water	5.1	0.62
Triethylamine (Note #2 and #3)	3.2	0.52

Formulating Notes:
1. 72-7231 must be at a minimum pH of 6.7 for dispersing.
2. The amount of triethylamine necessary to obtain the final pH of 8.0 to 8.3 will vary.
3. Adjust the paint to its final pH immediately after processing to prevent settling.

Paint Properties:
   % Nonvolatile (By Weight): 51.24
               (By Volume): 36.53
   Pigment to Binder Ratio: 1.00
   Pigment Volume Concentration: 22.18
   Weight Per Gallon: 10.82
   Theoretical VOC (Pounds/Gallon): 0.69
               (Grams/Liter): 83
   pH: 8.0-8.3
   Viscosity #2 Zahn (Seconds): 35-45

Typical Cured Film Properties:
   Cure Schedule: 10 Minutes at 350F
   Substrate: Cold Rolled Steel
   Gloss (60/20): 94/70
   Pencil Hardness: Pass H
   % Crosshatch Adhesion: 100
   Impact (In. Lbs.):
      Direct: 160
      Reverse: 80
   Humidity (Hours): 344
      Blisters: None
      % Crosshatch Adhesion: 100

SOURCE: Cargill, Inc.: CARGILL Formulary: Formula 1877-9B

Waterborne Acrylic Bake Enamel

The WR-17 study was conducted to demonstrate the improved corrosion resistance afforded a waterborne system by choosing a more hydrophobic catalyst, NACURE 155 (DNNDSA), over a more hydrophilic catalyst, K-CURE 1040 (p-TSA). Free acid catalyst, rather than the blocked versions, were used in this study, since the formula already contained a neutralizing amine. A summary of the formulation, it's physical properties and test results are listed below:

Grind:
TiO2 2020	150.7
Acrysol WS-68	39.7
2-Butoxyethanol	6.7
n-Butanol	6.3
DMEA	0.9
D.I. Water	46.9

Pebble mill grind to 7+ on Hegman

Let Down:
Acrysol WS-68	436.3
n-Butanol	5.4
Cymel 303LF	45.1
2-Butoxyethanol	39.0
*DMEA (10% in water)	22.0
*D.I. Water	201.0

Pigment/Binder: 40/60
Acrylic/Melamine: 80/20
Total Solids, %: 37.7
Solids 1 hr/110C: 37.4
Total Resin Solids: 21.1
Wt/gal @ 25C (lb/gal): 9.65
V.O.C.: 1.84

* Used to adjust pH and viscosity.
Film Properties:
Substrate: B1000
Cure: 15 min./149C
Thickness: 0.7-0.8 mil

	Control	*K-1040 p-TSA	*N-155 DNNDSA
Pencil Hardness	H-2H	H-2H	H-2H
Tukon KNH25	13.2	13.2	13.0
MEK Rub (2X)	24	42	50
Impact Reverse, in/lbs	90-100	120-130	90-100
Direct, in/lbs	40-50	30-40	30-40
Crosshatch, % Adhesion	100	100	100
Salt Spray, 150 hours	10/6-7mm	10/6-7mm	10/3-4mm

* 0.5% on TRS

SOURCE: King Industries: Formulation WR-17

Water Reducible, Force Cure Black Gloss Enamel

Formulation 10-0091-G shows how CHEMPOL 10-0091 may be used with a thermoplastic acrylic latex to gain exceptional adhesion to untreated steel surfaces. This blend provides cured films with outstanding adhesion, hardness, and corrosion resistance, especially when coating untreated steel substrate.

Formulation 10-0091-G was designed primarily for curing in force dry application, but it may also be used in air dry situations.

10-0091-G is formulated to be sprayed "as is". When used in this manner, the volatile organic compounds at point of application are 2.93 pounds per gallon, excluding water. These "VOC" emissions conform to most existing and pending state and federal Clean Air regulations.

It is strongly recommended that the mixing procedure be followed precisely. Changes in the raw material or mixing procedure may cause wide variations in the finished product.

The pH of the finish should be monitored during manufacture and maintained at a value of 7.5-8.0. If the pH is allowed to go beyond 8.5 "seeding" of the acrylic latex may occur.

Ingredients:	Pounds	Gallons
Chempol 10-0091	53.83	6.12
DMAMP-80	3.23	0.40
Nalco 955-815	2.42	0.24
Polycompound W-953	1.61	0.16
EA-1075	1.61	0.16
Water	96.85	11.62
Bykumen	2.02	0.24
Carbon black (Sterling R)	16.14	1.05

Charge to steel ball mill and grind 7-1/2 to 8 Hegman gage.
Add the following to mill and grind 30 minutes.

	Pounds	Gallons
Chempol 10-0091	53.83	6.12
DMAMP-80	2.66	0.32
Water	57.39	6.86

Let Down With:

	Pounds	Gallons
Chempol 10-0091	50.69	17.12
Cobalt Hydrocure 5%	1.94	0.16
Manganese Hydrocure 5%	1.94	0.16
VP-321	0.81	0.08
Texanol	28.25	3.35
Anti-skinning agent	1.61	0.16
Nalco 2309	0.56	0.08
Nalco 2302	0.24	0.08
DMAMP-80	10.33	1.29
Water	250.12	30.75
Ucar 4431X	117.84	13.48

Properties:
    Total solids, % by weight: 30.17
    Total solids, % by volume: 21.31
    Viscosity, #4 Ford Cup, seconds: 45-55

SOURCE: Freeman Polymers Division: CHEMPOL 10-0091-G Formulation

Water Reducible, Gloss White Baking Enamel

Ingredients:	Pounds	Gallons
Grind paste.		
Chempol 10-0503	152.73	16.97
Butyl Cellosolve	61.12	8.15
Byk P-104S	4.11	0.52
Dimethylethanolamine	8.92	1.05
Titanium Dioxide (R-902)	405.24	12.28
Water	48.64	5.86

Blend resin and ethylene glycol monobutyl ether. Add Byk P-104S and mix. Add DMEA and mix. Add TiO2 followed by water. Grind to 7 1/2 Hegman.

Stabilize:		
Chempol 10-0503	31.14	3.46
Cymel 303	118.20	11.82
DMEA	8.67	1.02
Water	13.03	1.57

Blend stabilizing solution and mix into grind.

Let down with:		
Chempol 10-0503	154.35	17.15
DMEA	16.83	1.98
Cycat 4040	2.77	0.37
Dipropylene glycol monomethyl ether	26.39	3.32
Byk 020	7.27	0.92
Water	12.54	13.56

Properties:
  Viscosity: 43 sec. #4 Ford cup
  pH: 8.7
  Nonvolatile, %: 66.3
  Polyester/Melamine: 68/32
  Pigment/binder weight ratio: 1.00/1.00
  Cure cycle: 15 min. @ 300F
  Gloss, 60: 88

SOURCE: Freeman Polymers Division: CHEMPOL 10-0503-A Formulation

Water Reducible, High Gloss White Baking Enamel

Suggested Starting Formulation 10-0503-B is formulated to illustrate the outstanding performance of CHEMPOL 10-0503 saturated polyester. Excellent paint stability and outstanding film hardness, gloss, flexibility, and adhesion are the hallmarks of this water reducible polyester.

Ingredients:	Pounds	Gallons
Chempol 10-0503	148.21	15.12
Butyl Cellosolve	55.61	7.41
DMEA	13.07	1.77
Hypermer PS2	2.78	0.33
Coroc A2678M	8.34	0.95
Deionized Water	185.47	22.29

Mix until clear, then add using a Cowles dissolver

Tiona RCL-535	333.69	9.56

Ball or pebble grind to Hegman 7-1/2+ and let down with the following pre-mixed resin solution:

Chempol 10-0503	111.23	11.35
Butyl Cellosolve	27.81	3.70
Cymel 303	83.42	8.34
DMEA	7.23	0.98
Deionized Water	48.38	5.82

Premix and slowly add under stirring the following:

Acrysol ASE60	1.11	0.13
Deionized Water	82.31	9.89

And, finally, premix and add:

Cycat 4040	2.78	0.35
DMEA	0.56	0.08
Deionized Water	16.13	1.94

Properties:
    % Weight Solids: 54.96
    % Volume Solids: 36.32
    VOC, #/gallon: 5.08
    VOC, #/gallon-water: 2.92
    P/B ratio: 1.2
    % PVC: 26.3
    Binder/Melamine ratio: 70/30
    Viscosity, #4 Ford Cup: 26 seconds

SOURCE: Freeman Polymers Division: CHEMPOL 10-0503-B Formulation

Water Soluble Baking Enamels

Curing Information:
   Suggested Water Reduction Schedule:

Ingredients:	Parts by Weight
Chempol 10-0501	44.4
Butoxyethanol	14.8
Triethylamine	4.6
Water	29.6
Resimene 745	6.6

Ingredients:	Parts by Weight
Chempol 10-0501	44.4
Butoxyethanol	14.8
Triethanolamine	4.1
Water	30.1
Resimene 745	6.6

Suggested Baking Schedule:

Time	Temperature
10 minutes	350F
20 minutes	325F

SOURCE: Freeman Polymers Division: CHEMPOL 10-0501 Formulation

White Baking Enamel

	Pounds	Gallons
Cargill Water Reducible Oil Free Polyester 72-7203	161.3	18.13
Triethylamine	7.9	1.31
Cargill High Solids Monomeric Methylated Melamine 23-2347	47.6	4.71
Deionized Water	56.7	6.80
Dow Corning 14 Additive	2.8	0.43
Titanium Dioxide Ti-Pure R-902	237.9	6.96

High Speed to a 7 Hegman

	Pounds	Gallons
Cargill Water Reducible Oil Free Polyester 72-7203	94.4	10.60
Triethylamine	13.4	2.21
Deionized Water	345.5	41.48

	Pounds	Gallons
Deionized Water (Hold for Viscosity Adjustment)	61.4	7.37

Paint Properties:
   % Nonvolatile (By Weight): 46.42
                (By Volume): 31.25
   Pigment to Binder Ratio: 0.99
   Pigment Volume Concentration: 22.26
   Weight Per Gallon: 10.29
   Theoretical VOC (Pounds/Gallon): 1.98
                (Grams/Liter): 237
   Water:Cosolvent Weight Ratio: 84.09:15.91
   pH: 8.0-8.5
   Viscosity #2 Zahn Cup (Seconds): 60-70

Typical Cured Film Properties:
   Cure Schedule: 10 Minutes at 350F
   Substrate: Cold Rolled Steel
   Gloss (60/20): 94/82
   Pencil Hardness: Pass H
   % Crosshatch Adhesion: 100
   Impact (In. Lbs.):
      Direct: 160
      Reverse: 60
   Humidity (Hours): 150
      Blisters: None
      % Crosshatch Adhesion: 100
   Salt Spray (Hours): 150
      Scribe Creep (mm): 3
      Scribe Blisters: Medium Dense #4
      Surface Blisters: None
      Surface Corrosion: None

SOURCE: Cargill, Inc.: CARGILL Formulary: Formula P1877-30B

White Baking Enamel

	Pounds	Gallons
Cargill Water Reducible Self Crosslinking		
Resin 73-7390	73.0	8.57
Dimethylethanolamine	4.5	0.61
Titanium Dioxide Ti-Pure R-900	224.9	6.43
Patcote 577	3.0	0.44
Byk-301	3.0	0.37
Diethylene Glycol Monobutyl Ether	1.5	0.19
Deionized Water	90.0	10.80

High Speed to a 7 Hegman

	Pounds	Gallons
Cargill Water Reducible Self Crosslinking		
Resin 73-7390	215.4	25.29
Dimethylethanolamine	14.0	1.90
Raybo 62-HydroFlo	5.0	0.59
Ethylene Glycol Monobutyl Ether	4.0	0.53
Deionized Water	368.9	44.28

Paint Properties:
    % Nonvolatile (By Weight): 42.62
            (By Volume): 29.04
    Pigment to Binder Ratio: 1.10
    Pigment Volume Concentration: 22.13
    Weight Per Gallon: 10.07
    Theoretical VOC (Pounds/Gallon): 2.65
              (Grams/Liter): 318
    Water:Cosolvent Weight Ratio: 79.40:20.60
    pH: 8.6
    Viscosity #2 Zahn Cup (Seconds): 30

Typical Cured Film Properties:
    Cure Schedule: 15 Minutes at 350F
    Substrate: Cold Rolled Steel
    Gloss (60): 85
    Pencil Hardness: Pass 3H
    Impact (In. Lbs.):
      Direct: 130
      Reverse: 40
    Salt Spray (Hours): 150
      Scribe Creep (mm): 1
      Scribe Blisters: Few, #4
      Scribe Corrosion (mm): 1
      Surface Blisters: None
      Surface Corrosion: None

SOURCE: Cargill, Inc.: CARGILL Formulary: Formula P1396-77

White Baking Enamel

Material:	Lbs.	Gals.
Titanox 2101	226.0	6.78
Arolon 580-W-42	113.0	12.71
Butoxy Ethanol	45.0	5.99
Water	22.0	2.64
Odorless Mineral Spirits	5.4	0.86
Pebble mill 18-24 hours.		
Letdown:		
Arolon 580-W-2	526.0	59.17
Cymel 373	38.6	3.68
n-Butanol	30.0	4.44
Water	31.0	3.73
Sodium Nitrite	1.0	.06

Analysis:
  Pigment Volume Concentration, Percent: 18.1
  Pigment/Binder Ratio: .75/1
  Percent Solids, Weight: 52.8
  Percent Solids, Volume: 37.5
  Viscosity @ 25C, #4 Ford, Secs.: 40-60
  Pounds/Gallon: 10.39
  pH: 7.0-7.5
  VOC (excluding water):
      Grams/Liter: 212
      Pounds/Gallon: 1.77

SOURCE: Reichhold Chemicals, Inc.: Waterborne Handbook:
        Formulation P-233A

White Baking Enamel

Material:	Lbs.	Gals.
Arolon 860-W-45	271.39	30.84
Surfynol 104H	2.45	0.31
Colloid 679	2.69	0.29
Water	57.56	6.91
Triethylamine	2.98	0.33
Ti-Pure R-902	215.80	6.50

Grind to a 7 Hegman on Cowles or pebble mill.

Letdown:		
Arolon 860-W-45	324.90	36.92
Cymel 303	53.30	5.33
Premix and add under agitation:		
Butoxy Ethanol	26.70	3.56
Butyl Carbitol	26.80	3.37
Byk 301	1.03	0.13
Water	41.15	4.94
Triethylamine	5.14	0.57

Analysis:
   Pigment Volume Concentration, Percent: 17.57
   Pigment/Binder Ratio: 0.80/1
   Percent Solids, Weight: 52.59
   Percent Solids, Volume: 42.66
   Viscosity @ 25C, #4 Ford, Secs.: 35-45
   Pounds/Gallon: 10.32
   pH: 8.2-8.5
   VOC (excluding water):
      Grams/Liter: 152.4
      Pounds/Gallon: 1.27

SOURCE: Reichhold Chemicals, Inc.: Waterborne Handbook:
        Formulation 2402-17D

White Baking Enamel

Material:	Lbs.	Gals.
Kelsol 4097-WG4-55	119.8	13.27
Butoxy Ethanol	9.9	1.32
Water	140.4	16.85
Titanox 2101	263.6	7.91

Pebble or sandmill grind to 7 Hegman.

Letdown:		
Kelsol 4097-WG4-55	329.4	36.48
Cymel 303	82.3	8.23
Byk 301	2.7	0.33
Water	130.0	15.61

Adjust final pH to 8.2-8.8, if necessary, with DMEA.

Analysis:
  Pigment Volume Concentration, Percent: 19.5
  Pigment/Binder Ratio: 0.8/1.0
  Percent Solids, Weight: 55.1
  Percent Solids, Volume: 40.7
  Viscosity @ 25C, KU: 70-80
  Pounds/Gallon: 10.8
  pH: 8.2-8.6
  VOC (excluding water):
    Grams/Liter: 186
    Pounds/Gallon: 1.6
  Formulation 2293-86

White Gloss Baking Enamel

Material:	Lbs.	Gals.
Water	261.5	31.40
Triethylamine	11.6	1.90
Kelsol 5293-B-75	129.9	14.62
Titanox 2101	303.9	9.12

Pebble mill to a 7 Hegman grind.

Letdown:		
Kelsol 5293-B-75	216.2	24.35
Cymel 303	78.1	7.81
Triethylamine	30.4	5.02
Butoxy Ethanol	20.6	2.74
Water	22.2	2.66
DC29	3.3	0.38

Analysis:
  Pigment Volume Concentration, Percent: 20.8
  Pigment/Binder Ratio: 0.9/1
  Percent Solids, Weight: 59.6
  Percent Solids, Volume: 43.8
  Viscosity @ 25C, #4 Ford, Secs.: 60-70
  Pounds/Gallon: 10.8
  pH: 8.2-8.6
  VOC (excluding water):
    Grams/Liter: 271
    Pounds/Gallon: 2.26
SOURCE: Reichhold Chemicals, Inc.: Waterborne Handbook:
        Formulation 2425-22

White Baking Enamel

Material:	Lbs.	Gals.
Water	86.0	10.32
DMEA	4.0	0.54
Synthemul 40-427	46.0	5.41
Foamaster R	1.0	0.14
Diethylene Glycol	30.0	3.22
Titanox 2020	300.0	8.77
High speed disperser, Hegman grind of 7 1/2.		
Super Beckamine 27-568	66.7	6.74
Water	35.3	4.24
Synthemul 40-431	428.0	49.77
DMEA	3.0	0.41

Analysis:
    Pigment Volume Concentration, Percent: 18.8
    Pigment/Binder Ratio: 1/1
    Percent Solids, Weight: 60.4
    Percent Solids, Volume: 46.7
    Viscosity @ 25C, #4 Ford, Secs.: 25-30
    Pounds/Gallon: 11.2
    pH: 8.3-8.8
    VOC (excluding water):
        Grams/Liter: 145.0
    Melamine, Percent: 19.8
Bonderite 37, 0.7-0.9 mils, 20 min. @ 300F cure.
    60 Gloss: 87
    20 Gloss: 68
    Mek, 30 Double Rubs: Exc.
    Impact Resistance:
        Forward: F8
        Reverse: F120
    150 Hours Salt Fog:
        Appearance: No rust, no blisters
        60 Gloss: 90
        20 Gloss: 72
    4-Hour Spot Test:
        Toluene: Exc., no blisters
        Varsol: Exc., no blisters
        Gasoline: Exc., no blisters
        5% NaOH Solution: Exc., no blisters
        5% H2SO4: Exc., no blisters

SOURCE: Reichhold Chemicals, Inc.: Waterborne Handbook:
        Formulation 431-01

White Baking Enamel

Material:	Pounds	Gallons
Rutile Titanium Dioxide	250.0	7.50
CMD 9012 Acrylic Resin	114.0	12.67
Hexamethoxymethylmelamine Resin (2)	45.8	4.58
Polyester Polyol Resin (3)	21.4	2.40
n-Butyl Alcohol	18.7	2.78
High Speed Disperse to 7-8 NS. Add		
CMD 9012 Acrylic Resin	91.3	10.14
Hexamethoxymethylmelamine Resin (2)	36.7	3.67
Polyester Polyol Resin (3)	17.1	1.91
n-Butyl Alcohol	15.0	2.22
Mix thoroughly and add SLOWLY under agitation:		
Dimethylethanolamine	17.5	2.24
Mix well and add slowly:		
Deionized Water	415.6	49.89
Adjust pH to 8.0-8.7 using Dimethylethanolamine		

(2) Cymel 303
(3) Niax PCP-0301

Typical Coating Properties:
    Total Weight Solids: 50.4%
    Total Volume Solids: 36.5%
    P.V.C.: 20.5%
    Viscosity: 115-120 Krebs Units
    pH: 8.0-9.7 (Adjusted)
    Reduced Viscosity: 40-50"
        (3:1 in DI Water)   #4 Ford Cup

Resin Composition (Solids):
    Acrylic Resin: 56%
    Polyester Polyol Resin: 14%
    Hexamethoxymethylmelamine Resin: 30%
Volatile Organic Compounds (VOC):
    Pounds/Gallon: 2.21
    Grams/Liter: 265

Typical Film Properties: Uncatalyzed:
    Pencil Hardness: 4H
    60 Gloss: 86
    Reverse Impact: Cracks 40 in/lbs.
    Direct Impact: Pass 100 in/lbs.
    1/4" Conical Mandrel: Fails 1/2"
    MEK Rubs: 100-B (P.H.)
Catalyzed (5):
    Pencil Hardness: 6-7H
    60 Gloss: 97
    Reverse Impact: 10 in/lbs.
    Direct Impact: Pass 40, Fails 60 in/lbs.
    (5) Catalyst is 0.25%--(Total Resin Solids) Syntex 3981

SOURCE: Shell Chemical Co.: 25-200 White Baking Enamel

White, High Gloss, Water Reducible Baking Enamel

Formulation 10-1105-A is a high gloss white baking enamel based on a water reducible coconut alkyd, CHEMPOL 10-1105. Coatings based on CHEMPOL 10-1105 demonstrate good hardness, humidity and overbake resistance. The product should be used for general purpose baking enamels over treated or untreated metal substrate where color retention is required.

Ingredients:	Pounds	Gallons
Titanium Dioxide (Ti-Pure R-902)	304.78	9.15
Chempol 10-1105	123.08	13.47
Byk P-104S	3.07	0.39
Byk 020	2.76	0.38
Dimethylethanolamine	36.74	0.91
Ethylene glycol monobutyl ether	37.78	5.03
Water	36.55	4.39

Grind in Pebble Mill to 7 1/2 Hegman gauge.

Add the following and redisperse for 30 minutes:

Chempol 10-1105	25.21	2.76
Cymel 303	121.77	12.18
Dimethylethanolamine	6.51	0.88
Water	8.05	0.97

Let down with:

Chempol 10-1105	131.58	14.40
Dimethylethanolamine	13.26	1.79
Cycat 4040	2.22	0.28
Ucar Solvent 2LM	20.61	2.59
Ethylene glycol monobutyl ether	10.27	1.37
Water	241.86	29.07

Properties:
  Total solids, % by weight: 57.03
  Total solids, % by volume: 41.07
  Reduction for spray application (pbw): 150 parts paint/
                                          7 parts water
  Reduction viscosity, #4 Ford cup, seconds: 55-65
  Pigment/binder weight ratio: 0.96/1.0
  VOC, lbs/gals, (calculated): 2.81
  PVC, %: 22.29
  Gloss (60 Gardner Glossmeter): 94
  Weight/gallon, lbs.: 10.58
  pH: 8.0-9.0
  Flash-off time, (3 mil wet film): 5-10 minutes
  Suggested cure schedule: 10 minutes @ 325F
  % Catalyst (based on vehicle solids): 0.55
  Acrylic/Melamine (based on vehicle solids): 62/38

SOURCE: Freeman Polymers Division: CHEMPOL 10-1105-A Formulation

White, High Gloss, Water Reducible Baking Enamel

Formulation 10-1706-B is a high gloss white baking enamel based on a water reducible acrylic copolymer, CHEMPOL 10-1706. CHEMPOL 10-1706-B demonstrates excellent humidity resistance, gloss retention in QUV exposure testing and good corrosion resistance. CHEMPOL 10-1706-B should be used for high perform- ance industrial baking finishes where exterior durability is extremely important.

Ingredients:	Pounds	Gallons
Titanium Dioxide (Ti-Pure R-902)	300.60	9.03
Chempol 10-1706	113.30	12.88
Byk P-104S	3.02	0.38
Byk 020	2.72	0.37
Dimethylethanolamine	6.65	0.90
Ethylene glycol monobutyl ether	45.35	6.04
Water	36.05	4.33

Grind in Pebble Mill to 7 1/2 Hegman gauge.
Add the following and redisperse for 10 minutes:

Chempol 10-1706	23.13	2.63
Cymel 303	120.10	12.01
Dimethylethanolamine	6.42	0.87
Water	9.67	1.16

Let down with:

Chempol 10-1706	120.33	13.67
Dimethylethanolamine	13.30	1.80
Cycat 4040	2.19	0.27
Ucar Solvent 2LM	20.48	2.58
Ethylene glycol monobutyl ether	18.90	2.58
Water	237.64	28.56

Properties:
    Total solids, % by weight: 57.04
    Total solids, % by volume: 41.96
    Reduction for spray application (pbw): 150 parts paint/
                                             7 parts water
    Reduction viscosity, #4 Ford cup, seconds: 55-60
    Pigment/binder weight ratio: 0.96/1.00
    VOC, lbs/gal (calculated), less water: 2.77
    PVC, %: 21.52
    Gloss (60 Gardner Glossmeter): 90
    Weight/gallon, lbs.: 10.86
    pH: 8.0-9.0
    Flash-off time, (3.0 mil wet film): 5-10 minutes
    Suggested cure schedule: 10 minutes @ 325F
    % Catalyst (based on vehicle solids): 0.7
    Acrylic/Melamine (based on vehicle solids): 62/38

SOURCE: Freeman Polymers Division: CHEMPOL 10-1706-B Formulation

White, High Gloss, Water Reducible Baking Enamel

10-1300-B white, high gloss, water reducible baking enamel was formulated to demonstrate the performance of CHEMPOL 10-1300 in a general purpose high gloss baking enamel over treated (Bonderite) and untreated metal surfaces. 10-1300-B is intended for spray applications.

10-1300-B can be prepared without first having to solubilize CHEMPOL 10-1300. It is important that the procedure for preparing this enamel be closely followed. Changes in pigment wetting agents or melamine resin may cause variation in gloss and performance.

When reduced as directed, 10-1300-B contains 19.5% organic solvent and amine, 80.5% water by volume which meets Rule 66 requirements for water reducible coatings containing organic solvents.

Ingredients:	Pounds	Gallons
Premix in order listed:		
Chempol 10-1300	75.19	9.09
Triethylamine	5.47	0.93
Polycompound W-953	3.20	0.42
Nalco 65J-769	1.68	0.17
Water	102.64	12.29
Bykumen	3.20	0.42
Titanium Dioxide (Ti-Pure R-900)	210.50	6.06

Grind Pebble Mill to 7 1/2 Hegman gauge.
Add to mill the following and disperse one hour:

Chempol 10-1300	75.19	0.09
Triethylamine	5.47	0.93

Let down with:
Premix in order listed and add while under agitation:

Chempol 10-1300	105.25	12.71
Resimene 1-720	35.11	3.54
Triethylamine	12.80	2.11
Exkin No. 2	1.68	0.17
Paint Additive #14	2.11	0.25
Catalyst 600	1.09	0.08
Water	347.66	41.74

Properties:
    Total solids, % by weight: 42.60
    Total solids, % by volume: 28.87
    Viscosity #4 Ford cup, seconds: 75-90
    Reduction for spray application: 10 parts 10-1300-B/
        by volume   1 part water
    Pigment/binder weight ratio: 1.0/1.0
    PVC, %: 21
    pH: 8.5-9.5

SOURCE: Freeman Polymers Division: CHEMPOL 10-1300-B Formulation

Water Reducible Gloss White Baking Enamel

Ingredients:	Pounds	Gallons
Chempol 10-0173	126.75	14.64
Dimethylaminoethanol	8.58	1.16
Byk VP020	2.47	0.33
Byk P-104S	4.17	0.53
Butyl Carbitol	19.80	2.49
Water	69.43	8.33
Ti-Pure R-900	207.31	6.04

Weigh in first four ingredients and mix well. Add water separately with agitation. Add pigment slowly with continued agitation.

Sandmill to #7 Hegman keeping grind temperature below 120F with water jacket.

Let down with:

Chempol 10-0173	118.86	13.72
Dimethylaminoethanol	6.17	0.84
Butyl Cellosolve	20.75	2.76
Byk 301	2.95	0.35
Cymel 303	57.30	5.72
Water	262.02	31.45
Water (hold to adjust viscosity)	96.67	11.64

In the "let down", preblend the first four ingredients and mix well. Add water to the mixture with continued agitation. Then add the "let down" portion to the grind slowly with good agitation.

Check the pH and adjust with more Dimethylaminoethanol if necessary. Use the "adjust viscosity" water to bring the viscosity of the coating to spec at room temperature.

Properties:
   Total solids, % by weight: 43-44
   Total solids, % by volume: 30-31
   Weight/gallon, lbs.: 10.0-10.5
   VOC (lbs/gal), less water: 2.7-2.8
   PVC, %: 20.003
   Viscosity, #2 Zahn: 30"-40"
   pH: 8.5-9.0
   Substrate; Bonderite 1000 CRS
   Bake: 10' @ 325F
   DFT: 1.0-1.2 mils
   Gloss: 60: 80
          20: 40
   Pencil hardness: F-2H
   MEK: Passed 50 Double Rubs
   Impact resistance: Passed 60 inch pounds
   Crosshatch: 100% retention

SOURCE: Freeman Polymers Division: CHEMPOL 10-0173-B Formula

Section V
Exterior Paints and Related

Accent-Base Trim Paint

Ingredients:	Pounds	Gallons
Pigment Grind:		
Water	302.5	36.32
Cellosize Hydroxyethyl Cellulose ER-4400	5.0	0.45
Propylene Glycol	51.8	6.00
2-Amino-2-methyl-1-propanol (1)	2.0	0.25
Dispersant (2)	2.5	0.25
Antifoam (3)	1.9	0.25
Preservative (4)	2.3	0.25
Triton Nonionic Surfactant N-101	2.2	0.25
Clay (5)	50.0	2.07
Nepheline Syenite (6)	25.0	1.13
Letdown:		
Ucar Latex 624	438.9	49.88
Ucar Filmer IBT	11.0	1.39
Antifoam (3)	1.9	0.25
Mildewcide (7)	7.3	0.75
Triton Nonionic Surfactant CA	2.1	0.25
Triton Anionic Surfactant GR-7M	1.1	0.13
Ammonium Hydroxide, 28% Aqueous Solution	0.9	0.13

Suppliers:
(1) 2-Amino-2-methyl-1-propanol - AMP-95 or equivalent
(2) Dispersant - "Tamol" 1124 or equivalent
(3) Antifoam - "Colloid" 653 or equivalent
(4) Preservative - "Nuosept" 95 or equivalent
(5) Clay - "Polygloss" 90 or equivalent
(6) Nepheline Syenite - "Minex" 10 or equivalent
(7) Mildewcide - "Troysan" Polyphase AF-1 or equivalent

Paint Properties:
   Pigment Volume Concentration, %: 12.0
   Volume Solids, %: 26.7
   Viscosity, equilibrated:
      Stormer, KU: 86
      ICI, poise: 0.8
   Weight per Gallon, lb: 9.08
   Freeze-thaw Stability, 3 cycles: Pass
   Heat Stability, 2 weeks at 120F: Pass

Film Properties:
   Gloss, 60: 50
   Gloss, 20: 13
   Flow and Leveling, brushed: Fair
   Wet Adhesion: Excellent
   Color Acceptance:
      Huls Color 64A-1A Nuance System: Excellent
      Huls Color 73A-1A Nuance System: Excellent

SOURCE: Ucar Emulsion Systems: UCAR Latex 624: Formulation
      Suggestion E-2421

Clear Wood Preservative

Material:	Lbs.	Gals.
Kelsol 3931-WG4-45	166.0	20.00
Butoxy Ethanol	7.5	1.00
Syloid 234	25.0	1.50

Grind the above to a 6 Hegman grind, then add the following and disperse 10-15 minutes:

Michemlube 743	28.7	3.50

Letdown:

Kelsol 3931-WG4-45	207.5	25.00

Premix the next two materials, then add under good agitation:

Manganese Hydrocure II	6.4	0.75
Butoxy Ethanol	7.5	1.00

Premix the next four materials, then add under good agitation:

Aerosol OT-75	2.3	0.25
Intercide T-O	7.3	0.75
Polyphase AF-1	9.8	1.00
Butoxy Ethanol	15.0	2.00

Add the following under good agitation:

Ammonium Hydroxide	1.9	0.25
Water	358.2	43.00

Analysis:
　　Pigment Volume Concentration, Percent: 6.2
　　Pigment/Binder Ratio: .2/1
　　Percent Solids, Weight: 26.0
　　Percent Solids, Volume: 24.1
　　Viscosity @ 25C, #2 Zahn, Secs.: 32
　　Pounds/Gallon: 8.4
　　pH: 8.5
　　VOC (excluding water):
　　　　Grams/Liter: 272
　　　　Pounds/Gallon: 2.3

SOURCE: Reichhold Chemicals, Inc.: Waterborne Handbook:
　　　　Formulation 2262-12C

Deep-Base Trim Paint

Ingredients:	Pounds	Gallons
Pigment Grind:		
Water	257.5	30.91
Cellosize Hydroxyethyl Cellulose ER-4400	4.5	0.41
Propylene Glycol	51.8	6.00
2-Amino-2-methyl-1-propanol (1)	2.0	0.25
Dispersant (2)	50.0	0.50
Antifoam (3)	1.9	0.25
Preservative (4)	2.3	0.25
Triton Nonionic Surfactant N-101	2.2	0.25
Titanium Dioxide (5)	75.0	2.15
Nepheline Syenite (6)	75.0	3.30
Letdown:		
Ucar Latex 624	464.3	52.76
Ucar Filmer IBT	11.6	1.46
Antifoam (3)	1.9	0.25
Mildewcide (7)	7.3	0.75
Triton Nonionic Surfactant CA	2.1	0.25
Triton Anionic Surfactant GR-7M	1.1	0.13
Ammonium Hydroxide, 28% Aqueous Solution	0.9	0.13

Suppliers:
(1) 2-Amino-2-methyl-1-propanol - AMP-95 or equivalent
(2) Dispersant - "Tamol" 1124 or equivalent
(3) Antifoam - "Colloid" 653 or equivalent
(4) Preservative - "Nuosept" 95 or equivalent
(5) Titanium Dioxide - "Tiona" RCL-535 or equivalent
(6) Nepheline Syenite - "Minex" 10 or equivalent
(7) Mildewcide - "Troysan" Polyphase AF-1 or equivalent

Paint Properties:
   Pigment Volume Concentration, %: 18.0
   Volume Solids, %: 30.3
   Viscosity, equilibrated:
      Stormer, KU: 90
      ICI, poise: 0.95
   Weight per Gallon, lb: 9.66
   Freeze-thaw Stability, 3 cycles: Pass
   Heat Stability, 2 weeks at 120F: Pass

Film Properties:
   Gloss, 60: 33
   Gloss, 20: 6
   Flow and Leveling, brushed: Fair
   Wet Adhesion: Very Good
   Color Acceptance:
      Huls 23B-3D Nuance System: Excellent
      Huls 73C-3D Nuance System: Very Good

SOURCE: Ucar Emulsion Systems: UCAR Latex 624: Formulation
         Suggestion E-2420

Economy White House Paint

Ingredients:	Pounds	Gallons
Pigment Grind:		
Water	492.7	59.15
Cellosize Hydroxyethyl Cellulose ER-30M	4.5	0.41
Preservative (1)	2.3	0.25
Dispersant (2)	7.4	0.75
Triton Nonionic Surfactant N-101	2.1	0.25
Antifoam (3)	1.9	0.25
Propylene Glycol	13.0	1.50
Titanium Dioxide (4)	150.0	4.31
Clay (5)	50.0	2.31
Talc (6)	150.0	6.32
Letdown:		
Ucar Latex 624	199.0	22.61
Ucar Filmer IBT	6.0	0.76
Antifoam (3)	1.9	0.25
Mildewcide (7)	4.9	0.50
Triton Nonionic Surfactant CA	2.1	0.25
Ammonium Hydroxide, 28% Aqueous Solution	0.9	0.13

Suppliers:
(1) Preservative - "Nuosept" 95 or equivalent
(2) Dispersant - "Tamol" 1124 or equivalent
(3) Antifoam - "Colloid" 650 or equivalent
(4) Titanium Dioxide - "Tiona" RCL-535 or equivalent
(5) Clay - "Kaopaque" 105 or equivalent
(6) Talc - "Nytal" 300 or equivalent
(7) Mildewcide - "Troysan" Polyphase AF-1 or equivalent

Paint Properties:
   Pigment Volume Concentration, %: 55.0
   Volume Solids, %: 23.7
   Viscosity, equilibrated:
      Stormer, KU: 86
      ICI, poise: 0.4
   Weight per Gallon, lb: 10.89
   Freeze-thaw Stability, 3 cycles: Pass
   Heat Stability, 2 weeks at 120F: Pass

Film Properties:
   Contrast Ratio, 6-mil wet drawdown (1): 0.927
   Reflectance, %: 88.6
   Sheen, 85: 3.9
   Flow and Leveling, brushed: Good
   Wet Adhesion: Excellent

   (1) Opacity could be increased by toning.

SOURCE: UCAR Emulsion Systems: UCAR Latex 624: Formulation
      Suggestion E-2414

Exterior Flat House Paint-White

Ingredients:	Pounds	Gallons
Pigment Grind:		
Water	301.2	36.16
Cellosize Hydroxyethyl Cellulose ER-30M	4.0	0.36
Preservative (1)	2.3	0.25
Dispersant (2)	7.7	0.75
Antifoam (3)	1.9	0.25
Triton Nonionic Surfactant N-101	2.1	0.25
Propylene Glycol	25.9	3.00
Titanium Dioxide (4)	225.0	6.59
Nepheline Syenite (5)	100.0	4.61
Talc (6)	100.0	4.21
Letdown:		
Ucar Latex 625	357.2	40.36
Ucar Filmer IBT	8.9	1.12
Antifoam (3)	1.9	0.25
Mildewcide (7)	15.0	1.46
Triton Nonionic Surfactant CA	2.1	0.25
Ammonium Hydroxide, 28% Aqueous Solution	0.9	0.13

Suppliers:
(1) Preservative - "Nuosept" 95 or equivalent
(2) Dispersant - "Colloid" 270 or equivalent
(3) Antifoam - "Colloid" 640 or equivalent
(4) Titanium Dioxide - "Tronox" CR-828 or equivalent
(5) Nepheline Syenite - "Minex" 4 or equivalent
(6) Talc - "Nytal" 300 or equivalent
(7) Mildewcide - "NuoCide" 404-D or equivalent

Paint Properties:
   Pigment Volume Concentration, %: 44.7
   Volume Solids, %: 34.5
   Viscosity, equilibrated:
      Stormer, KU: 93
      ICI, poise: 0.5
   Weight per Gallon, lb: 11.56
   Freeze-thaw Stability, 3 cycles: Pass
   Heat Stability, 2 weeks at 120F: Pass

Film Properties:
   Contrast Ratio:
      6-mil wet drawdown: 0.981
   Reflectance, %: 89.8
   Sheen, 85: 4.0
   Wet Adhesion, gloss alkyd:
      4-hour dry: Excellent
     24-hour dry: Excellent

SOURCE: Ucar Emulsion Systems: UCAR Latex 625: Formulation
       Suggestion E-2437

Exterior Flat Paint

Natrosol 250HR, 2 1/2%	110
Ethylene Glycol	25
Propylene Glycol	34
Texanol	11
Dispersant (4.5 lbs. active)	18
KTPP	1.5
Triton CF-10	2.5
Drew L-475	1.0
Horsehead RF-30	250
St. Joe #40 ZnO	50
Asbestine 3X	179
Nopcocide N-96	5
Natrasol 250HR 2-1/2%	30
Water	34
Drew L-475	3
NH4OH (28%)	2
Rhoplex AC-388	464

Weight per Gallon: 12.08   Volume Solids: 40.2%
Weight Solids: 58.3%   PVC: 39

Formulation 1203-7

Exterior Flat Paint

Natrosol 250HR 2-1/2%	110
Ethylene Glycol	25
Propylene Glycol	34
Texanol	11
Dispersant (4.5 lbs. active)	12.6
KTPP	1.5
Triton CF-10	2.5
Drew L-475	1.0
Zopaque RCL-9	250
St. Joe #40 ZnO	50
Asbestine 3X	179
Skane M8	2
Natrosol 250HR 2-1/2%	30
Water	34
Drew L-475	3
NH4OH (28%)	2
Rhoplex AC-388	464

Weight per Gallon: 12.08   Volume Solids: 40.5%
Weight Solids: 58.7%   PVC: 39

Formulation 1203-8

SOURCE: Uniroyal Chemical Co., Inc.: POLYWET RC-54

Exterior Flat Paint
E 13998, Federal Specification TT-P-19C, Amendment 1

Materials:	Lbs	Gal
Water	100.0	12.0
Tamol 850	6.0	0.5
Potassium tripolyphosphate	1.0	----
Dowicil 75	0.5	----
Igepal CO-610	3.0	0.3
Ethylene glycol	20.0	2.2
Skane M-8	2.0	0.2
Hercules 501 defoamer	3.0	0.3
Texanol	12.0	1.5
Ti-Pure R-900	250.0	7.5
Horsehead XX503	50.0	1.1
Nytal 300	125.0	5.3
Rhoplex AC-829	401.0	43.6
Ammonium hydroxide, 28%	2.0	0.2
3% Natrosol Plus solution/water	208.0	25.3

Paint Properties:
    Pigment volume concentration, %: 37.2
    Solids, wt %: 56
    Solids, volume %: 39
    Pigment, wt %: 35.7
    Wt/gal, lbs: 11.9

Properties exhibited when paint is thickened with Natrosol Plus:
    Viscosity, KU: 98
    Brushing viscosity, poises: 1.5
    Spatter: 8
    Leveling: 2
    Sag resistance: 21.75
    Scrub cycles to failure: 283
    Contrast ratio: 0.92

SOURCE: Aqualon Co.: NATROSOL PLUS: Formulation 2

Exterior Flat White House Paint with Zinc Oxide

Ingredients:	Pounds	Gallons
Pigment Grind:		
Water	307.1	36.87
Cellosize Hydroxyethyl Cellulose ER-30M	3.5	0.32
Preservative (1)	2.3	0.25
Dispersant (2)	10.3	1.00
Antifoam (3)	1.9	0.25
Triton Nonionic Surfactant N-101	2.1	0.25
Propylene Glycol	25.9	3.00
Potassium Tripolyphosphate (KTPP)	1.0	0.05
Titanium Dioxide (4)	225.0	6.59
Zinc Oxide (5)	25.0	0.54
Nepheline Syenite (6)	100.0	4.61
Talc (7)	75.0	3.16
Letdown:		
Ucar Latex 625	357.2	40.36
Ucar Filmer IBT	8.9	1.12
Antifoam (3)	1.9	0.25
Mildewcide (8)	10.2	1.00
Triton Nonionic Surfactant CA	2.1	0.25
Ammonium Hydroxide, 28% Aqueous Solution	0.9	0.13

Suppliers:
(1) Preservative - "Nuosept" 95 or equivalent
(2) Dispersant - "Colloid" 270 or equivalent
(3) Antifoam - "Colloid" 640 or equivalent
(4) Titanium Dioxide - "Tronox" CR-828 or equivalent
(5) Zinc Oxide - "XX-503R" or equivalent
(6) Nepheline Syenite - "Minex" 4 or equivalent
(7) Talc - "Nytal" 300 or equivalent
(8) Mildewcide - "Nuocide" 404-D or equivalent

Paint Properties:
   Pigment Volume Concentration, %: 43.8
   Volume Solids, %: 34.0
   Viscosity, equilibrated:
      Stormer, KU: 87
      ICI, poise: 0.4
   Weight per Gallon, lb: 11.60
   Freeze-thaw Stability, 3 cycles: Pass

Film Properties:
   Contrast Ratio: 6-mil wet drawdown: 0.981
   Reflectance, %: 89.9
   Sheen, 85: 5.7
   Wet Adhesion, gloss alkyd:
      4-hour dry: Excellent
      24-hour dry: Excellent
   Chalk Adhesion, ASTM #5: Excellent

SOURCE: Ucar Emulsion Systems: UCAR Latex 625: Formulation
      Suggestion E-2441

Exterior House Paint-Tint Base

Ingredients:	Pounds	Gallons
Pigment Grind:		
Water	267.0	32.05
Cellosize Hydroxyethyl Cellulose ER-30M	3.5	0.32
Preservative (1)	2.3	0.25
Dispersant (2)	5.2	0.50
Antifoam (3)	1.9	0.25
Triton Nonionic Surfactant N-101	2.1	0.25
2-Amino-2-methyl-1-propanol (4)	2.0	0.25
Propylene Glycol	25.9	3.00
Titanium Dioxide (5)	150.0	4.39
Nepheline Syenite (6)	200.0	9.23
Letdown:		
Ucar Latex 625	415.5	46.95
Ucar Filmer IBT	10.4	1.31
Antifoam (3)	1.9	0.25
Mildewcide (7)	7.3	0.75
Triton Nonionic Surfactant CA	2.1	0.25

Suppliers:
(1) Preservative - "Nuosept" 95 or equivalent
(2) Dispersant - "Colloid" 270 or equivalent
(3) Antifoam - "Colloid" 640 or equivalent
(4) 2-Amino-2-methyl-1-propanol - AMP-95 or equivalent
(5) Titanium Dioxide - "Tronox" CR-808 or equivalent
(6) Nepheline Syenite - "Minex" 4 or equivalent
(7) Mildewcide - "Troysan Polyphase" AF-1 or equivalent

Paint Properties:
    Pigment Volume Concentration, %: 38.0
    Volume Solids, %: 35.8
    Viscosity, equilibrated:
        Stormer, KU: 94
        ICI, poise: 0.55
    Weight per Gallon, lb: 10.97
    Freeze-thaw Stability, 3 cycles: Pass
    Heat Stability, 2 weeks at 120F: Pass

Film Properties:
    Contrast Ratio: 6-mil wet drawdown: 0.955
    Reflectance, %: 89.7
    Sheen, 85: 4.5
    Wet Adhesion, gloss alkyd:
        4-hour dry: Excellent
        24-hour dry: Excellent
    Color Acceptance, 4 fl. oz./gal:
        Huls 888-7214E: Excellent
        Huls 888-1045F: Excellent

SOURCE: Ucar Emulsion Systems: UCAR Latex 625: Formulation
        Suggestion E-2438

Exterior Semi-Gloss

Material:	Lbs.	Gals.
Water	116.6	14.00
Triton X-102	2.0	0.26
Colloid 226	8.0	0.78
Colloid 643	2.0	0.26
Butyl Carbitol	9.6	1.21
AMP 95	1.6	0.18
Nopcocide N-40D	12.0	1.31

Premix the following, then add:

	Lbs.	Gals.
Propylene Glycol	25.1	2.90
Natrosol 250MR	1.5	0.13
Titanox 2101	250.0	7.31
ASP 170	40.0	1.86

High speed disperser, 15 minutes maximum.

	Lbs.	Gals.
Water	10.6	1.27
Colloid 643	2.0	0.26
Synthemul 40-412	439.0	50.46
Triton N-57	1.5	0.16

Premix the following before adding:

	Lbs.	Gals.
Propylene Glycol	25.1	2.90
SCT 270	24.0	2.75
Water	100.0	12.00

Analysis:
   Pigment Volume Concentration, Percent: 27.5
   Percent Solids, Weight: 47.6
   Percent Solids, Volume: 33.3
   Initial Viscosity, KU: 80-85
   Pounds/Gallon: 10.70
   pH: 8.5
   VOC (excluding water):
      Grams/Liter: 175.0
   Blister Resistance: Exc.
   Wet Adhesion (Alkyd): Exc.
   Contrast Ratio (3 mils wet): 0.975
   Gloss @ 60: 25

SOURCE: Reichhold Chemicals, Inc.: Waterborne Handbook:
        Formulation 412-01

Exterior Sheen House Paint-White

Ingredients:	Pounds	Gallons
Pigment Grind:		
Water	199.9	24.00
Ucar Polyphobe 104 Rheology Modifier	3.3	0.37
2-Amino-2-methyl-1-propanol (1)	3.0	0.37
Preservative (2)	2.3	0.25
Dispersant (3)	7.7	0.75
Antifoam (4)	1.9	0.25
Triton Nonionic Surfactant N-101	2.1	0.25
Propylene Glycol	34.6	4.00
Titanium Dioxide (5)	225.0	6.67
Nepheline Syenite (6)	75.0	3.46
China Clay (7)	25.0	1.14
Letdown:		
Water	65.7	7.89
Water)Premix	16.7	2.00
Ucar Polyphobe 102 Rheology Modifier)	17.4	2.00
Ucar Latex 625	391.4	44.23
Ucar Filmer IBT	9.8	1.24
Antifoam (4)	1.9	0.25
Mildewcide (8)	7.3	0.75
Surfactant (9)	1.1	0.13

Suppliers:
(1) 2-Amino-2-methyl-1-propanol - AMP-95 or equivalent
(2) Preservative - "Nuosept" 95 or equivalent
(3) Dispersant - "Colloid" 270 or equivalent
(4) Antifoam - "Colloid" 643 or equivalent
(5) Titanium Dioxide - "Tioxide" TR92 or equivalent
(6) Nepheline Syenite - "Minex" 10 or equivalent
(7) China Clay - "Burgess" No. 98 or eqiuivalent
(8) Mildewcide - "Troysan Polyphase" AF-1 or equivalent
(9) Surfactant - "Pentex" 99 Anionic Surfactant or equivalent

Paint Properties:
   Pigment Volume Concentration, %: 35.0
   Volume Solids, %: 32.2
   Viscosity, equilibrated:
      Stormer, KU: 90
      ICI, poise: 1.1
   Weight per Gallon, lb: 10.91
   Freeze-thaw Stability, 3 cycles: Pass
   Heat Stability, 2 weeks at 120F: Pass
Film Properties:
   Contrast Ratio: 6-mil wet drawdown: 0.971
   Reflectance, %: 90.9
   Sheen, 85: 53.3
   Gloss, 60: 11.7
   Flow and Leveling, brushed: Good
   Wet Adhesion, gloss alkyd: 4-hour dry: Excellent
                     24-hour dry: Excellent

SOURCE: Ucar Emulsion Systems: UCAR Latex 625: Formulation
      Suggestion E-2439

Exterior Trim House Paint-White

Ingredients:	Pounds	Gallons
Pigment Grind:		
Water	166.6	20.00
Ucar Polyphobe 102 Rheology Modifier	8.8	1.00
2-Amino-2-methyl-1-propanol (1)	3.0	0.37
Preservative (2)	2.3	0.25
Propylene Glycol	60.5	7.00
Dispersant (3)	7.4	0.62
Antifoam (4)	1.9	0.25
Triton Nonionic Surfactant N-57	2.1	0.25
Titanium Dioxide (5)	250.0	7.50
Attapulgite Clay (6)	2.0	0.10
Letdown:		
Water	79.4	9.53
Water)Premix	12.5	1.50
Ucar Polyphobe 102 Rheology Modifier)	13.2	1.50
Ucar Latex 625	420.8	47.55
Ucar Filmer IBT	10.5	1.33
Antifoam (4)	2.8	0.37
Mildewcide (7)	7.3	0.75
Triton Anionic Surfactant GR-7M	1.1	0.13

Suppliers:
(1) 2-Amino-2-methyl-1-propanol - AMP-95 or equivalent
(2) Preservative - "Nuosept" 95 or equivalent
(3) Dispersant - "Colloid" 286 or equivalent
(4) Antifoam - "Colloid" 653 or equivalent
(5) Titanium Dioxide - "Ti-Pure" R-702 or equivalent
(6) Attapulgite Clay - "Attagel" 50 or equivalent
(7) Mildewcide - "Troysan Polyphase" AF-1 or equivalent

Paint Properties:
   Pigment Volume Concentration, %: 25.2
   Volume Solids, %: 30.1
   Viscosity, equilibrated:
      Stormer, KU: 90
      ICI, poise: 1.1
   Weight per Gallon, lb: 10.52
   Freeze-thaw Stability, 3 cycles: Pass
   Heat Stability, 2 weeks at 120F: Pass
Film Properties:
   Contrast Ratio: 6-mil wet drawdown: 0.967
   Reflectance, %: 92.3
   Gloss, 60: 74.1
   Gloss, 20: 26.4
   Flow and Leveling, brushed: Very Good
   Wet Adhesion, gloss alkyd: 4-hour dry: Excellent
                       24-hour dry: Excellent

SOURCE: Ucar Emulsion Systems: UCAR Latex 625: Formulation
       Suggestion E-2440

Exterior White

	Pounds
Natrosol 250 HR, 2%	100
Tamol 731	7
Igepal CO-610	3
Colloid 677	3
Ethylene Glycol	25
Texanol	8
Troysan Polyphase AF-1	5
Tronox CR 820	250
Nytal 300	100
Water	150
Natrosol 250 HR, 2%	175
Flexbond 325	300

    Yield, gallons: 101
    Solids, %: 46
    PVC, %: 40
    Viscosity, KU: 90
    VOC, gm/l - water: 125
    Formulation No. 32514

Lower Cost White and Tint

	Pounds
Methocel J-12-HS, 3%	100
Tamol 731	6
AMP-95	2
Witco 3056-A	2
Igepal CO-610	3
Ethylene Glycol	20
Texanol	8
Tronox CR-812	175
Satintone #1	100
Camel-Carb	125
Celite 499	50
Water	276
Witco 3056-A	1
Merbac 35	1
Methocel J-12-HS, 3%	100
Flexbond 325	180

    Yield, gallons: 100
    Solids, %: 48
    PVC, %: 63
    VOC, gm/l - water: 117
    Formulation No. 32513

SOURCE: Air Products and Chemicals, Inc.: FLEXBOND 325 Vinyl-Acrylic Emulsion: Typical Starting Formulations

Exterior White House Paint

Natrosol Plus, Grade 330, 3%	100
Tamol 731	7
AMP 95	2
Surfynol TG	2
Drew L-475	2
Propylene Glycol	25
Texanol	10
Ti-Pure R-902	250
Nytal 300	100
Troysan Polyphase AF-1	5
Nuosept 95	1
Water	165
Natrosol Plus, Grade 330, 3%	150
Drew L-475	2
Aerosol OT, 75%	1
Flexbond 380 DEV	300

    Yield, gallons: 101
    Solids, %: 46
    PVC, %: 40
    Viscosity, KU: 100
    VOC, gm/l-water: 139

SOURCE: Air Products and Chemicals, Inc.: FLEXBOND 380 DEV:
       Formulation 38014

Wood Panel Filler

	Pounds
Water	247
Tamol 850	14
Triton CF-10	3
Foamaster R	2
Ti-Pure R-901	95
ASP-400	616
Novacite S-325	327
Disperse on high shear disperser	
Merbac 35	0.35
Vinac 885	236

    Viscosity: 73 KU
    Yield: 100 gallons
    Solids by Weight: 73.2%
    Solids by Volume: 53.7%
    PVC: 79%
    Weight/Gallon: 14.5 pounds

SOURCE: Air Products and Chemicals, Inc.: VINAC 885 Emulsion:
       Formulation No. 2794-76-1

Flexible, Premium-Performance, High-Solids Caulk

	Pounds	Gallons	Parts by Weight
UCAR Latex 154 (60% N.V.)	488.0	54.3	38.0
Surfactant (1)	9.0	0.8	0.7
Dispersant (2)	10.0	0.5	0.8
Plasticizer (3)	115.5	12.5	9.0
Tooling Agent (4)	19.5	2.9	1.5
Ethylene Glycol	13.0	1.4	1.0
Filler (5)	616.0	27.2	48.0
Titanium Dioxide (6)	12.5	0.4	1.0

Suppliers:
(1) Surfactant - "Triton" X-405 or equivalent
(2) Dispersant - Composition T or equivalent
(3) Plasticizer - "Santicizer" 160 or equivalent
(4) Tooling Agent - "Varsol" No. 1 or equivalent
(5) Filler - "Atomite" or equivalent
(6) Titanium Dioxide - "Ti-Pure" R-901 or equivalent

Properties:
    Total Solids: 83.3% by wt
    Weight per Gallon, Theoretical: 12.8 lb
    Filler/Latex Solids Ratio: 2.2/1.0
    Plasticizer (on total binder solids): 28.5% by wt
    Shore "A" Hardness (4 weeks 73F, 50% RH): 26
    Tensile (2 weeks, 73F, 50% RH): 9 psi
    Elongation (2 weeks, 73F, 50% RH): 185%
    180 Peel Adhesion (per inch):
        Pinewood: 11 lb (cohesive/adhesive)
        Aluminum: 23 lb (cohesive)
        Glass: 20 lb (cohesive/adhesive)
    Extrusion: Good gunnability
    ASTM Shrinkage (28 days, 73F, 50% RH): 25% or less
    Slump at Room Temperature: None
    Recovery (25% Extension RT for 1/2 hr.): 70% or greater
    XW Accelerated Weathering, Wood Channel, 1500+ hours: Excel-
        lent

SOURCE: Union Carbide Corp.: UCAR Latex 154: Table 8

General Purpose Mastic

Ingredients:	Pounds	Gallons
UCAR Latex 154	685.8	76.2
Antifoam (1)	7.4	1.0
Potassium Tripolyphosphate (KTPP)	1.0	0.1
Dispersant (2)	4.9	0.5
TERGITOL Nonionic Surfactant NP-40, 70%		
Aqueous Solution	8.7	1.0
Plasticizer (3)	23.3	2.5
Odorless Mineral Spirits	13.2	2.0
Preservative (4)	2.0	0.3
Zinc Oxide	25.0	0.5
Titanium Dioxide	25.0	0.7
Mica	150.0	6.4
Talc	100.0	4.4
Calcium Carbonate	100.0	4.4

Suppliers:
(1) Antifoam - "Colloid" 679 or equivalent
(2) Dispersant - "Tamol" 850 or equivalent
(3) Plasticizer - "Santicizer" 160 or equivalent
(4) Preservative - "Troysan" 364 or equivalent

Properties:
   Weight per Gallon: 11.4 lb
   Pigment/Binder: 1/1
   Total Solids:
     by Volume: 60.6%
     by Weight: 70.7%
   Pigment Volume Concentration (PVC): 27%
   Consistency: Paste-like

SOURCE: UCAR Latex 154: Table 6

Quality Deep-Base House Paint

Ingredients:	Pounds	Gallons
Pigment Grind:		
Water	277.0	33.25
Cellosize Hydroxyethyl Cellulose ER-30M	4.0	0.36
Preservative (1)	2.3	0.25
Antifoam (2)	1.9	0.25
Dispersant (3)	5.0	0.50
Triton Nonionic Surfactant N-101	2.2	0.25
2-Amino-2-methyl-1-propanol (4)	2.0	0.25
Propylene Glycol	17.3	2.00
Titanium Dioxide (5)	75.0	2.15
Nepheline Syenite (6)	250.0	11.33
Letdown:		
Ucar Latex 624	411.2	46.73
Ucar Filmer IBT	10.3	1.30
Antifoam (2)	1.9	0.25
Mildewcide (7)	7.3	0.75
Triton Nonionic Surfactant CA	2.1	0.25
Ammonium Hydroxide, 28% Aqueous Solution	0.9	0.13

Suppliers:
(1) Preservative - "Nuosept" 95 or equivalent
(2) Antifoam - "Colloid" 650 or equivalent
(3) Dispersant - "Tamol" 1124 or equivalent
(4) 2-Amino-2-methyl-1-propanol - AMP-95 or equivalent
(5) Titanium Dioxide - "Tiona" RCL-535 or equivelent
(6) Nepheline Syenite - "Minex" 4 or equivalent
(7) Mildewcide - "Troysan" Polyphase AF-1 or equivalent

Paint Properties:
   Pigment Volume Concentration, %: 38.0
   Volume Solids, %: 35.5
   Viscosity, equilibrated:
      Stormer, KU: 92
      ICI, poise: 1.0
   Weight per Gallon, lb: 10.70
   Freeze-thaw Stability, 3 cycles: Pass
   Heat Stability, 2 weeks at 120F: Pass

Film Properties:
   Sheen, 85: 2.2
   Flow and Leveling, brushed: Good
   Wet Adhesion: Excellent
   Color Acceptance:
      Huls 23B-3D Nuance System: Excellent
      Huls 73C-3D Nuance System: Very Good
   Color Match to Chip:
      Huls 23B-3D Nuance System: Excellent
      Huls 73C-3D Nuance System: Very Good

SOURCE: Ucar Emulsion Systems: UCAR Latex 624: Formulation
      Suggestion E-2416

Quality Deep-Base House Paint

Ingredients:	Pounds	Gallons
Pigment Grind:		
Water	277.0	33.25
Cellosize Hydroxyethyl Cellulose ER-30M	4.0	0.36
Preservative (1)	2.3	0.25
Antifoam (2)	1.9	0.25
Dispersant (3)	5.0	0.50
Triton Nonionic Surfactant N-101	2.2	0.25
2-Amino-2-methyl-1-propanol (4)	2.0	0.25
Propylene Glycol	17.3	2.00
Titanium Dioxide (5)	75.0	2.15
Nepheline Syenite (6)	250.0	11.33
Letdown:		
Ucar Latex 624	411.2	46.73
Ucar Filmer IBT	10.3	1.30
Antifoam (2)	1.9	0.25
Mildewcide (7)	7.3	0.75
Triton Nonionic Surfactant CA	2.1	0.25
Ammonium Hydroxide, 28% Aqueous Solution	0.9	0.13

Suppliers:
(1) Preservative - "Nuosept" 95 or equivalent
(2) Antifoam - "Colloid" 650 or equivalent
(3) Dispersant - "Tamol" 1124 or equivalent
(4) 2-Amino-2-methyl-1-propanol - AMP-95 or equivalent
(5) Titanium Dioxide - "Tiona" RCL-535 or equivelent
(6) Nepheline Syenite - "Minex" 4 or equivalent
(7) Mildewcide - "Troysan" Polyphase AF-1 or equivalent

Paint Properties:
    Pigment Volume Concentration, %: 38.0
    Volume Solids, %: 35.5
    Viscosity, equilibrated:
        Stormer, KU: 92
        ICI, poise: 1.0
    Weight per Gallon, lb: 10.70
    Freeze-thaw Stability, 3 cycles: Pass
    Heat Stability, 2 weeks at 120F: Pass

Film Properties:
    Sheen, 85: 2.2
    Flow and Leveling, brushed: Good
    Wet Adhesion: Excellent
    Color Acceptance:
        Huls 23B-3D Nuance System: Excellent
        Huls 73C-3D Nuance System: Very Good
    Color Match to Chip:
        Huls 23B-3D Nuance System: Excellent
        Huls 73C-3D Nuance System: Very Good

SOURCE: Ucar Emulsion Systems: UCAR Latex 624: Formulation
        Suggestion E-2416

Quality Sheen White House Paint

Ingredients:	Pounds	Gallons
Pigment Grind:		
Water	295.5	35.48
Cellosize Hydroxyethyl Cellulose ER-30M	3.5	0.32
Preservative (1)	2.3	0.25
Propylene Glycol	34.6	4.00
Dispersant (2)	7.4	0.75
Triton Nonionic Surfactant N-101	2.2	0.25
Antifoam (3)	1.9	0.25
Titanium Dioxide (4)	225.0	6.75
Nepheline Syenite (5)	100.0	4.61
Letdown:		
Ucar Latex 624	394.6	44.84
Ucar Filmer IBT	9.9	1.25
Mildewcide (6)	7.3	0.75
Antifoam (3)	1.9	0.25
Ammonium Hydroxide, 28% Aqueous Solution	2.0	0.25

Suppliers:
(1) Preservative - "Nuosept" 95 or equivalent
(2) Dispersant - "Tamol" 1124 or equivalent
(3) Antifoam - "Colloid" 653 or equivalent
(4) Titanium Dioxide - "Ti-Pure" R-902 or equivalent
(5) Nepheline Syenite - "Minex" 10 or equivalent
(6) Mildewcide - "Troysan" Polyphase AF-1 or equivalent

Paint Properties:
   Pigment Volume Concentration, %: 35.0
   Volume Solids, %: 32.5
   Viscosity, equilibrated:
      Stormer, KU: 92
      ICI, poise: 0.8
   Weight per Gallon, lb: 10.88
   Freeze-thaw Stability, 3 cycles: Pass
   Heat Stability, 2 weeks at 120F: Pass

Film Properties:
   Contrast Ratio, 6-mil wet drawdown: 0.962
   Reflectance, %: 91.4
   Gloss, 60: 8.2
   Gloss, 85: 45.0
   Flow and Leveling, brushed: Good
   Wet Adhesion: Excellent

SOURCE: Ucar Emulsion Systems: UCAR Latex 624: Formulation
      Suggestion E-2422

Quality Tint-Base House Paint

Ingredients:	Pounds	Gallons
Water	267.6	32.13
Cellosize Hydroxyethyl Cellulose ER-30M	3.0	0.27
Preservative (1)	2.3	0.25
Antifoam (2)	1.9	0.25
Dispersant (3)	7.4	0.75
Triton Nonionic Surfactant N-101	2.2	0.25
2-Amino-2-methyl-1-propanol (4)	2.0	0.25
Propylene Glycol	21.6	2.50
Titanium Dioxide (5)	150.0	4.30
Nepheline Syenite (6)	225.0	10.19
Letdown:		
Ucar Latex 624	406.5	46.19
Ucar Filmer IBT	10.2	1.29
Antifoam (2)	1.9	0.25
Mildewcide (7)	7.3	0.75
Triton Nonionic Surfactant CA	2.1	0.25
Ammonium Hydroxide, 28% Aqueous Solution	0.9	0.13

Suppliers:
(1) Preservative - "Nuosept" 95 or equivalent
(2) Antifoam - "Colloid" 650 or equivalent
(3) Dispersant - "Tamol" 1124 or equivalent
(4) 2-Amino-2-methyl-1-propanol - AMP-95 or equivalent
(5) Titanium Dioxide - "Tiona" RCL-535 or equivalent
(6) Nepheline Syenite - "Minex" 4 or equivalent
(7) Mildewcide - "Troysan" Polyphase AF-1 or equivalent

Paint Properties:
    Pigment Volume Concentration, %: 40.0
    Volume Solids, %: 36.2
    Viscosity, equilibrated:
        Stormer, KU: 94
        ICI, poise: 1.2
    Weight per Gallon, lb: 11.12
    Freeze-thaw Stability, 3 cycles: Pass
    Heat Stability, 2 weeks at 120F: Pass

Film Properties:
    Contrast Ratio, 6-mil wet drawdown: 0.946
    Reflectance, %: 88.9
    Sheen, 85: 3.5
    Flow and Leveling, brushed: Good
    Wet Adhesion: Excellent
    Color Acceptance:
        Huls 32B-3T Nuance System: Excellent
        Huls 73C-2T Nuance System: Excellent

SOURCE: Ucar Emulsion Systems: UCAR Latex 624: Formulation
        Suggestion E-2415

Quality Tint Base House Paint (Conventional)

Ingredients:	Pounds	Gallons
Pigment Grind:		
Water	237.2	28.47
Cellosize Hydroxyethyl Cellulose ER-30M	4.0	0.36
Preservative (1)	2.3	0.25
Dispersant (2)	10.2	1.00
Triton Nonionic Surfactant N-101	2.2	0.25
Propylene Glycol	25.9	3.00
Antifoam (3)	1.9	0.25
Titanium Dioxide (4)	175.0	5.09
Nepheline Syenite (5)	225.0	10.37
Letdown:		
Opaque Polymer (6)	-----	-----
Ucar Latex 376	435.1	48.08
Butyl Carbitol Solvent	11.9	1.50
Antifoam (3)	1.9	0.25
Mildewcide (7)	7.3	0.75
Nonionic Surfactant (8)	2.1	0.25
Ammonium Hydroxide 28% Aqueous Solution	1.0	0.13

Suppliers:
(1) Preservative - "Nuosept" 95 or equivalent
(2) Dispersant - "Colloid" 226/35 or equivalent
(3) Antifoam - "Colloid" 640 or equivalent
(4) Titanium Dioxide - "Ti-Pure R-902" or equivalent
(5) Nepheline Syenite - "Minex" 4 or equivalent
(6) Opaque Polymer - "Ropaque" OP-62 or equivalent
(7) Mildewcide - "Troysan Polyphase" AF-1 or equivalent
(8) Nonionic Surfactant - "Color-Sperse" 188-A or equivalent

Paint Properties:
   Pigment Volume Concentration (PVC), %: 38.5
   Solids, %:
      by Volume: 40.2
      by Weight: 56.9
   Viscosity (initial):
      Stormer, KU: 85
      ICI, poise: 1.0
   pH (initial): 8.2
   Weight per Gallon, lb.: 11.43
   Freeze-Thaw Stability, 3 cycles: Pass (+4 KU)
   Heat Stability, 2 weeks at 120F: Pass (+2 KU)

Film Properties:
   Contrast Ratio, 3 mil film: 0.966
   Reflectance, %: 90.5
   Sheen, 85: 5.5
   Color Acceptance (1): Excellent
   (1) 4 fl. oz./gallon Huls 888-1045"F" Colorant

SOURCE: Ucar Emulsion Systems: UCAR Latex 367/UCAR Latex 376:
        Formulation Suggestion E-2211 (Conventional)

Quality Tint Base House Paint (Reduced TiO2)

Ingredients:	Pounds	Gallons
Pigment Grind:		
Water	203.6	24.44
Cellosize Hydroxyethyl Cellulose ER-30M	4.2	0.39
Preservative (1)	2.3	0.25
Dispersant (2)	7.7	0.75
Triton Nonionic Surfactant N-101	2.2	0.25
Propylene Glycol	25.9	3.00
Antifoam (3)	1.9	0.25
Titanium Dioxide (4)	125.0	3.63
Nepheline Syenite (5)	150.0	6.91
Letdown:		
Opaque Polymer (6)	105.8	12.27
Ucar Latex 376	406.6	44.93
Butyl Carbitol Solvent	13.9	1.75
Antifoam (3)	1.9	0.25
Mildewcide (7)	7.3	0.75
Nonionic Surfactant (8)	2.1	0.25
Ammonium Hydroxide 28% Aqueous Solution	1.0	0.13

Suppliers:
(1) Preservative - "Nuosept" 95 or equivalent
(2) Dispersant - "Colloid" 226/35 or equivalent
(3) Antifoam - "Colloid" 640 or equivalent
(4) Titanium Dioxide - "Ti-Pure" R-902 or equivalent
(5) Nepheline Syenite - "Minex" 4 or equivalent
(6) Opaque Polymer - "Ropaque" OP-62 or equivalent
(7) Mildewcide - "Troysan Polyphase" AF-1 or equivalent
(8) Nonionic Surfactant - "Color-Sperse" 188-A or equivalent

Paint Properties:
   Pigment Volume Concentration (PVC), %: 42.4
   Solids, %:
      by Volume: 42.4
      by Weight: 51.7
   Viscosity (initial):
      Stormer, KU: 85
      ICI, poise: 0.9
   pH (initial): 8.1
   Weight per Gallon, lb.: 10.61
   Freeze-Thaw Stability, 3 cycles: Pass (+3 KU)
   Heat Stability, 2 weeks at 120F: Pass (+2 KU)

Film Properties:
   Contrast Ratio, 3 mil film: 0.952
   Reflectance, %: 91.6
   Sheen, 85: 14.0
   Color Acceptance (1): Excellent
   (1) 4 fl. oz./gallon Huls 888-1045F Colorant

SOURCE: Ucar Emulsion Systems: UCAR Latex 367/UCAR Latex 376:
       Formulation Suggestion E-2211-A (Reduced TiO2)

Quality White House Paint

Ingredients:	Pounds	Gallons
Pigment Grind:		
Water	252.1	30.27
Cellosize Hydroxyethyl Cellulose ER-30M	4.0	0.36
Preservative (1)	2.3	0.25
Dispersant (2)	10.5	1.00
Potassium Tripolyphosphate (KTPP)	2.0	0.10
Triton Nonionic Surfactant N-101	2.2	0.25
2-Amino-2-methyl-1-propanol (3)	1.0	0.13
Propylene Glycol	34.5	4.00
Antifoam (4)	1.9	0.25
Titanium Dioxide (5)	225.0	6.54
Zinc Oxide (6)	25.0	0.54
Mica (7)	25.0	1.07
Talc (8)	175.0	7.37
Letdown:		
Ucar Latex 379	408.3	45.12
Ucar Filmer IBT	17.8	2.25
Mildewcide (9)	2.2	0.25
Antifoam (4)	1.9	0.25

Suppliers:
(1) Preservative - "Nuosept" 145 or equivalent
(2) Dispersant - "Tamol" 960 or equivalent
(3) 2-Amino-2-methyl-1-propanol - AMP-95 or equivalent
(4) Antifoam - "Colloid" 640 or equivalent
(5) Titanium Dioxide - "Ti-Pure" R-900 or equivalent
(6) Zinc Oxide - "Pasco" 311 or equivalent
(7) Mica - Mica 325WG or equivalent
(8) Talc - IT-325 or equivalent
(9) Mildewcide - Skane M-8 or equivalent

Comparative Properties:
   Pigment Volume Concentration, %: 40.1
   Solids, %:
      by Volume: 39.3
      by Weight: 57.7
   Weight per Gallon, lb: 11.9
Paint Properties:
   Viscosity, initial:
      Stormer, KU: 100
      ICI, poise: 0.9
   pH, initial: 8.9
   Freeze-thaw Stability, 3 cycles: Pass
   Heat Stability, 2 weeks at 120F: Pass
Film Properties:
   Contrast Ratio, 6-mil bar: 0.966
   Reflectance, %: 90.5
   Sheen, 85: 5.1

SOURCE: Ucar Emulsion Systems: UCAR Latex 379: Formulation
      Suggestion E-1992

Quality White House Paint

Ingredients:	Pounds	Gallons
Pigment Grind:		
Water	224.0	26.89
Cellosize Hydroxyethyl Cellulose QP-15,000	3.5	0.32
Dispersant (1)	9.1	1.00
Nonionic Surfactant (2)	2.0	0.25
Antifoam (3)	0.9	0.13
Potassium Tripolyphosphate (KTPP)	1.0	0.05
Propylene Glycol	25.9	3.00
Titanium Dioxide (4)	250.0	7.26
Nepheline Syenite (5)	150.0	6.91
Clay (6)	50.0	2.73
Let Down:		
Ucar Acrylic 525	427.9	47.54
Ucar Filmer IBT	16.9	2.13
Antifoam (3)	2.6	0.37
Mildewcide (7)	9.0	1.17
Ammonium Hydroxide, 28% Aqueous Solution	1.8	0.25

Suppliers:
(1) Dispersant - "Byk" VP-155 or equivalent.
(2) Nonionic Surfactant - "Triton" N-101 or equivalent
(3) Antifoam - "Colloid" 643 or equivalent
(4) Titanium Dioxide - "Zopaque" RCL-9 or equivalent
(5) Nepheline Syenite - "Minex" 4 or equivalent
(6) Clay - "Optiwhite" or equivalent
(7) Mildewcide - "Super-Ad-It" or equivalent

Paint Properties:
   Pigment Volume Concentration, %: 42.0
   Total Solids, %:
      by Volume: 40.2
      by Weight: 58.0
   Viscosity, initial:
      Stormer, KU: 95-100
      ICI, poise: 1.7
   pH, initial: 9.1
   Weight per Gallon, lb: 11.75
   Freeze-Thaw, 3 cycles: Pass
   Heat Stability, 2 wk at 120F: Pass

SOURCE: Union Carbide Corp.: UCAR Acrylic 525: Formulation
      Suggestion E-2076

Quality White House Paint

Ingredients:	Pounds	Gallons
Pigment Grind:		
Water	294.7	35.38
Cellosize Hydroxyethyl Cellulose ER-30M	3.0	0.27
Preservative (1)	2.3	0.25
Propylene Glycol	25.9	3.00
Dispersant (2)	7.4	0.75
Triton Nonionic Surfactant N-101	2.2	0.25
Antifoam (3)	1.9	0.25
Titanium Dioxide (4)	225.0	6.75
Clay (5)	50.0	2.28
Nepheline Syenite (6)	150.0	6.91
Letdown:		
Ucar Latex 624	364.2	41.39
Ucar Filmer IBT	9.1	1.15
Mildewcide (7)	7.3	0.75
Antifoam (3)	2.8	0.37
Ammonium Hydroxide, 28% Aqueous Solution	2.0	0.25

Suppliers:
(1) Preservative - "Nuosept" 95 or equivalent
(2) Dispersant - "Tamol" 1124 or equivalent
(3) Antifoam - "Colloid" 653 or equivalent
(4) Titanium Dioxide - "Ti-Pure" R-902 or equivalent
(5) Clay - "Iceberg" or equivalent
(6) Nepheline Syenite - "Minex" 4 or equivalent
(7) Mildewcide - "Troysan" Polyphase AF-1 or equivalent

Paint Properties:
   Pigment Volume Concentration, %: 45.0
   Volume Solids, %: 35.4
   Viscosity, equilibrated:
      Stormer, KU: 92
      ICI, poise: 1.2
   Weight per Gallon, lb: 11.48
   Freeze-thaw Stability, 3 cycles: Pass
   Heat Stability, 2 weeks at 120F: Pass

Film Properties:
   Contrast Ratio, 6-mil wet drawdown: 0.956
   Reflectance, %: 90.6
   Sheen, 85: 3.0
   Flow and Leveling, brushed: Good
   Wet Adhesion: Excellent

SOURCE: Ucar Emulsion Systems: UCAR Latex 624: Formulation
       Suggestion E-2413

Quality White House Paint (Conventional)

Ingredients:	Pounds	Gallons
Pigment Grind:		
Water	271.9	32.64
Cellosize Hydroxyethyl Cellulose ER-30M	4.0	0.36
Dispersant (1)	10.5	1.00
Potassium Tripolyphosphate (KTPP)	2.0	0.10
Triton Nonionic Surfactant N-101	2.2	0.25
2-Amino-2-methyl-1-propanol (2)	1.0	0.13
Propylene Glycol	25.9	3.00
Antifoam (3)	1.9	0.25
Titanium Dioxide (4)	225.0	6.54
Zinc Oxide (5)	25.0	0.54
Mica (6)	25.0	1.07
Talc (7)	175.0	7.37
Letdown:		
Opaque Polymer (8)	-----	----
Ucar Latex 367	408.3	45.12
Ucar Filmer IBT	8.9	1.13
Mildewcide (9)	2.2	0.25
Antifoam (3)	1.9	0.25

Suppliers:
(1) Dispersant - "Tamol" 960 or equivalent
(2) 2-Amino-2-methyl-1-propanol - AMP-95 or equivalent
(3) Antifoam - "Colloid" 640 or equivalent
(4) Titanium Dioxide - "Ti-Pure" R-900 or equivalent
(5) Zinc Oxide - "Pasco" 311 or equivalent
(6) Mica - "Mica" 325WG or equivalent
(7) Talc - "IT-325" or equivalent
(8) Opaque Polymer - "Ropaque" OP-62 or equivalent
(9) Mildewcide - "Skane" M8 or equivalent

Paint Properties:
   Pigment Volume Concentration (PVC), %: 40.1
   Solids, %:
      by Volume: 38.7
      by Weight: 57.7
   Viscosity (initial):
      Stormer, KU: 85
      ICI, poise: 0.8
   pH (initial): 8.5
   Weight per Gallon, lb.: 11.91
   Freeze-Thaw Stability, 3 cycles: Pass (+4 KU)
   Heat Stability, 2 weeks at 120F: Pass (+3 KU)
Film Properties:
   Contrast Ratio, 3 mil film: 0.964
   Reflectance, %: 91.9
   Sheen 85: 5.9

SOURCE: Ucar Emulsion Systems: UCAR Latex 367/UCAR Latex 376:
        Formulation Suggestion E-2210 (Conventional)

Quality White House Paint (Reduced TiO2)

Ingredients:	Pounds	Gallons
Pigment Grind:		
Water	234.2	28.12
Cellosize Hydroxyethyl Cellulose ER-30M	4.2	0.39
Dispersant (1)	10.5	1.00
Potassium Tripolyphosphate (KTPP)	2.0	0.10
Triton Nonionic Surfactant N-101	2.2	0.25
2-Amino-2-methyl-1-propanol (2)	1.0	0.13
Propylene Glycol	25.9	3.00
Antifoam (3)	1.9	0.25
Titanium Dioxide (4)	175.0	5.09
Zinc Oxide (5)	25.0	0.54
Mica (6)	25.0	1.07
Talc (7)	100.0	4.21
Letdown:		
Opaque Polymer (8)	101.9	11.85
Ucar Latex 367	382.4	42.25
Ucar Filmer IBT	9.9	1.25
Mildewcide (9)	2.2	0.25
Antifoam (3)	1.9	0.25

Suppliers:
(1) Dispersant - "Tamol" 960 or equivalent
(2) 2-Amino-2-methyl-1-propanol - AMP-95 or equivalent
(3) Antifoam - "Colloid" 640 or equivalent
(4) Titanium Dioxide - "Ti-Pure" R-900 or equivalent
(5) Zinc Oxide - "Pasco" 311 or equivalent
(6) Mica - "Mica" 325WG or equivalent
(7) Talc - IT-325 or equivalent
(8) Opaque Polymer - "Ropaque" OP-62 or equivalent
(9) Mildewcide - "Skane" M8 or equivalent

Paint Properties:
   Pigment Volume Concentration (PVC), %: 44.0
   Solids, %:
      by Volume: 38.8
      by Weight: 53.0
   Viscosity (initial):
      Stormer, KU: 83
      ICI, poise: 0.8
   pH (initial): 8.7
   Weight per Gallon, lb: 11.05
   Freeze-Thaw Stability, 3 cycles: Pass (+4 KU)
   Heat Stability, 2 weeks at 120F: Pass (+4 KU)
Film Properties:
   Contrast Ratio, 3 mil film: 0.962
   Reflectance, %: 92.6
   Sheen, 85: 16.2
Reduced TiO2 formulation not "fine-tuned" for optical properties.

SOURCE: Ucar Emulsion Systems: UCAR Latex 367/UCAR Latex 376:
      Formulation Suggestion E-2210-A (Reduced TiO2)

Tint-Base Trim Paint

Ingredients:	Pounds	Gallons
Pigment Grind:		
Water	219.2	26.31
Cellosize Hydroxyethyl Cellulose ER-4400	3.0	0.27
Propylene Glycol	60.5	7.00
2-Amino-2-methyl-1-propanol (1)	2.0	0.25
Dispersant (2)	6.1	0.62
Antifoam (3)	1.9	0.25
Preservative (4)	2.3	0.25
Triton Nonionic Surfactant N-101	2.2	0.25
Titanium Dioxide (5)	150.0	4.30
Nepheline Syenite (6)	50.0	2.20
Letdown:		
Ucar Latex 624	486.2	55.25
Ucar Filmer IBT	12.2	1.54
Antifoam (3)	1.9	0.25
Mildewcide (7)	7.3	0.75
Triton Nonionic Surfactant CA	2.1	0.25
Triton Anionic Surfactant GR-7M	1.1	0.13
Ammonium Hydroxide, 28% Aqueous Solution	0.9	0.13

Suppliers:
(1) 2-Amino-2-methyl-1-propanol - AMP-95 or equivalent
(2) Dispersant - Tamol 1124 or equivalent
(3) Antifoam - "Colloid" 653 or equivalent
(4) Preservative - "Nuosept" 95 or equivalent
(5) Titanium Dioxide - "Tiona" RCL-535 or equivalent
(6) Nepheline Syenite - "Minex" 10 or equivalent
(7) Mildewcide - "Troysan" Polyphase AF-1 or equivalent

Paint Properties:
   Pigment Volume Concentration, %: 20.0
   Volume Solids, %: 32.5
   Viscosity, equilibrated:
      Stormer, KU: 92
      ICI, poise: 1.05
   Weight per Gallon, lb: 10.09
   Freeze-thaw Stability, 3 cycles: Pass
   Heat Stability, 2 weeks at 120F: Pass

Film Properties:
   Contrast Ratio, 6-mil wet drawdown: 0.942
   Reflectance, %: 89.5
   Gloss, 60: 45
   Gloss, 20: 10
   Flow and Leveling, brushed: Good
   Wet Adhesion: Excellent
   Color Acceptance:
      Huls 32B-3T Nuance System: Excellent
      Huls 73C-2T Nuance System: Excellent

SOURCE: Ucar Emulsion Systems: UCAR Latex 624: Formulation E-2419

TT-P-19D Acrylic House Paint

Ingredients:	Pounds	Gallons
Pigment Grind:		
Water	203.7	24.45
Cellosize Hydroxyethyl Cellulose ER-30M	3.5	0.32
Preservative (1)	2.3	0.25
Dispersant (2)	12.8	1.25
Triton Nonionic Surfactant N-101	2.2	0.25
Antifoam (3)	1.9	0.25
Potassium Tripolyphosphate (KTPP)	1.0	0.05
Propylene Glycol	25.9	3.00
Titanium Dioxide (4)	260.0	7.59
Calcium Carbonate (5)	200.0	8.86
Letdown:		
Ucar Latex 624	448.4	50.95
Ucar Filmer IBT	11.2	1.41
Antifoam (3)	2.8	0.37
Mildewcide (6)	7.3	0.75
Ammonium Hydroxide, 28% Aqueous Solution	2.0	0.25

Suppliers:
(1) Preservative - "Nuosept" 95 or equivalent
(2) Dispersant - "Colloid" 226/35 or equivalent
(3) Antifoam - "Colloid" 653 or equivalent
(4) Titanium Dioxide - "Ti-Pure" R-902 or equivalent
(5) Calcium Carbonate - "Snowflake" or equivalent
(6) Mildewcide - "Troysan" Polyphase AF-1 or equivalent

Specifications:	TT-P-19D	E-2423 Results
Total NV, % by weight	50 (minimum)	58.8
Total NV, % by volume	40 (minimum)	40.4
Nonvolatile Organic Content		
(% by weight)	19 (minimum)	19.2
Stormer Viscosity, KU	80-100	96
Fineness of Dispersion	4 (minimum)	4
Reflectance, %	90 (minimum)	91.4
Gloss, 60	20 (maximum)	5.8
Drying Time, set to touch,		
minutes	10 (minimum)	10
Drying Through, hours	2 (maximum)	<2
VOC, grams per liter	250 (maximum)	96
Contrast Ratio, 6-mil wet		
drawdown	0.95 (minimum)	0.96
Lead, % by weight	0.06 (maximum)	0
Accelerated Storage	30 days at 125F	Pass
Freeze-thaw Stability	3 cycles, no coagulation,	Pass
	KU change 8 max	-3
Flexibility	No cracking or flaking	Pass
Alkali Resistance	No blistering or	
	remulsification	Pass
Compatibility	Uniform color and gloss	Pass
Biological Growth	8+ surface disfigurement	10
Accelerated Weathering	No chalking: trace chalking	
	Lightness difference index 1.5:	-1.32

SOURCE: Ucar Emulsion Systems: UCAR Latex 624: Formulation E-2423

White Trim Paint (Conventional)

Ingredients:	Pounds	Gallons
Pigment Grind:		
Water	241.2	28.96
Cellosize Hydroxyethyl Cellulose ER-4400	1.0	0.09
Preservative (1)	2.3	0.25
Propylene Glycol	60.5	7.00
Dispersant (2)	7.4	0.75
2-Amino-2-methyl-1-propanol (3)	2.0	0.25
Triton Nonionic Surfactant N-57	2.1	0.25
Antifoam (4)	1.9	0.25
Titanium Dioxide (5)	250.0	7.16
Letdown:		
Ucar Polyphobe Thickener 102	13.1	1.50
Ucar Latex 624	447.7	50.88
Opaque Polymer (6)	-----	-----
Ucar Filmer IBT	11.2	1.41
Antifoam (4)	2.8	0.37
Mildewcide (7)	7.3	0.75
Triton Anionic Surfactant GR-7M	1.1	0.13

Suppliers:
(1) Preservative - "Nuosept" 95 or equivalent
(2) Dispersant - "Tamol" 1124 or equivalent
(3) 2-Amino-2-methyl-1-propanol - AMP-95 or equivalent
(4) Antifoam - "Colloid" 653 or equivalent
(5) Titanium Dioxide - "Tiona" RCL-535 or equivalent
(6) Opaque Polymer - "Ropaque" OP-62 or equivalent
(7) Mildewcide - "Troysan" Polyphase AF-1 or equivalent

Paint Properties:
   Pigment Volume Concentration, %: 23.0
   Volume Solids, %: 31.1
   Viscosity, equilibrated:
      Stormer, KU: 100
      ICI, poise: 0.95
   Weight per Gallon, lb: 10.52
   Freeze-Thaw Stability, 3 cycles: Pass
   Heat Stability, 2 weeks at 120F: Pass

Film Properties:
   Contrast Ratio, 6-mil wet drawdown: 0.957
   Reflectance, %: 90.6
   Gloss, 60: 65
   Gloss, 20: 20
   Flow and Leveling, brushed: Good
   Wet Adhesion: Excellent

SOURCE: Ucar Emulsion Systems: UCAR Latex 624: Formulation
        Suggestion E-2418A (Conventional)

White Trim Paint (Opaque-Polymer Modified)

Ingredients:	Pounds	Gallons
Pigment Grind:		
Water	239.6	28.76
Cellosize Hydroxyethyl Cellulose ER-4400	1.0	0.09
Preservative (1)	2.3	0.25
Propylene Glycol	60.5	7.00
Dispersant (2)	6.9	0.70
2-Amino-2-methyl-1-propanol (3)	2.0	0.25
Triton Nonionic Surfactant N-57	2.1	0.25
Antifoam (4)	1.9	0.25
Titanium Dioxide (5)	231.7	6.64
Letdown:		
Ucar Polyphobe Thickener 102	13.1	1.50
Ucar Latex 624	428.4	48.67
Opaque Polymer (6)	25.6	2.98
Ucar Filmer IBT	11.2	1.41
Antifoam (4)	2.8	0.37
Mildewcide (7)	7.3	0.75
Triton Anionic Surfactant GR-7M	1.1	0.13

Suppliers:
(1) Preservative - "Nuosept" 95 or equivalent
(2) Dispersant - "Tamol" 1124 or equivalent
(3) 2-Amino-2-methyl-1-propanol - AMP-95 or equivalent
(4) Antifoam - "Colloid" 653 or equivalent
(5) Titanium Dioxide - "Tiona" RCL-535 or equivalent
(6) Opaque Polymer - "Ropaque" OP-62 or equivalent
(7) Mildewcide - "Troysan" Polyphase AF-1 or equivalent

Paint Properties:
  Pigment Volume Concentration, %: 26.4
  Volume Solids, %: 31.1
  Viscosity, equilibrated:
    Stormer, KU: 100
    ICI, poise: 0.95
  Weight per Gallon, lb: 10.38
  Freeze-thaw Stability, 3 cycles: Pass
  Heat Stability, 2 weeks at 120F: Pass

Film Properties:
  Contrast Ratio, 6-mil wet drawdown: 0.956
  Reflectance, %: 90.6
  Gloss, 60: 50
  Gloss, 20: 11
  Flow and Leveling, brushed: Good
  Wet Adhesion: Very Good

SOURCE: Ucar Emulsion Systems: UCAR Latex 624: Formulation
        Suggestion E-2418B (Opaque-Polymer Modified)

Section VI
Interior Paints and Related

Architectural Interior Flat

	Lbs/100 Gal.
Water	591.4
Cellosize QP-15,000	6.9

Add Cellosize under agitation; <u>mix 5 minutes</u>

KTPP	1.0

Add and <u>mix 2 minutes</u>

Triton N101	5.2
Tamol 731	5.0
DrewPlus L475	0.3
Ethylene Glycol	23.4

Add and <u>mix 2 minutes</u>

TiO2	130.7
Snow*Tex 45	112.7
Min-U-Sil 40	56.4

Add slowly; disperse to a 4+ Hegman

Wallpol 40-136	104.5
Exxate 1300	8.3
DrewPlus L475	0.7
Dowicil 75	1.4

Add slowly under agitation, <u>mix 10 minutes</u>

    Viscosity: 90 KU at 25C
    Wt/gal: 10.48 lbs.
    N.V.W.: 35.6%
    N.V.V.: 19.2%
    PVC: 0.612
    Pigment: 28.6%
    Contrast Ratio: 0.95
    Reflectance (Hunter Green): 90.5
    Gloss 60: 1.9
    Gloss 20: 1.3

SOURCE: U.S. Silica Co.: Formula AC-106

Best Quality Flat Wall Paint

Ingredients:	Pounds	Gallons
Pigment Grind:		
Water	348.2	41.80
Cellosize Hydroxyethyl Cellulose ER-30M	4.5	0.41
Preservative (1)	2.3	0.25
Dispersant (2)	5.7	0.62
Potassium Tripolyphosphate (KTPP)	1.0	0.05
Triton Nonionic Surfactant N-101	2.2	0.25
2-Amino-2-methyl-1-propanol (3)	1.0	0.13
Propylene Glycol	21.6	2.50
Antifoam (4)	1.9	0.25
Titanium Dioxide (5)	175.0	5.47
Clay (6)	175.0	7.99
Calcium Carbonate (7)	100.0	4.43
Nepheline Syenite (8)	75.0	3.40
Letdown:		
Ucar Latex 376	282.7	31.20
Butyl Carbitol Solvent	7.9	1.00
Antifoam (4)	1.9	0.25

Suppliers:
(1) Preservative - "Nuosept" 95 or equivalent
(2) Dispersant - "Tamol" 731 or equivalent
(3) 2-Amino-2-methyl-1-propanol - AMP-95 or equivalent
(4) Antifoam - "Colloid" 640 or equivalent
(5) Titanium Dioxide - "Tronox" CR-813 or equivalent
(6) Clay - "Satintone" W or equivalent
(7) Calcium Carbonate - "#1 White" or equivalent
(8) Nepheline Syenite - "Minex" 4 or equivalent

Paint Properties:
   Pigment Volume Concentration (PVC), %: 57.0
   Solids, %: by Volume: 37.4
             by Weight: 57.2
   Viscosity (initial): Stormer, KU: 90
                        ICI, poise: 1.4
   pH (initial): 8.2
   Weight per Gallon, lb.: 12.06
   Freeze-Thaw Stability, 3 cycles: Pass (+8 KU)
   Heat Stability, 2 weeks at 120F: Pass (+5 KU)

Film Properties:
   Contrast Ratio, 3 mil film: 0.980
   Reflectance, %: 93.6
   Sheen, 85: 4.2
   ASTM Scrub Cycles, with shim:
      Initial Break: 335
      Failure, 100%: 1100
   Porosity, ASTM D-3258: 23.8
   Color (1): (1) Tinted 1% by weight with Huls 888-7214E
      Color Acceptance: Excellent
      R.T. Drawdown, R.T. Touchup (2): Excellent
      R.T. Drawdown, C.T. Touchup (2): Very Good
   (2) R.T.=72F/50% relative humidity C.T.=40F/50% relative hum.

SOURCE: UCAR Emulsion Systems: UCAR Latex 367/UCAR Latex 376:
        Formulation Suggestion I-2215

Better Quality Flat Wall Paint

Ingredients:	Pounds	Gallons
Pigment Grind:		
Water	419.4	50.35
Cellosize Hydroxyethyl Cellulose ER-15M	5.5	0.50
Preservative (1)	2.3	0.25
Dispersant (2)	4.6	0.50
Potassium Tripolyphosphate (KTPP)	1.0	0.05
Triton Nonionic Surfactant N-101	2.2	0.25
2-Amino-2-methyl-1-propanol (3)	1.0	0.13
Propylene Glycol	17.3	2.00
Antifoam (4)	1.9	0.25
Titanium Dioxide (5)	150.0	4.80
Clay (6)	125.0	5.81
Calcium Carbonate (7)	200.0	8.86
Letdown:		
Ucar Latex 367	226.3	25.00
Ucar Filmer IBT	7.9	1.00
Antifoam (4)	1.9	0.25

Suppliers:
(1) Preservative - "Nuosept" 95 or equivalent
(2) Dispersant - "Tamol" 731 or equivalent
(3) 2-Amino-2-methyl-1-propanol - AMP-95 or equivalent
(4) Antifoam - "Colloid" 640 or equivalent
(5) Titanium Dioxide - "Ti-Pure" R-931 or equivalent
(6) Clay - "Satintone" W or equivalent
(7) Calcium Carbonate - "#1 White" or equivalent

Paint Properties:
   Pigment Volume Concentration (PVC), %: 60.2
   Solids, %:
      by Volume: 32.3
      by Weight: 52.3
   Viscosity (initial):
      Stormer, KU: 95
      ICI, poise: 1.1
   pH (initial): 8.0
   Weight per Gallon, lb.: 11.66
   Freeze-Thaw Stability, 3 cycles: Pass (+6 KU)
   Heat Stability, 2 weeks at 120F: Pass (+4 KU)

Film Properties:
   Contrast Ratio, 3 mil film: 0.969
   Reflectance, %: 92.3
   Sheen, 85: 1.9
   ASTM Scrub Cycles, with shim:
      Initial Break: 190
      Failure, 100%: 575
   Porosity, ASTM D-3258: 22.9

SOURCE: Ucar Emulsion Systems: UCAR Latex 367/UCAR Latex 376:
      Formulation Suggestion I-2216

Dripfree Acrylic Latex Paint
Pigment dispersion, Cowles Dissolver

	Pounds	Gallons
1.5% Sodium CMC* water gel	200	23.9
Vancide TH	1	0.1
Ethylene glycol	25	3.0
Darvan No. 7	10	1.0
Antifoamer	4	0.4
Ammonium hydroxide 28%	2	0.2
Titanium dioxide	200	5.8
Nytal 300	225	9.5
Reduction:		
Rhoplex AC-33 or equivalent	327	37.7
4% Veegum T water gel	136	16.0
Wetting agent	20	2.4

* Hercules 7H3SF or equivalent

Consistency, immediate KU: 90   Consistency, one day KU: 94
Consistency, one week KU: 94   Consistency, one month KU: 95
Pigment volume concentration: 49 Percent solids by weight: 51

Formula No. 1246

Dripfree Acrylic Latex Paint
Pigment dispersion, Cowles Dissolver

	Pounds	Gallons
1.5% Sodium CMC* water gel	200	23.9
Vancide TH	1	0.1
Ethylene glycol	25	3.0
Darvan No. 7	10	1.0
Antifoamer	4	0.4
Ammonium hydroxide 28%	2	0.2
Titanium dioxide	200	5.8
Nytal 300	225	9.5
Reduction:		
Rhoplex AC-33 or equivalent	260	30.0
4% Veegum T water gel	202	23.7
Wetting agent	20	2.4

* Hercules 7H3SF or equivalent

Consistency, immediate KU: 93   Consistency, one day KU: 96
Consistency, one week KU: 96   Consistency, one month KU: 96
Pigment volume concentration: 55 Percent solids by weight: 49

Formula No. 1247

SOURCE: R.T. Vanderbilt Co., Inc.: VEEGUM T in Latex Paints

Economy Flat Wall Paint

Ingredients:	Pounds	Gallons
Pigment Grind:		
Water	496.0	59.54
Cellosize Hydroxyethyl Cellulose ER-30M	5.5	0.50
Preservative (1)	2.3	0.25
Dispersant (2)	4.6	0.50
Potassium Tripolyphosphate (KTPP)	1.0	0.05
Triton Nonionic Surfactant N-101	2.2	0.25
2-Amino-2-methyl-1-propanol (3)	1.0	0.13
Propylene Glycol	13.0	1.50
Antifoam (4)	1.0	0.13
Titanium Dioxide (5)	75.0	2.34
Clay (6)	200.0	9.30
Calcium Carbonate (7)	150.0	6.65
Silica (8)	25.0	1.30
Letdown:		
Ucar Latex 376	147.8	16.31
Butyl Carbitol Solvent	7.9	1.00
Antifoam (4)	1.9	0.25

Suppliers:
(1) Preservative - "Nuosept" 95 or equivalent
(2) Dispersant - "Tamol" 731 or equivalent
(3) 2-Amino-2-methyl-1-propanol - AMP-95 or equivalent
(4) Antifoam - "PA-454" or equivalent
(5) Titanium Dioxide - "Tronox" CR-813 or equivalent
(6) Clay - "ASP-NC2" or equivalent
(7) Calcium Carbonate -"#1 White" or equivalent
(8) Silica - "Celite" 281 or equivalent
Paint Properties:
   Pigment Volume Concentration (PVC), %: 70.0
   Solids, %:
      by Volume: 28.0
      by Weight: 47.7
   Viscosity (initial):
      Stormer, KU: 95
      ICI, poise: 0.6
   pH (initial): 8.5
   Weight per Gallon, lb.: 11.34
   Freeze-Thaw Stability, 3 cycles: Pass (+3 KU)
   Heat Stability, 2 weeks at 120F: Pass

Film Properties:
   Contrast Ratio, 3 mil film: 0.974
   Reflectance, %: 91.5
   Sheen, 85: 2.4
   ASTM Scrub Cycles, with shim:
      Initial Break: 25
      Failure, 100%: 40
   Porosity, ASTM D-3258: 27.3

SOURCE: Ucar Emulsion Systems: UCAR Latex 367/UCAR Latex 376:
      Formulation Suggestion I-2218

Economy Flat Wall Paint

Ingredients:	Pounds	Gallons
Pigment Grind:		
Water	484.6	58.18
Cellosize Hydroxyethyl Cellulose ER-30M	5.5	0.50
Preservative (1)	2.3	0.25
Dispersant (2)	4.6	0.50
Triton Nonionic Surfactant N-101	2.2	0.25
Tergitol Nonionic Surfactant NP-40	2.2	0.25
2-Amino-2-methyl-1-propanol (3)	1.0	0.13
Propylene Glycol	13.0	1.50
Antifoam (4)	1.0	0.13
Titanium Dioxide (5)	75.0	2.34
Clay (6)	200.0	9.30
Calcium Carbonate (7)	150.0	6.65
Silica (8)	25.0	1.30
Letdown:		
Ucar Latex 379	147.8	16.31
Butyl Carbitol Solvent	7.9	1.00
Antifoam (4)	1.9	0.25
Propylene Glycol	10.0	1.16

Suppliers:
(1) Preservative - "Nuosept" 145 or equivalent
(2) Dispersant - "Tamol" 731 or equivalent
(3) 2-Amino-2-methyl-1-propanol - AMP-95 or equivalent
(4) Antifoam - "PA-454" or equivalent
(5) Titanium Dioxide - "Tronox" CR-813 or equivalent
(6) Clay - ASP-NC2 or equivalent
(7) Calcium Carbonate - "#1 White" or equivalent
(8) Silica - "Celite" 281 or equivalent

Comparative Properties:
   Pigment Volume Concentration, %: 70.0
   Solids, %:
      by Volume: 28.5
      by Weight: 47.4
   Weight per Gallon, lb: 11.34
Paint Properties:
   Viscosity, initial:
      Stormer, KU: 89
      ICI, poise: 0.6
   pH, initial: 8.7
   Freeze-thaw Stability, 3 cycles: Pass
   Heat Stability, 2 weeks at 120F: Pass
Film Properties:
   Contrast Ratio, 6-mil bar: 0.966
   Reflectance, %: 89.5
   Sheen, 85: 2.1
   ASTM Scrub Cycles, with shim:
      Initial Break: 153
      Failure, 10%: 189

SOURCE: Ucar Emulsion Systems: UCAR Latex 379: Formulation I-1988

Flat Interior Latex Paint

	Pounds	Gallons
Dispersion:		
Water	250	30.0
Van Gel B	3	0.2
Hydroxyethyl Cellulose (1)	5	0.3
Vancide TH	1	0.1
Darvan No. 7	15	1.5
Surfactant (3)	2	0.2
Ethylene Glycol	25	2.7
Coalescent (4)	12	1.5
Defoamer (5)	1	0.1
Titanium Dioxide (6)	150	4.7
Nytal 300	310	13.1
Reduction:		
Water	185	22.2
Defoamer	1	0.1
Vinyl-acrylic Latex (7)	210	23.3

Pigment Volume Concentration, %: 60
Solids by Weight, %: 50
Consistency, KU: 87
Brookfield Viscosity, Poises:
    At 10 RPM: 68
    At 100 RPM: 15
Thixotropic Index: 4.5

Raw materials or equivalents:
1) Cellosize QP-4400
3) Triton CF-10
4) Texanol
5) Nopco NDW
6) Ti-Pure R-931
7) Everflex E

SOURCE: R.T. Vanderbilt Co., Inc.: Van Gel B: Formula No. F-103

Flat Interior Latex Paint

	Pounds	Gallons
Dispersion:		
Water	350	42.0
Van Gel B	4.5	0.2
Hydroxyethyl Cellulose (2)	4.5	0.3
Vancide TH	1	0.1
Darvan No. 7	15	1.5
Surfactant (3)	2	0.2
Ethylene Glycol	25	2.7
Coalescent (4)	12	1.5
Defoamer (5)	1	0.1
Titanium Dioxide (6)	150	4.7
Nytal 300	310	13.1
Reduction:		
Water	85	10.2
Defoamer	1	0.1
Vinyl-acrylic Latex (7)	210	23.3

Pigment Volume Concentration, %: 60
Solids by Weight, %: 50
Consistency, KU: 94
Brookfield Viscosity, Poises:
    At 10 RPM: 102
    At 100 RPM: 20
Thixotropic Index: 5.1

(2) Cellosize QP-15,000
(3) Triton CF-10
(4) Texanol
(5) Nopco NDW
(6) Ti-Pure R-931

SOURCE: R.T. Vanderbilt Co., Inc.: VAN GEL B: Formula No. F-104

Flat Interior Latex Paint

	Pounds	Gallons
Dispersion:		
Water	250	30.0
Van Gel B	3	0.2
Hydroxyethyl Cellulose (1)	5	0.3
Vancide TH	1	0.1
Darvan No. 7	15	1.5
Surfactant (3)	2	0.2
Ethylene Glycol	25	2.7
Coalescent (4)	12	1.5
Defoamer (5)	1	0.1
Titanium Dioxide (6)	150	4.7
Nytal 300	310	13.1
Reduction:		
Water	170	20.4
Defoamer	1	0.1
Acrylic Latex (8)	221	25.1

```
Pigment Volume Concentration, %: 60
Solids by Weight, %: 50
Consistency, KU: 78
Brookfield Viscosity, Poises:
    At 10 RPM: 64
    At 100 RPM: 11
Thixotropic Index: 5.8
```

Raw Materials or equivalents:
1) Cellosize QP-4400
3) Triton CF-10
4) Texanol
5) Nopco NDW
6) TiPure R-931
8) Rhoplex AC-388

SOURCE: R.T. Vanderbilt Co., Inc.: VAN GEL B: Formula No. F-105

Flat Interior Latex Paint

	Pounds	Gallons
Dispersion:		
Water	350	42.0
Van Gel B	4.5	0.2
Hydroxyethyl Cellulose (2)	4.5	0.3
Vancide TH	1	0.1
Darvan No. 7	15	1.5
Surfactant (3)	2	0.2
Ethylene Glycol	25	2.7
Coalescent (4)	12	1.5
Defoamer (5)	1	0.1
Titanium Dioxide (6)	150	4.7
Nytal 300	310	13.1
Reduction:		
Water	70	8.4
Defoamer	1	0.1
Acrylic Latex (8)	221	25.1

Pigment Volume Concentration, %: 60
Solids by Weight, %: 50
Consistency, KU: 85
Brookfield Viscosity, Poises:
    At 10 RPM: 124
    At 100 RPM: 16
Thixotropic Index: 7.8

Raw Materials or Equivalents:
2) Cellosize QP-15,000
3) Triton CF-10
4) Texanol
5) Nopco NDW
6) Ti-Pure R-931
8) Rhoplex AC-388

SOURCE: R.T. Vanderbilt Co., Inc.: VAN GEL B: Formula No. F-106

Flat Interior Vinyl-Acrylic Latex Paint

Dispersion:	Pounds	Gallons
Water	350	42.0
Van-Gel	4.5	.2
Hydroxyethyl cellulose (1)	4.5	.3
Vancide TH	1	.1
Darvan No. 7	15	1.5
Ethylene glycol	25	2.7
Coalescent (2)	12	1.5
Defoamer (3)	2	.2
Titanium dioxide (4)	150	4.7
Nytal 300	310	13.1
Reduction:		
Water	67	8.0
Surfactant - 5% solution (5)	20	2.4
Vinyl-acrylic latex (6)	210	23.3

(1) QP15,000 or equivalent
(2) Texanol or equivalent
(3) Nopco NDW or equivalent
(4) R-931 or equivalent
(5) Aerosol OT-B or equivalent
(6) Everflex E or equivalent

Paint Properties:
    Consistency, KU: 98
    Pigment volume concentration, %: 60
    Solids by weight, %: 50

This formulation exhibits greater body and brush pick-up while maintaining the smooth and even flow of Development Formulation No. F-101.

Paint Preparation:
    VAN-GEL is added to initial water charge at slow agitation and then dispersed at high speed until a 7 Hegman is obtained (5-10 minutes). Hydroxyethyl cellulose and other ingredients for pigment dispersion are then added in the usual manner. After dispersion step, the paint is reduced at slow agitation with remaining ingredients.

SOURCE: R.T. Vanderbilt Co., Inc.: Development Formulation No. F-102

Flat Interior Vinyl-Acrylic Latex Paint

Dispersion:	Pounds	Gallons
Water	250	30.0
Van Gel	3	0.2
Hydroxyethyl cellulose (1)	5	0.3
Vancide TH	1	0.1
Darvan No. 7	15	1.5
Ethylene glycol	25	2.7
Coalescent (2)	12	1.5
Defoamer (3)	2	0.2
Titanium dioxide (4)	150	4.7
Nytal 300	310	13.1
Reduction:		
Water	167	20.0
Surfactant - 5% solution (5)	20	2.4
Vinyl-acrylic latex (6)	210	23.3

(1) QP4400 or equivalent
(2) Texanol or equivalent
(3) Nopco NDW or equivalent
(4) R-931 or equivalent
(5) Aerosol OT-B or equivalent
(6) Everflex E or equivalent

Paint Properties:
    Consistency, KU: 87
    Pigment volume concentration, %: 60
    Solids by weight, %: 50

    This formulation is the first in a series intended to provide
a range of thixotropic properties for the paint formulator.

Paint Preparation:
    VAN GEL is added to initial water charge at slow agitation
and then dispersed at high speed until a 7 Hegman is obtained
(5 - 10 minutes). Hydroxyethyl cellulose and other ingredients
for pigment dispersion are then added in the usual manner. After
dispersion step, the paint is reduced at slow agitation with
remaining ingredients.

SOURCE: R.T. Vanderbilt Co., Inc.: Developmental Formulation
        No. F-101

Good Quality Flat Wall Paint

Ingredients:	Pounds	Gallons
Pigment Grind:		
Water	427.7	51.35
Cellosize Hydroxyethyl Cellosize ER-30M	5.0	0.45
Preservative (1)	2.3	0.25
Dispersant (2)	5.7	0.62
Potassium Tripolyphosphate (KTPP)	1.0	0.05
Triton Nonionic Surfactant N-101	2.2	0.25
Propylene Glycol	17.3	2.00
Antifoam (3)	1.9	0.25
Titanium Dioxide (4)	125.0	3.91
Clay (5)	175.0	9.55
Calcium Carbonate (6)	175.0	7.75
Letdown:		
Ucar Latex 367	201.0	22.19
Ucar Filmer IBT	7.9	1.00
Antifoam (3)	1.9	0.25
Ammonium Hydroxide 28% Aqueous Solution	1.0	0.13

Suppliers:
(1) Preservative - "Nuosept" 95 or equivalent
(2) Dispersant - "Tamol" 731 or equivalent
(3) Antifoam - "Colloid" 643 or equivalent
(4) Titanium Dioxide - "Tronox" CR-813 or equivalent
(5) Clay - "Optiwhite" or equivalent
(6) Calcium Carbonate - "#1 White" or equivalent

Paint Properties:
   Pigment Volume Concentration (PVC), %: 65.0
   Solids, %:
      by Volume: 32.6
      by Weight: 51.8
   Viscosity (initial):
      Stormer, KU: 92
      ICI, poise: 1.2
   pH (initial): 8.0
   Weight per Gallon, lb: 11.50
   Freeze-Thaw Stability, 3 cycles: Pass (+8 KU)
   Heat Stability, 2 weeks at 120F: Pass (+3 KU)

Film Properties:
   Contrast Ratio, 3 mil film: 0.979
   Reflectance, %: 93.1
   Sheen, 85: 2.3
   ASTM Scrub Cycles, with shim:
      Initial Break: 115
      Failure, 100%: 350
   Porosity, ASTM D-3258: 28.0

SOURCE: Ucar Emulsion Systems: UCAR Latex 367/UCAR Latex 376:
      Formulation Suggestion I-2217

Good Quality Interior Flat

Natrosol Plus, Grade 330, 3%	100
Propylene Glycol	24
Colloid 226/35	5
Texanol	8
Surfynol TG	3
Colloid 643	2
Ti-Pure R-901	150
Satintone W	150
Duramite	150
Celite 499	25
Tektamer 38	0.5
Water	193
Colloid 643	2
Natrosol Plus, Grade 330, 3%	120
Aerosol OT, 75%	1
Flexbond 380 DEV	240

    Yield, gallons: 100
    Solids, %: 51.8
    PVC, %: 58.5
    Viscosity, KU: 98
    VOC, gm/l - water: 121
    Reflectance: 91.8
    Contrast Ratio: 0.96
    85 Sheen: 1.5
    Formulation 38012

68% PVC Interior Flat

Natrosol 250 HR, 2%	100
Colloid 226/35	5
AMP-95	2
Surfynol TG	2
Propylene Glycol	24
Texanol	8
Drew L-475	3
Ti-Pure R-900	125
Samhide 583	12
Satintone W	150
Snowflake	150
Celite 499	25
Tektamer 38	0.5
Water	200
Natrosol 250 HR, 2%	190
Flexbond 380 DEV	160

    Yield, gallons: 102   VOC, gm/l-water: 124
    Solids, %: 47.6   Reflectance: 92.8
    PVC, %: 68   Contrast Ratio: 0.97
    Viscosity, KU: 90   85 Sheen: 2
    Formulation 38013

SOURCE: Air Products and Chemicals, Inc.: FLEXBOND 380 DEV

Green Floor Paint

Material:	Weight (Lbs.)	Volume (Gals.)
Water	14.0	1.68
Methyl Carbitol	50.0	5.80
Exp. Dispersant QR-681M (35%)	10.4	1.14
NH4OH (28%)	1.0	0.13
Triton CF-10	2.0	0.23
Drew L-405	1.0	0.14
Chrome Oxide Green X-1134	242.1	5.69
Letdown:		
Maincote HG-54	632.6	73.70
Butyl Propasol	31.5	4.25
Ektasolve EEH	5.3	0.71
Sodium Nitrite (15%)	9.0	1.08
Drew Y-250	2.5	0.35
NH4OH (28%)	5.0	0.68
Rheology Modifier QR-1001 (20.8%)	25.0	2.82
Water	13.3	1.60

Formulation Constants:
    Pigment Volume Content, %: 16.3
    Volume Solids, %: 34.9
    Viscosity, KU:
        Initial: 72
        Equilibrated: 74
    Viscosity, ICI, poise:
        Initial: 1.5
        Equilibrated: 1.5
    pH:
        Initial: 9.2
        Equilibrated: 9.2

SOURCE: Rohm and Haas Co.: Maintenance Coatings: MAINCOTE HG-54:
      Formulation FP-54-2

High-Build Interior Flat Wall Paint Based on a Vinyl Acrylic Copolymer Thickened with Acrysol RM-825

Materials:	Weight/Ratio	Volume/Gallons
Water	120.0	14.40
Tamol 960 (40%)	10.0	0.94
Ethylene Glycol	25.0	2.69
Dowicil 75	1.0	0.08
PAG-188	2.0	0.25

Add the following at low speed:

Attagel 50	5.0	0.25
Ti-Pure R-900	200.0	5.84
Optiwhite	100.0	5.45
Imsil A-15	75.0	3.40

Grind the above on a high-speed impeller mill at 3800-4500 FPM for 20 minutes. At a slower speed let down as follows:

Water	140.0	16.80
Acrysol RM-825	33.6	3.86
Butyl Carbitol	14.0	1.75
Vinyl Acrylic Copolymer (55%)	350.0	38.68
PAG-188	4.0	0.50
Ammonium Hydroxide (28%)	1.5	0.20
Water	40.2	4.83
Toners*	0.7	0.08

Formulation Constants:
  PVC: 43%
  Volume Solids: 35%
  Initial Viscosity: 80-90 KU
  Equilibrated Viscosity: 75-110 KU
  ICI Viscosity: 1.9-2.7 P
  CARB (Organic Volatile Level) g/L: 167
  pH: 7.8-8.2

  ** Tenneco 8800 Line Lampblack 0.5
     Tenneco 8800 Line Raw Umber 0.2

SOURCE: Rohm and Haas Co.: Trade Sales Coatings: ACRYSOL RM-825: Formulation F-VA-2

High Build Interior Flat Wall Paint Based on a Vinyl Acrylic Copolymer Thickened with Experimental Rheology Modifier QR-708

Materials:	Parts Per Hundred Weight Ratio*	(Volume Basis)
Water	120.0	14.40
Tamol 960 (40%)	10.0	0.94
Ethylene Glycol	25.0	2.69
Dowicil 75	1.0	0.08
PAG-188	2.0	0.25

Add the following at low speed:

Attagel 50	5.0	0.25
Ti-Pure R-900	200.0	5.84
Optiwhite	100.0	5.45
Imsil A-15	75.0	3.40

Grind the above on a high speed impeller at 3800-4500 FPM for 20 minutes. At a slower speed let down as follows:

Water	140.0	16.80
Experimental Rheology Modifier QR-708 (35%)	24.0	2.67
Butyl Carbitol	20.0	2.50
Vinyl Acrylic Copolymer (55%)	350.0	38.68
PAG-188	4.0	0.50
Ammonium Hydroxide (28%)	1.5	0.20
Water	43.9	5.27
Toners**	0.7	0.08

Formulation Constants:
  PVC: 43%
  Volume Solids: 35%
  Initial Stormer Viscosity: 80-90 K.U.
  Equilibrated Stormer Viscosity: 75-110 K.U.
  ICI Viscosity: 1.9-2.7 P
  CARB (Organic Volatile Level) g/L: 167
  pH: 7.8-8.2

  * Using weight ratio in pound units will yield approximately 100 gallons of paint, while with kilograms, 833 liters will result.

  ** Tenneco 8800 Line Lampblack 0.5
  ** Tenneco 8800 Line Raw Umber 0.2

SOURCE: Rohm and Haas Co.: Experimental Rheology Modifier QR-708: Formulation XF-VA-1

High-quality Flat Wall Paint

Ingredients:	Pounds	Gallons
Pigment Grind:		
Water	330.6	39.69
Cellosize Hydroxyethyl Cellulose ER-30M	4.5	0.41
Preservative (1)	2.3	0.25
Dispersant (2)	5.7	0.62
Triton Nonionic Surfactant N-101	2.2	0.25
Tergitol Nonionic Surfactant NP-40	2.2	0.25
2-Amino-2-methyl-1-propanol (3)	1.0	0.13
Propylene Glycol	21.6	2.50
Antifoam (4)	1.9	0.25
Titanium Dioxide (5)	175.0	5.47
Clay (6)	175.0	7.99
Calcium Carbonate (7)	100.0	4.43
Nepheline Syenite (8)	75.0	3.40
Letdown:		
Ucar Latex 379	282.7	31.20
Butyl Carbitol Solvent	13.8	1.75
Antifoam (4)	1.9	0.25
Propylene Glycol	10.0	1.16

Suppliers:
(1) Preservative - "Nuosept" 145 or equivalent
(2) Dispersant - "Tamol" 731 or equivalent
(3) 2-Amino-2-methyl-1-propanol - AMP-95 or equivalent
(4) Antifoam - "Colloid" 640 or equivalent
(5) Titanium Dioxide - "Tronox" CR-813 or equivalent
(6) Clay - Satintone W or equivalent
(7) Calcium Carbonate - "#1 White" or equivalent
(8) Nepheline Syenite - "Minex" 4 or equivalent
Comparative Properties:
   Pigment Volume Concentration, %: 57.0
   Solids, %:
      by Volume: 37.8
      by Weight: 56.9
   Weight per Gallon, lb: 12.05
Paint Properties:
   Viscosity, initial:
      Stormer, KU: 92
      ICI, poise: 1.0
   pH, initial: 8.4
   Freeze-thaw Stability, 3 cycles: Pass
   Heat Stability, 2 weeks at 120F: Pass
Film Properties:
   Contrast Ratio, 6-mil bar: 0.979
   Reflectance, %: 90.5
   Sheen, 85: 3.8
   ASTM Scrub Cycles, with shim:
      Initial Break: 881
      Failure, 10%: 1053

SOURCE: Ucar Emulsion Systems: UCAR Latex 379: Formulation
      Suggestion I-1985

Interior Gloss White

Material	Lbs.	Gals.
Water	100.0	12.00
Propylene Glycol	40.0	4.63
Colloid 226	8.0	0.78
Colloid 643	1.0	0.13
AMP 95	1.5	0.18
Proxel GXL	1.5	0.17
Premix the following, then add:		
Propylene Glycol	48.8	5.65
Natrosol 250MR	3.0	0.26
Tiona RCL-9	250.0	7.31
High speed disperser, 15 minutes maximum.		
Synthemul 40-430	525.0	60.34
Water	56.0	6.72
Igepal CO-610	3.0	0.34
Butyl Carbitol	10.0	1.35
Colloid 643	1.0	0.13

Analysis:
    Pigment Volume Concentration, Percent: 20.2
    Percent Solids, Weight: 46.9
    Percent Solids, Volume: 38.2
    Initial Viscosity, KU: 85-95
    Pounds/Gallon: 10.49
    pH: 8.3-8.8
    VOC (excluding water):
        Grams/Liter: 238.0
    60 Gloss: 65-75
    20 Gloss: 20-25

SOURCE: Reichhold Chemicals, Inc.: Waterborne Formulations:
        Formulation 430-04

Interor Semigloss - Acrylic

	Pounds
Water	30.0
Cellosize QP4400, Thickener (2.5%)	92.0
Polywet ND-2, Dispersant (25%)	7.2
Propylene Glycol	20.0
Colloids 600, Defoamer	2.0
TiPure R-900, Rutile Titanium Dioxide	170.0

Disperse at high speed. Add

Rhoplex AC-490, Acrylic Emulsion	382.0
Texanol, Coalescing Solvent	10.0
Triton CF-10, Wetting Agent	2.0
Super Ad-It, Preservative	0.5
Water	20.0
Ethylene Glycol	20.0
Water	101.0
Cellosize QP4400, Thickener (2.5%)	112.5
Colloid 600, Defoamer	1.0
Triton N-57, Wetting Agent	8.0

Weight per Gallon: 9.58 pounds
Solids: 36.1%
PVC: 20.9%
Viscosity: 80 KU
Gloss, 60: 68

Paint Dispersions:
POLYWET ND-2 and ND-35 exhibit the following outstanding properties in latex paints: superior resistance to foaming in package and applied, absence of cratering, improved scrub resistance and superior adhesion to chalked paint and enamel surfaces. POLYWET ND-2 is the lower solids version of ND-35 (25% vs. 35%).

SOURCE: Uniroyal Chemical Co., Inc.: POLYWET ND-2 and ND-35: Formulation

Interior Semigloss White

Propylene Glycol	50
Tamol 731	12
Texanol	10
AMP 95	2
Triton CF10	2
Drew L-484	1
Ti-Pure R-900	240
Attagel 50	3
Tektamer 38	0.5
Water	182
Aerosol OT-75%	1
Natrosol Plus, Grade 330, 2%	100
Drew L-484	3
Flexbond 380 DEV	430
Acrysol RM-825	22

Yield, gallons: 100.7
Solids, %: 45.5
PVC, %: 22.3
Viscosity, KU: 95
ICI Viscosity, poise: 1.55
VOC, gm/l - water: 222
60 gloss: 68
20 gloss: 24

SOURCE: Air Products and Chemicals, Inc.: FLEXBOND 380 DEV:
Formulation 38018

Latex Interior Satin

	Lbs./100 Gal.
Water	200.00
Tamol 731	1.60
KTPP	0.40
Triton N-101	0.80
QP-15,000	1.40

Add all to water under agitation, mix 5 minutes

R-900	120.00
Min-U-Sil 15	46.60
Snow*Tex 45	30.00
Dowicil 75	0.80
Drew Plus L-475	1.00
Exxate 1300	10.20

Add all under agitation, disperse to 6+ Hegman

Water	108.40
Ucar SCT-275	18.40

Mix 5 minutes

28% Ammonia	0.80
Wallpol 40-136	464.00

```
60 Gloss: 28.5
20 Gloss:  6.0
Viscosity: 88 KU at 25C
PVC: 20.8%
Pigment: 19.6%
Contrast Ratio: 98.0
Reflectance (Y): 90.8%
Lbs./gal: 10.0
N.V.V.: 34.1%
N.V.W.: 47.5%
```

Gloss, contrast ratio, and reflectance measured at 2 mil dry film thickness.

SOURCE: U.S. Silica Co.: Formula AC-108

Latex Interior Semi-Gloss

	Lbs./100 Gal.
Water	200.00
Tamol 731	1.60
KTPP	0.40
Triton N-101	0.80
QP-15,000	1.40

Add all to water under agitation, mix 5 minutes

R-900	110.00
Min-U-Sil 10	28.00
Dowicil 75	0.80
Drew Plus L-475	1.00
Exxate 1300	10.20

Add all under agitation, disperse to 7+ Hegman

Water	106.40
Ucar SCT-275	18.00

Mix 5 minutes

28% Ammonia	0.80
Wallpol 40-136	490.20

    60 Gloss: 58.0
    20 Gloss: 20.3
    Viscosity: 87 KU at 25C
    PVC: 13.8%
    Pigment: 14.2%
    Contrast Ratio: 97.3
    Reflectance (Y): 90.9%
    Lbs./gal.: 9.7
    N.V.V.: 33.0%
    N.V.W.: 42.8%

Gloss, contrast ratio, and reflectance measured at 2 mil dry film thickness.

SOURCE: U.S. Silica Co.: Formula AC-105

Low Cost, High Hide Interior Flat

	Lbs./100 Gal.
Water	566.8
Cellosize QP 15,000	7.0

Add Cellosize under agitation, mix 5 minutes

KTPP	1.0

Add and mix 2 minutes

Triton N101	5.2
Tamol 731	5.0
Drewplus L475	0.4
Ethylene Glycol	23.7

Add and mix 2 minutes

R900	100.0
Snow*Tex 45	250.0

Add slowly; disperse to a 4+ Hegman

Wallpol 40-136	104.4
Exxate 1300	8.2
Drewplus L475	0.7
Dowcil 75	1.5

Add slowly under agitation, mix 10 minutes

    Viscosity: 92 KU at 25C
    Wt/gal: 10.74 lbs.
    N.V.W.: 39.4%
    N.V.V.: 22.1%
    PVC: .704
    Pigment: 32.59%
    P/B: 4.79
    Contrast Ratio: 0.97
    Reflectance (Hunter Green): 92.5
    Gloss 60: 2.0
    Gloss 20: 1.3

SOURCE: U.S. Silica Co.: Formula AC-110

Low-Cost Interior Flat Tint Base

	lbs/99.7 gal	g/l
Premix on Kadyzolver at 2,000 rpm for 10 min:		
Water	200.0	240.4
KTPP	2.0	2.4
Ross & Rowe 551	2.0	2.4
Tamol 731 (25%)	5.0	6.0
Nopco NXZ	2.0	2.4
Ethylene glycol	20.0	24.0
Carbitol acetate	10.0	12.0
Thickener (2.5% solution)	101.0	121.4
Add and grind at 3,500 rpm for 15 min:		
Ti-Pure R-901	175.0	210.3
Camel Carb	150.0	180.3
Iceburg clay	125.0	150.3
1160 silica	25.0	30.0
Let down (low speed on an air stirrer):		
Makon 10	3.0	3.6
Polyco 2161 (55%)	200.0	240.4
Merbac 35	0.5	0.6
Water	40.0	48.1
Thickener (2.5% solution)	110.0	132.2
NH4OH, 28% ammonium hydroxide	1.0	1.2

Tint with Cal/Ink 8800 series colorants at 2 oz/gal

Physical Properties:
   PVC, %: 62.7
   Volume solids, %: 30.3
   Weight solids, %: 49.9
   Density, lbs/gal (kg/l): 11.75 (1.41)
   Stormer viscosity (overnight), KU: 88+-3
   pH: 8.0-8.5

SOURCE: Aqualon Co.: NATROSOL B Hydroxyethylcellulose: Formula
        Table II

Low VOC Paint 65PVC Flat

	Pounds
Water	100
Natrosol 250HR	2
Kathon LX	1
Surfynol TG	2
Potassium Carbonate	3
Colloid 226	7
Colloid 640	3
TiPure R-900	125
Huber 70C	150
Hubercarb Q-6	175
Colloid 640	3
Water	193
Natrosol 250HR 2%	200
Airflex 738	190

Yield, gallons: 100.9
Solids: 48%
PVC: 65%
Viscosity: 90 KU
Contrast Ratio, 3 mil: 0.9714
Reflectance: 90.76
85 Sheen: 1.9
Formulation 12776-16

Low VOC 60 PVC Flat

	Pounds
Water	300
Natrosol Plus, Grade 330	6.7
Surfynol TG	2
Kathon LX	1
Colloid 226	7
Colloid 640	2
TiPure R-901	125
Huber 70C	150
Hubercarb Q-325	200
Colloid 643	3
Water	136
Airflex 738	250

Yield, gallons: 100.8
Solids: 51.10%
PVC: 60%
Formulation 12061-76

SOURCE: Air Products and Chemicals, Inc.: Low VOC Paint
Formulations Based on AIRFLEX 738 Emulsion

Low VOC Semi Gloss

	Pounds
Water	75
Tamol 731	8
AMP 95	2
Surfynol TG	1
Kathon LX	1
Colloid 640	3
Ti-Pure R-900	250
Water	100
Airflex 738	220
Rhoplex SG10	140
Colloid 640	2
Acrysol RM-5	50
Water	184
AMP95	6
Acrysol TT-935	5

Yield, gallons: 99.5
Solids: 43%
PVC: 27.50%
Viscosity: 91 KU
ICI Viscosity: 1.35 Poise
60 Gloss: 61
Formulation 11975-30A

Low VOC Eggshell Enamel

	Pounds
Water	200
Tamol 731	10
Surfynol TG	2
Kathon LX	1
Colloid 640	3
Potassium Carbonate	5
Ti-Pure R-900	225
Spacerite S-11	20
Atomite	110
Colloid 640	2
Water	170
Polyphobe 111	20
Airflex 738	350

Yield, gallons: 100
Solids: 48.50%
PVC: 41.20%
Viscosity: 100 KU
60 Gloss: 13
85 Sheen: 29
Reflectance: 91.89
Contrast Ratio, 3 mil: 0.971
Formulation 12776-39

SOURCE: Air Products and Chemicals, Inc.: Low VOC Paint Formulas

Polyvinyl Acetate Latex Paint
Pigment dispersion, Cowles Dissolver

	Pounds	Gallons
1.5% Sodium CMC* water gel	134	16.0
Vancide TH	1	0.1
Water	116	13.9
Ethylene glycol	25	3.0
Darvan No. 7	10	1.0
Antifoamer	4	0.4
Titanium dioxide	200	5.8
Nytal 300	150	6.3
Reduction:		
Water	130	15.6
Everflex BG or equivalent	264	29.6
4% Veegum T water gel	50	5.9
Wetting agent	20	2.4

* Hercules 7H3SF or equivalent
Consistency, immediate KU: 67   Consistency, one day KU: 75
Consistency, one week KU: 79   Consistency, one month KU: 81
Pigment volume concentration: 46 Percent solids by weight: 45

Formula No. 1242

Stipple Finish PVA Latex Paint
Pigment dispersion, Cowles Dissolver

	Pounds	Gallons
1.5% Sodium CMC* water gel	134	16.0
Vancide TH	1	0.1
Water	116	13.9
Ethylene glycol	25	3.0
Darvan No. 7	10	1.0
Antifoamer	4	0.4
Titanium dioxide	200	5.8
Nytal 300	300	12.6
Reduction:		
Water	78	9.3
Everflex BG or equivalent	264	29.6
4% Veegum T water gel	50	5.9
Wetting agent	20	2.4

* Hercules 7H3SF or equivalent
Consistency, immediate KU: 93   Consistency, one day KU: 97
Consistency, one week KU: 97   Consistency, one month KU: 97
Pigment volume concentration: 57 Percent solids by weight: 54

Formula No. 1243

SOURCE: R.T. Vanderbilt Co., Inc.: VEEGUM T in Latex Paints

Premium Quality Interior White Eggshell Wall Paint
Based on Rhoplex AC-64 Thickened with Acrysol RM-825

Materials:	Weight/Pounds	Volume/Gallons
Water	83.2	10.00
Dowicil 75	1.0	0.09
Tamol 731 (25%)	11.5	1.25
Propylene Glycol	25.0	2.91
Patcote 888	2.0	0.25
Ti-Pure R-900	250.0	7.30
Optiwhite	100.0	5.45
Attagel 50	5.0	0.26

Grind the above on a high-speed impeller mill at 3800-4500 FPM
for 20 minutes. At a slower speed let down as follows:

Water	83.2	10.00
Rhoplex AC-64	350.0	39.55
Texanol	10.6	1.33
Patcote 888	4.0	0.50
Water	83.2	10.00
Butyl Carbitol) Premix	31.0	3.88
Acrysol RM-825)	26.6	3.06
Natrosol MR (2.5%)	12.0	1.45
Ammonium Hydroxide (28%)	1.5	0.21
Water	20.9	2.51

Formulation Constants:
   PVC: 36%
   Volume Solids: 35.4%
   Initial Viscosity:
      KU: 85+-5 KU
      ICI: 1.6+-0.2P
   Equilibrated Viscosity:
      KU: 95+-5KU
      ICI: 1.7+-0.2P
   Gloss 85: 20
        60: 6
   CARB (Organic Volatile Level) g/L: 188

SOURCE: Rohm and Haas Co.: Trade Sales Coatings: ACRYSOL RM-825:
      Formulation F-64-7

Quality Flat Tint Base

Ingredients:	Pounds	Gallons
Pigment Grind:		
Water	175.0	20.98
Cellosize Hydroxyethyl Cellulose QP-15,000H	4.5	0.39
Preservative (1)	2.3	0.26
Dispersant (2)	10.2	1.02
Nonionic Surfactant (3)	3.0	0.34
Propylene Glycol	25.9	2.99
Antifoam (4)	0.9	0.13
Titanium Dioxide (5)	175.0	5.24
Nepheline Syenite (6)	200.0	9.20
Letdown:		
Ucar Latex 376 (7)	183.2	20.24
Ucar Acrylic 522 (7)	199.5	22.42
Ucar Filmer IBT	14.2	1.79
Antifoam (4)	2.6	0.37
Mildewcide (8)	5.0	0.33
Wetting Aid (9)	2.3	0.20
Ammonium Hydroxide 28% Aqueous Solution	1.8	0.24

Suppliers:
(1) Preservative - "Cosan" 145 or equivalent
(2) Dispersant - "Colloid" 226/35 or equivalent
(3) Nonionic Surfactant - "Igepal" CO-630 or equivalent
(4) Antifoam - "Colloid" 643 or equivalent
(5) Titanium Dioxide - "Tronox" CR-822 or equivalent
(6) Nepheline Syenite - "Minex" 4 or equivalent
(7) Ucar Latex 376: Ucar Acrylic 522 in 50:50 volume solids blend, respectively.
(8) Mildewcide - "Nopcocide" N-96 or equivalent
(9) Wetting Agent - "Dextrol OC-50" or equivalent

Paint Properties:
   Pigment Volume Concentration (PVC), %: 40.9
   Solids, %:
      by Volume: 37.2
      by Weight: 53.3
   Viscosity (initial):
      Stormer, KU: 104
      ICI, poise: 1.3
   pH (initial): 8.4
   Weight per Gallon, lb: 11.24

Film Properties:
   Contrast Ratio, 3 mil film: 0.962
   Reflectance, %: 90.5

SOURCE: Ucar Emulsion Systems: UCAR Latex 367/UCAR Latex 376: Formulation Suggestion E-2212

Quality Interior White

	Pounds
Methocel J-12-HS, 3%	75
Tamol 731	6
AMP-95	3
Igepal CO-610	4
Ethylene Glycol	25
Texanol	8
Witco 3056-A	2
Tronox CR-812	250
Satintone #5	100
Zeolex 80	20
Gold Bond R	75
Water	215
Merbac 35	1
Witco 3056-A	1
Methocel J-12-HS, 3%	110
Flexbond 325	300

    Yield, gallons: 103
    Solids, %: 51
    PVC, %: 49
    VOC, gm/l - water: 130
    Formulation No. 32511

Interior Semigloss White

	Pounds
Propylene Glycol	40
Texanol	13
Tamol 731	12
Drew L-484	2
AMP-95	3
Water	25
Ti-Pure R-900	250
Drew L-484	3
Merbac 35	1
Aerosol OT	1
Flexbond 325	400
Natrosol Plus, Grade 330, 2%	100
Rheolate 278	18
Water	185

    Yield, gallons: 99.7
    Solids, %: 45
    PVC, %: 24
    VOC, gm/l-water: 195
    Viscosity, KU: 82
    ICI Viscosity, poise: 1.24
    60 Gloss: 64
    20 Gloss: 21
    Leneta Leveling: 7
    Formulation No. 32528

SOURCE: Air Products and Chemicals, Inc.: Flexbond 325 Vinyl-Acrylic Emulsion: Typical Starting Formulations

Quality Latex Flat
Optiwhite Approach

	Lbs./100 Gal.
Water	340
PA-328	3.00
Nuosept 95	2.00
Ethylene Glycol	18.60
KTPP	1.00
Tamol 731	6.90
Triton X-100	4.50
Propylene Glycol	9.30
Cellosize QP-4400	4.00
Ammonia	1.00
CR-813	120
Optiwhite	200
#1 White (CaCO3)	150
Celite 281	35
3011	265
Texanol	11.88
PA-328	3.00

Evaluation Results:
Hiding 3 Mil Bird Blade on Opacity Panel:
    Reflectance: .903
    Contrast Ratio: .972
Tint Strength .5 Blue/200 grams:
    Reflectance: .727
Sheen: 85: 1.9
       60: 2.3
Viscosity KU: 95

SOURCE: Burgess Pigment Co.: Formula 95 ICE/OPT

Quality Latex Interior Flat
Optiwhite Approach

	Lbs./100 Gal.
Water	261
Troysan 174	2.25
Natrosol Plus 330	3.50
KTPP	1.68
Propylene Glycol	61
AMP-95	3.00
Tamol 1124	8.00
Igepal CO-630	3.95
Colloid 697	4.00
RCL-9	160
#10 White	200
Optiwhite	200
Unocal 3084	268
Colloid 697	4.00
Texanol	13.00
Triton GR-7	3.90
RM-825	9.75

Evaluation Results:
Hiding 3 Mil Bird Blade on Opacity Panel:
    Reflectance: .901
    Contrast Ratio: .970
Tint Strength .5 Blue/200 grams:
    Reflectance: .715
Sheen:
    85: 1.7
    60: 2.7
Viscosity KU: 95

SOURCE: Burgess Pigment Co.: Formula 148 OP-62/OPT

Quality Latex Semi-Gloss
Burgess No. 98 Approach

	Lbs./100 Gal.
Water	294
Tamol 731	5.00
AMP-95	3.00
Tergitol NP-10	2.00
Drew L-464	2.00
Troysan 186	2.00
CR-800	250
Burgess No. 98	75
Attagel 50	4.00
Texanol	16.00
Ucar 379	280
Ucar 525	180
Propylene Glycol	40
Triton GR-7M	1.00
Drew L-464	4.00
Ucar SCT-275	18

Evaluation Results:
Hiding 3 Mil Bird Blade on Opacity Panel:
    Reflectance: .920
    Contrast Ratio: .981
Tint Strength .5 Blue/200 grams:
    Reflectance: .737
Sheen:
    85: 82.8
    60: 33.4
    20: 3.8
Viscosity KU: 96

SOURCE: Burgess Pigment Co.: Formula 100 OP-62/98

Red Floor Paint

Material:	Weight (Lbs.)	Volume (Gals.)
Water	14.0	1.68
Methyl Carbitol	50.0	5.80
Exp. Dispersant QR-681M (35%)	10.4	1.14
NH4OH (28%)	1.0	0.13
Triton CF-10	2.0	0.23
Drew L-405	1.0	0.14
Chrome Oxide Green X-1134	41.6	0.98
Red Iron Oxide R-2899	201.3	4.71

Grind on the high-speed cowles for twenty minutes; then, add the
    following letdown:

Maincote HG-54	632.6	73.70
Butyl Propasol	31.5	4.25
Ektasolve EEH	5.3	0.71
Sodium Nitrite (15%)	9.0	1.08
Drew Y-250	2.5	0.35
NH4OH (28%)	5.0	0.68
Rheology Modifier QR-1001 (20.8%)	21.0	2.36
Water	7.2	2.06

Formulation Constants:
    Pigment Volume Content, %: 16.3
    Volume Solids, %: 34.9
    Viscosity, KU:
        Initial: 82
        Equilibrated: 85
    Viscosity, ICI, poise:
        Initial: 1.5
        Equilibrated: 1.5
    pH:
        Initial: 9.2
        Equilibrated: 9.2

SOURCE: Rohm and Haas Co.: Maintenance Coatings: MAINCOTE HG-54:
        Formulation FP-54-1

Semigloss Interior White Paint

Materials:	Lbs	Gal
Propylene glycol	80.0	9.30
Tamol SG-1, 35%	8.5	0.89
Hercules SGL defoamer	2.0	0.26
Rutile titanium dioxide	240.0	7.03
Silica (small-particle, amorphous)	25.0	1.13
Water	25.0	3.00

Mill the above in a high-speed mill (Cowles) for 20 minutes and
   let down at a slower speed, as follows:

Rhoplex AC-417, 48.0%	500.0	56.00
Hercules SGL defoamer	2.7	0.37
Propylene glycol	10.0	1.16
Texanol	21.6	2.73
Super Ad-It) Premix	1.0	0.12
Triton GR-7M)	0.5	0.06
Water and/or 3% Natrosol Plus	169.0	20.20

Formulation Constants:
   Wt/gal, lbs: 10.65
   Pigment volume concentration, %: 24.8
   Volume solids, %: 32.9

Properties exhibited when paint is thickened with Natrosol Plus:
   Wt%polymer: 0.33
   Viscosity, KU: 92
   Brushing viscosity, poises: 1.2
   Spatter: 8
   Leveling: 5
   Sag resistance, mils: 19.5
   Scrub cycles to failure: 600
   60 gloss: 42.9

SOURCE: Aqualon Co.: NATROSOL PLUS: Formula SG-41-3

Semi-Gloss White Formulation

Materials:	Weight Ratio*	Parts Per Hundred (Volume Basis)
Charge to a Cowles tank and mix for 5 minutes at a low speed of about 1,500 feet per minute.		
Rhoplex AC-604 (46%)	160.0	17.73
Tamol 731 (25%)	14.0	1.70
Triton CF-10	2.0	0.24
Nopco NXZ	0.5	0.06
Dimethylaminoethanol)Premix	1.1	0.13
Water)	2.2	0.27

Slowly add the titanium dioxide and silicas while increasing mixer speed to 4,200 feet per minute.

Grind for 20 minutes:		
Ti-Pure R-960	200.2	6.06
Syloid 978	13.3	0.80
Syloid 308	6.7	0.40

Gradually add the following and blend at a low speed of about 1,500 feet per minute:

Rhoplex AC-604 (46%)	490.0	54.30
Water	10.0	1.21
Butyl Cellosolve (60% in water)	50.0	6.00
Cellosize QP-40 (5%)	8.4	1.40

If a lower temperature curing is required, replace some of the water in the formulation with catalyst.

Formulation Constants:
Pigment volume content, %: 19
TiO_2/binder ratio: 40/60
Approximate solids, %:
  Weight: 54
  Volume: 45

* Using weight ratio in pound units will yield approximately 100 gallons of paint, while with kilograms, 833 liters will result.

SOURCE: Rohm and Haas Co.: Building Products: RHOPLEX AC-604: Formulation Based on RHOPLEX AC-604 Emulsion

Styrene-Butadiene Latex Paint
Pigment dispersion, Cowles Dissolver

	Pounds	Gallons
1.5% Sodium CMC* water gel	200	23.9
Vancide TH	1	0.1
Ethylene glycol	25	3.0
Darvan No. 7	10	1.0
Antifoamer	4	0.4
Ammonium hydroxide, 28%	2	0.2
Titanium dioxide	200	5.8
Nytal 300	225	9.5
Reduction:		
Dow 762-W or equivalent	308	36.6
4% Veegum T water gel	146	17.1
Wetting agent	20	2.4

* Hercules 7H3SF or equivalent

Consistency, immediate KU: 80   Consistency, one day KU: 85
Consistency, one week KU: 86   Consistency, one month KU: 88
Pigment volume concentration: 47 Percent solids by weight: 52

Formula No. 1244

Acrylic Latex Paint
Pigment dispersion, Cowles Dissolver

	Pounds	Gallons
1.5% Sodium CMC* water gel	134	16.0
Vancide TH	1	0.1
Water	105	12.6
Ethylene glycol	25	3.0
Darvan No. 7	10	1.0
Antifoamer	4	0.4
Ammonium hydroxide 28%	2	0.2
Titanium dioxide	200	5.8
Nytal 300	225	9.5
Reduction:		
Rhoplex AC-33 or equivalent	322	37.2
4% Veegum T water gel	100	11.8
Wetting Agent	20	2.4

* Hercules 7H3SF or equivalent

Consistency, immediate KU: 79   Consistency, one day KU: 82
Consistency, one week KU: 83   Consistency, one month KU: 84
Pigment volume concentration: 50 Percent solids by weight: 51

Formula No. 1245

SOURCE: R.T. Vanderbilt Co., Inc.: VEEGUM T in Latex Paints

White Floor Paint

Material	Weight (Lbs.)	Volume (Gals.)
Water	4.0	0.48
Methyl Carbitol	50.0	5.80
Exp. Dispersant QR-681M	8.4	0.92
NH4OH (28%)	1.0	0.13
Triton CF-10	2.0	0.23
Drew L-405	1.0	0.14
TiPure R-900 titanium dioxide	195.0	5.70
Letdown:		
Maincote HG-54	632.6	73.70
Butyl Propasol	31.5	4.25
Ektasolve EEH	5.3	0.72
Sodium Nitrite (15%)	6.6	0.79
Drew Y-250	2.5	0.35
NH4OH (28%)	5.0	0.68
Rheology Modifier QR-1001/Water	50.9	6.11

Formulation Constants:
    Pigment Volume Content, %: 16.3
    Volume Solids, %: 34.9
    Viscosity, KU:
        Initial: 75
        Equilibrated: 77
    Viscosity, ICI, poise:
        Initial: 1.1
        Equilibrated: 1.2
    pH:
        Initial: 9.2
        Equilibrated: 9.1

SOURCE: Rohm and Haas Co.: Maintenance Coatings: MAINCOTE HG-54: Formulation FP-54-3

Section VII
Lacquers

Clear Air-Dry Lacquer

Material:	Lbs.	Gals.
Water	205.0	24.61
Methyl Carbitol	58.0	6.82
AMP 95	1.0	0.11
Dimethyl Phthalate	18.2	1.85
Aerosol TR-70	2.0	0.25
Foamaster G	1.0	0.15
Proxel GXL	1.0	0.10
Blend above in order; add to below slowly, under agitation.		
Synthemul 97-603	581.8	66.11

Analysis:
    Pigment Volume Concentration, Percent: 0.0
    Percent Solids, Weight: 30.2
    Percent Solids, Volume: 27.7
    Viscosity @ 25C, #4 Ford, Secs.: 25-30
    pH: 8.5-9.0
    Pounds/Gallon: 8.68
    VOC (excluding water):
        Grams/Liter: 191.0
    60 Gloss: 80-85
    MFFT, C: 0

SOURCE: Reichhold Chemicals, Inc.: Waterborne Handbook:
        Formulation T603-01

Clear, Full Gloss Waterborne Lacquer Based On Rhoplex CL-103 Emulsion for Air-Dry or Low-Force-Dry Application

	Weight	Volume
Materials:	Pounds	Gallons
Rhoplex CL-103 emulsion	668.0	77.5
Dee Fo 3000	1.4	0.2

Premix and add under agitation:

Butyl Carbitol	26.7	3.4
Butyl Cellosolve	40.1	5.4
Water	33.4	4.0

Add:

Dow Corning #14	26.7	4.0
Michem Emulsion 39235	22.9	2.8
14% ammonia	2.4	0.3
(adjust viscosity with water to a range of 20 to 26 seconds on a #2 Zahn cup)		
Water	20.3	2.4

Formulation Constants:
    Approximate solids, % (wt/vol): 33.6/31.4
    pH: 7.5-8.0
    Viscosity, #2 Zahn cup, seconds: 20 to 26
    VOC, g/liter: 249
        lb/gallon: 2.08
    Coalescent, wt % on polymer solids:
        Butyl Cellosolve: 15
        Butyl Carbitol: 10
    Heat/age Stability: Passes
    Freeze/thaw stability: Protect from freezing

SOURCE: Rohm and Haas Co.: RHOPLEX CL-103 Acrylic Emulsion: Formulation KC-53-1

Clear, Full-Gloss Waterborne Lacquer Based on Rhoplex CL-104 Emulsion for Air-Dry or Low-Force-Dry Application

Materials:	Weight Pounds	Volume Gallons
Rhoplex CL-104 emulsion	642.8	75.2
Dee Fo 3000	1.3	0.2

Premix and add under agitation:

Dow Corning #14	24.7	3.7
Butyl Carbitol	24.7	3.1
Hexyl Cellosolve	37.6	5.1
Water	29.2	3.5

Add:

Michem Emulsion 39235	21.8	2.6
14% Ammonia	3.0	0.4
(adjust viscosity with water to a range of 20 to 26 seconds on a #2 Zahn cup)	51.5	6.2

Formulation Constants:
    Approximate solids, % (wt/vol): 30.7/29.1
    pH: 7.5-8.0
    Viscosity, #2 Zahn cup, seconds: 20 to 26
    VOC, g/liter: 249
        lb/gallon: 2.07
    Coalescent, wt % on polymer solids:
        Butyl Cellosolve: 15
        Butyl Carbitol: 10
    Rheology modifier, wt % on polymer solids: 1.0
    Freeze/thaw stability: Protect from freezing
    Heat/age stability (140F/10 days): Passes

SOURCE: Rohm and Haas Co.: RHOPLEX CL-104 Acrylic Emulsion: Formulation WR-74-7

Clear Lacquer

| | Parts per Hundred | |
Materials:	Weight Ratio*	(Volume Basis)
Rhoplex WL-81	650.0	75.6
Water	110.0	13.2

Premix and add under agitation:

Ethylene glycol monobutyl ether	103.8	13.8
Diethylene glycol monobutyl ether	10.8	1.4
BYK-301	1.3	0.2
Patcote 519	0.9	0.1

To increase the lacquer viscosity add dilute ammonia solution consisting of 1 part concentrated ammonia (28%) and 1 part water. To decrease the lacquer viscosity thin with water.

Formulation Constants:
   pH: 7.5+-0.3
   Viscosity, cps.: 100 to 400
   Approximate Solids, % (wt., vol.): 31, 29
   Density, lbs./gal.: 8.4
   Freeze/Thaw Stability: Protect from freezing
   Mechanical Stability (5' Waring Blender): Satisfactory
   Heat Stability (10 days/140F): Satisfactory
   VOC, g/liter: 304
        lbs./gal.: 2.53
   % on Polymer Solids:
      Ethylene glycol monobutyl ether: 38.5
      Diethylene glycol monobutyl ether: 4.0

 * Using weight ratio in pound units will yield approximately
   100 gallons of lacquer while with kilograms, 833 liters
   will result.

This clear formulation can be used with Colanyl predispersed colorants to achieve various colors.

SOURCE: Rohm and Haas Co.: Industrial Coatings: RHOPLEX WL-81:
        Formulation WL-81-3

Clear Lacquer Based on Rhoplex WL-92 Polymer
Wood Formulation

Materials	Weight* (Pounds)	Volume (Gallons)
Rhoplex WL-92 (42 wt. % solids)	705.9	82.19
Patcote 519	1.7	0.24
Butyl Carbitol	59.3	7.41
Dowanol PNB	29.6	4.05
MichemLUBE 110 (25 wt. % solids)	28.0	3.38
Ammonia (14%) to pH of 7.9-8.1	----	----
Water to Reduce Viscosity	22.7	2.72

Formulation Constants:
  pH: 7.9-8.1
  Viscosity, cps: 400-600
  Approximate Solids, % (wt./vol.): 36/34
  Density, lbs./gal.: 8.5
  VOC, g/liter: 234
       lbs./gallon: 1.95
  Weight % on Polymer Solids:
     Butyl Carbitol: 20
     Dowanol PNB: 10

* Using weight in pound units will yield approximately 100
  gallons of lacquer. With kilograms, 833 liters will result.

SOURCE: Rohm and Haas Co.: Industrial Coatings: RHOPLEX WL-92:
        Formulation WL-92-1

Clear Lacquer Formulation for Low Force-Dry Applications

Materials:	Parts Per Hundred Weight*(Volume Basis)	
Rhoplex WL-96 (42%)	630.8	73.18
Patcote 519	1.7	0.24
Butyl Cellosolve	93.0	12.37
MichemLUBE 110 (25% solids)	27.7	3.34
Water	86.3	10.36

Formulation Constants:
    pH: 8.0
    Viscosity, cps: 600
    Approximate Solids, % (wt.,vol.): 33,30
    Density, lbs./gal.: 8.4
    Freeze/Thaw Stability: Protect from freezing
    Heat/Age Stability (140F/10 days): Satisfactory
    VOC, g/liter: 261
        lbs./gal.: 2.18
    % on Polymer Solids:
        Butyl Cellosolve: 35

    * Using weight in pound units will yield approximately 100
      gallons of lacquer. Weight units of kilograms yield 833
      liters.

SOURCE: Rohm and Haas Co.: Industrial Coatings: RHOPLEX WL-96:
        Formulation WL-96-1

Clear Waterborne Lacquer, Flatted to 40 Based on Rhoplex CL-103 Emulsion for Air-Dry or Low-Force-Dry Application

Materials:	Weight Pounds	Volume Gallons
Rhoplex CL-103 emulsion	655.0	75.9
Dee Fo 3000	1.3	0.2

Premix and add under agitation:

Butyl Carbitol	52.4	6.6
Butyl Propasol	26.2	3.6
Degussa TS-100	10.5	0.6
Water	29.6	3.6

Add:

Dow Corning #14	26.2	3.9
Michem Emulsion 39235	22.4	2.7
14% Ammonia	2.0	0.2

(adjust viscosity with water to a range of 20 to 26 seconds on a #2 Zahn cup)

Water	22.5	2.7

Formulation Constants:
    Approximate solids, % (wt/vol): 34.0/31.4
    pH: 7.5-8.0
    Viscosity, #2 Zahn cup, seconds: 20 to 26
    VOC, g/liter: 272
        lb/gallon: 2.27
    Coalescent, wt % on polymer solids:
        Butyl Propasol: 10
        Butyl Carbitol: 20
    Heat/age stability (140F/10 days): Passes
    Freeze/thaw stability: Protect from freezing

SOURCE: Rohm and Haas Co.: RHOPLEX CL-103 Acrylic Emulsion: Formulation KC-53-2

Clear Waterborne Lacquer, Flatted to 40 Gloss at 60 Based on Rhoplex CL-104 Emulsion for Low-Force-Dry Application

Materials:	Weight Pounds	Volume Gallons
Rhoplex CL-104 emulsion	638.8	74.7
Dee Fo 3000	1.2	0.2

Premix and add under agitation:

	Weight Pounds	Volume Gallons
Butyl Carbitol	49.2	6.2
Butyl Propasol	24.6	3.3
Degussa TS-100	9.8	0.5
Water	26.1	3.1

Add:

	Weight Pounds	Volume Gallons
Dow Corning #14	24.6	3.7
Michem Emulsion 39235	21.0	2.5
14% Ammonia	1.9	0.2

(adjust viscosity with water to a range of 20 to 26 seconds on a #2 Zahn cup)

	Weight Pounds	Volume Gallons
Water	46.3	5.6

Formulation Constants:
   Approximate solids, % (wt/vol): 31.5/29.5
   pH: 7.5-8.0
   Viscosity, #2 Zahn cup, seconds: 20 to 26
   VOC, g/liter: 271
       lb/gallon: 2.25
   Coalescent, wt % on polymer solids:
     Butyl Cellosolve: 10
     Butyl Carbitol: 20
   Freeze/thaw stability: Protect from freezing
   Heat/age stability (140F/10 days): Passes

SOURCE: Rohm and Haas Co.: RHOPLEX CL-104 Acrylic Emulsion: Formulation WR-74-8

Flat Black Lacquer

Materials:	Parts per Hundred Weight Ratio*	(Volume Basis)
Add the following with good agitation:		
Rhoplex WL-81 (41.5 wt% solids)	612.1	71.19
Water	65.9	7.91
Colanyl Black PR-A	24.5	2.35
Mix well, then sift in:		
Syloid 166	18.8	1.13
Add the following with good agitation:		
Ethylene glycol monobutyl ether	97.9	13.09
Texanol	10.4	1.32
Ammonium Benzoate, 10 wt% in water	23.5	2.82
Deefo 806-102 (or Patcote 519)	1.4	0.19

Note: If viscosity is too low, adjust up with 14% Ammonia.
      If viscosity is too high, adjust down with water.

Formulation Constants:
    pH: 7.9-8.4
    Viscosity, cps: 400-1000
    Approximate Solids, % (wt/vol): 31.0/28.7
    Density, lbs/gallon: 8.5
    Freeze/Thaw Stability: Protect from freezing
    VOC, g/liter: 330
        lbs/gallon: 2.66
    Weight % on Polymer Solids:
        Ethylene glycol monobutyl ether: 38.6
        Texanol: 4.1

* Using weight ratio in pound units will yield approximately
  100 gallons of lacquer while with kilograms, 833 liters
  will result.

SOURCE: Rohm and Haas Co.: Industrial Coatings: RHOPLEX WL-81:
        Formulation WL-81-8

Lacquer Type: Busan Primer
Substrate: Ferrous Metal

Materials:	Weight Ratio*	Volume Ratio*
Add the following with good agitation:		
Butyl Cellosolve	46.0	6.12
Water	137.0	16.44
Patcote 519	1.0	0.14
Igepal CTA-639	3.4	0.39
Tamol 165	14.8	1.68
Red Iron Oxide	82.5	1.93
Busan 11M-1	68.4	2.49
499 Talc	84.8	3.77
290 Lo-Micron Barytes	119.0	3.17

Pebble mill grind to 7 to 7.5 N.S. Hegman, then letdown in the
  following order with good agitation:

Rhoplex WL-91 (41.5 wt %)	420.0	48.72
Triton X-405 (reduced to 35 wt % solids)	9.9	1.08
Butyl Carbitol	26.2	3.27
Dibutylphthalate	17.4	1.99
Butyl Cellosolve	41.5	5.52
14% Ammonia to pH 9.0-9.3	6.8	0.85
Ammonium Benzoate (10 wt % in water)	20.9	2.51

  Adjust viscosity with water (200-600 cps)

Formulation Constants:
  Approximate Solids, % (wt/vol): 51/34
  Pigment/Binder Ratio: 65/35
  Pigment Volume Content, %: 34.8
  pH: 9.0-9.3
  Viscosity, cps: 200-600
  Density, lbs/gal: 11.0
  VOC, g/liter: 276
    , lbs/gallon: 2.30
  Coalescent, Wt % on Polymer Solids:
    Butyl Cellosolve: 50
    Butyl Carbitol: 15
    Dibutylphthalate: 10
  Freeze-Thaw Stability: Protect from freezing

* Using weight ratio in pounds will yield approximately 100
  gallons. With kilograms, approximately 833 liters will be
  obtained.

SOURCE: Rohm and Haas Co.: RHOPLEX WL for Acrylic Lacquers:
        Formulation WL-91-2

Lacquer Type: Chromate-Free
Substrate: Metal and Other Substrates

Materials:	Weight Ratio*	Volume Ratio*
Add the following with good agitation:		
Epotuf 38-690	27.5	3.27
Triethylamine	2.4	0.40
Aquacat	1.0	0.12
Magnacat	1.0	0.13
Butyl Cellosolve	13.9	1.85

Add in order and mix well:

Water	129.2	15.50
Tamol 165	2.7	0.31
Patcote 550	0.5	0.07
Raven 2000	33.8	2.32

Pebble mill grind to 7 to 7.5 N.S. Hegman, then letdown in the
  following order with good agitation:

Epotuf 38-690	)	148.1	17.63
Triethylamine	) Premix	12.1	2.00
Water	)	129.2	15.50

Mix well (recheck grind), then add the following in order with
  good agitation:

Rhoplex WL-81 (41.5 wt %)	287.4	33.26
Triton X-405 (reduced to 35 wt %)	4.8	0.55
Santicizer 160	23.7	2.53
Ammonium Benzoate (10 wt % in water)	24.0	2.88
Water	8.3	1.00

Formulation Constants:
    Approximate Solids, % (wt/vol): 36/33
    pH: 8.9
    Viscosity, cps: 300
    Epotuf 38-690/Rhoplex WL-81 (wt): 51/49
    Density, lbs/gal: 8.6
    VOC, g/liter: 224
       , lbs/gallon: 1.87
    Freeze-Thaw Stability: Protect from freezing

  * Using weight ratio in pounds will yield approximately 100
    gallons. With kilograms, approximately 833 liters will be
    obtained.

SOURCE: Rohm and Haas Co.: RHOPLEX WL for Acrylic Lacquers:
    Formulation WL-81-5

Lacquer Type: Chromate Primer
Substrate: Ferrous Metal

Materials:	Weight Ratio*	Volume Ratio*
Add the following with good agitation:		
Butyl Cellosolve	50.6	6.73
Water	148.7	17.84
Patcote 519	1.2	0.17
Igepal CTA-639	3.7	0.42
Tamol 165	16.1	1.89
Red Iron Oxide	89.5	2.09
Strontium Chromate	10.9	0.35
Zinc Yellow	3.3	0.12
499 Talc	32.5	1.45
290 Lo-Micron Barytes	44.7	1.19
Pebble mill grind to 7 to 7.5 N.S. Hegman, then letdown in the following order with good agitation:		
Rhoplex WL-91 (41.5 wt %)	455.8	52.88
Triton X-405 (reduced to 35 wt % solids)	21.5	2.46
Butyl Carbitol	28.4	3.55
Dibutylphthalate	19.0	2.17
Butyl Cellosolve	22.2	2.95
14% Ammonia to pH 9.0-9.3	7.4	0.92
Ammonium Benzoate, (10 wt % in water)	22.7	2.72
Adjust viscosity with water (200-600 cps)		

Formulation Constants:
   Approximate Solids, % (wt/vol): 42/30
   pH: 9.0-9.3
   Pigment Volume Content, %: 18.4
   Density, lbs/gal: 9.8
   VOC, g/liter: 279
      , lbs/gallon: 2.33
   Coalescent, Wt % on Polymer Solids:
      Butyl Cellosolve: 38.5
      Butyl Carbitol: 15
      Dibutylphthalate: 10
   Freeze-Thaw Stability: Protect from freezing

* Using weight ratio in pounds will yield approximately 100 gallons. With kilograms, approximately 833 liters will be obtained.

SOURCE: Rohm and Haas Co.: RHOPLEX WL for Acrylic Lacquers: Formulation WL-91-5

<u>Lacquer Type: Clear</u>
<u>Substrate: Metal and Other Substrates</u>

Materials:	Weight Ratio*	Volume Ratio*
Add the following with good agitation:		
Rhoplex WL-91 (41.5 wt %)	637.6	73.97
Water	107.9	12.95
Patcote 519	1.5	0.21

Premix and add under agitation:

Butyl Cellosolve	79.5	10.57
Butyl Carbitol	13.2	1.65
Dibutylphthalate	13.2	1.51
BYK-301	1.3	0.16

Formulation Constants:
   Approximate Solids, % (wt/vol): 33/31
   pH: 7.2-7.4
   Viscosity, cps: 200-450
   Density, lbs/gal: 8.5
   VOC, g/liter: 258
     , lbs/gallon: 2.15
   Coalescent, Wt % on Polymer Solids:
     Butyl Cellosolve: 30
     Butyl Carbitol: 5
     Dibutylphthalate: 5
   Freeze-Thaw Stability: Protect from freezing
   Heat Stability (10 days/140F): Satisfactory
   Mechanical Stability (5 minutes Waring Blender): Satisfactory

   Note: This clear formulation can be used with Colanyl predis-
       persed colorants to obtain various colors.

   * Using weight ratio in pounds will yield approximately 100
     gallons. With kilograms, approximately 833 liters will be
     obtained.

SOURCE: Rohm and Haas Co.: RHOPLEX WL for Acrylic Lacquers:
       Formulation WL-91-6

Lacquer Type: Clear
Substrate: Plastics

Materials:	Weight Ratio*	Volume Ratio*
Add the following with good agitation:		
Rhoplex WL-51 (41.5 wt %)	537.4	62.48
Water	138.1	16.57
Butyl Cellosolve	111.5	14.83
Butyl Carbitol	22.3	2.79
Dibutylphthalate	17.8	2.03
Ammonia (14%)	4.8	0.60

Formulation Constants:
    Approximate Solids, % (wt/vol): 29/27
    pH: 9.0
    Viscosity, cps: 1000
    Density, lbs/gal: 8.4
    VOC, g/liter: 361
      , lbs/gallon: 3.1
    Coalescent, Wt % on Polymer Solids:
      Butyl Cellosolve: 50
      Butyl Carbitol: 10
      Dibutylphthalate: 8
    Freeze-Thaw Stability: Protect from freezing

* Using weight ratio in pounds will yield approximately 100
  gallons. With kilograms, approximately 833 liters will be
  obtained.

Red Lacquer for Plastics Based on Rhoplex WL-51:

Materials:	Weight Ratio
Clear for Plastics (above)	783.3
Colanyl Red FRLL-A	105.0

SOURCE: Rohm and Haas Co.: RHOPLEX WL for Acrylic Lacquers:
       Formulation WL-51-9

Lacquer Type: Clear
Substrate: Wood

Materials:	Weight Ratio*	Volume Ratio*
Add the following with good agitation:		
Rhoplex WL-92 (42.0 wt %)	705.9	82.19
Patcote 519	1.7	0.24
Butyl Carbitol	59.3	7.41
Dowanol PNB	29.6	4.05
MichemLUBE 110 (25.0% solids)	28.0	3.38
Ammonia (14%) to pH of 7.9-8.1	----	----
Water to Reduce Viscosity	22.7	2.72

Formulation Constants:
   Approximate Solids, % (wt/vol): 36/34
   pH: 7.9-8.1
   Viscosity, cps: 400-600
   Density, lbs/gal: 8.5
   VOC, g/liter: 234
      , lbs/gallon: 1.95
   Coalescent, Wt % on Polymer Solids:
      Butyl Carbitol: 20
      Dowanol PNB: 10
   Freeze-Thaw Stability: Protect from freezing

* Using weight ratio in pounds will yield approximately 100
  gallons. With kilograms, approximately 833 liters will be
  obtained.

SOURCE: Rohm and Haas Co.: RHOPLEX WL for Acrylic Lacquers:
        Formulation WL-92-1

Lacquer Type: Clear
Substrate: Wood

Materials:	Weight Ratio*	Volume Ratio*
Add the following with good agitation:		
Rhoplex WL-96 (42.0 wt %)	624.0	72.39
Patcote 519	1.5	0.21
Butyl Carbitol	52.4	6.55
Butyl Propasol	26.2	3.58
MichemLUBE 110 (25 wt % solids)	24.7	2.98
Water	115.5	13.86
14% Ammonia to pH approx. 8.0	0.5	0.06

Formulation Constants:
   Approximate Solids, % (wt/vol): 32/30
   pH: 8.0
   Viscosity, cps: 600
   Density, lbs/gal: 8.5
   VOC, g/liter: 235
     , lbs/gallon: 1.96
   Coalescent, Wt % on Polymer Solids:
     Butyl Carbitol: 20
     Butyl Propasol: 10
   Freeze-Thaw Stability: Protect from freezing

  * Using weight in pounds will yield approximately 100 gallons.
    With kilograms, approximately 833 liters will be obtained.

SOURCE: Rohm and Haas Co.: RHOPLEX WL for Acrylic Lacquers:
      Formulation WL-96-7

Lacquer Type: Clear (and for use with predispersed colorants)
Substrate: Metal

Materials:	Weight Ratio*	Volume Ratio*
Add the following with good agitation:		
Kelsol 3905	240.0	28.07
Aquacat	1.9	0.23
Magnacat	3.8	0.47
Patcote 519	0.9	0.13
Patcote 577	0.9	0.13
28% Ammonia	5.1	0.66
Water	321.7	38.60
Mix well (% neutralization=58%, pH=5.8-6.0)		
Add in order, with agitation:		
Rhoplex WL-71 (41.5 wt %)	186.7	21.68
Water	32.2	3.86
Butyl Carbitol	20.0	2.50
14% Ammonia (to pH=8.4)	----	----
Water	21.4	2.57

Formulation Constants:
    Approximate Solids, % (wt/vol): 31.3/28.2
    pH: 8.4
    Viscosity, cps: 300-500
    Density, lbs/gal: 8.4
    VOC, g/liter: 246
      , lbs/gallon: 2.05
    Freeze-Thaw Stability: Protect from Freezing

    Note: Reduce for spray with 87/13 water/Butyl Cellosolve.

The following Colanyl series pigment dispersions are compatible
    with this formulation:

Code No.	Description	Pigment/Binder	Add x Lbs/100 Gallons
16-2010	Green 8G	13/87	70
18-1004	Black PR-A	5/95	40
11-1109	Yellow OT	13/87	70

    * Using weight ratio in pounds will yield approximately 100
      gallons. With kilograms, approximately 833 liters will be
      obtained.

SOURCE: Rohm and Haas Co.: RHOPLEX WL for Acrylic Lacquers:
        Formulation WL-71-5

Lacquer Type: Clear for Corrosion Resistance
Substrate: Copper and Brass

Materials:		Weight Ratio*	Volume Ratio*
Add the following with good agitation:			
Rhoplex WL-51 (41.5 wt %)		508.9	59.03
Water		164.9	19.78
Butyl Cellosolve	)	105.6	14.04
Butyl Carbitol	) Premix	21.2	2.65
Benzotriazole	)	1.6	----
Dibutylphthalate		16.9	1.94
Patcote 519		0.5	0.15
Byk-301		4.1	0.50
Ammonia (14%)		15.2	1.92

Formulation Constants:
    Approximate Solids, % (wt/vol): 29/27
    pH: 9.0
    Viscosity, cps: 1000**
    Density, lbs/gal: 8.4
    VOC, g/liter: 347
        , lbs.gallon: 2.9
    Coalescent, Wt % on Polymer Solids:
        Butyl Cellosolve: 50
        Butyl Carbitol: 10
        Dibutylphthalate: 8
    Freeze-Thaw Stability: Protect from freezing
    Heat Stability (10 days/140F): Satisfactory

 * Using weight ratio in pounds will yield approximately 100
   gallons. With kilograms, approximately 833 liters will be
   obtained.
** Viscosity may be adjusted with water prior to application.

SOURCE: Rohm and Haas Co.: RHOPLEX WL for Acrylic Lacquers:
        Formulation WL-51-4

Lacquer Type: Dark Walnut Wiping Stain
Substrate: Wood

Materials:	Weight Ratio*	Volume Ratio*
Add the following with good agitation:		
Rhoplex WL-93 (34 wt %)	342.6	39.84
Butyl Cellosolve	70.0	9.33
Santicizer 160	8.8	0.94
Deefo 806-102	1.0	0.12
Water	69.5	8.34
Water	330.6	39.69
Calco Nigrosine O2P PDR	21.8	1.03
Calco CID Orange Y Ex. Conc.	7.5	0.37
Calco CID Crocein Scarlet MOS Conc.	3.4	0.17
Calco CID Tartazine DB1 Conc.	3.4	0.17

Formulation Constants:
   Approximate Solids, % (wt/vol): 18.6/16.2
   pH: 7.3
   Viscosity, cps: 20
   Density, lbs/gal: 8.6
   VOC, g/liter: 338
      , lbs/gallon: 2.82
   Coalescent, Wt % on Polymer Solids:
      Butyl Cellosolve: 60.1
      Santicizer 160: 7.6
   Freeze-Thaw Stability: Protect from freezing
   Heat Stability (8 days/140F): Satisfactory

   * Using weight ratio in pounds will yield approximately 100
     gallons. With kilograms, approximately 833 liters will be
     obtained.

Procedure: Wipe directly onto wood, or spray (as is) onto wood;
           immediately wipe off excess.

SOURCE: Rohm and Haas Co.: RHOPLEX WL for Acrylic Lacquers:
        Formulation WL-93-1DW

Lacquer Type: Flat Black
Substrate: Metal and Other Substrates

Materials:	Weight Ratio*	Volume Ratio*
Add the following with good agitation:		
Rhoplex WL-81 (41.5 wt %)	612.1	70.84
Water	65.9	7.91
Colanyl Black PR-A	24.5	2.31
Mix well, then sift in:		
Syloid 166	18.8	1.14
Add the following with good agitation:		
Butyl Cellosolve	97.9	13.02
Texanol	10.4	1.31
Ammonium Benzoate, (10 wt % in water)	23.5	2.82
Deefo 806-102 (or Patcote 519)	1.4	0.19

Note: If viscosity is too low, adjust up with 14% ammonia.
If viscosity is too high, adjust down with water.

Formulation Constants:
  Approximate Solids, % (wt/vol): 34/30
  pH: 7.9-8.4
  Viscosity, cps: 400-1000
  Density, lbs/gal: 8.6
  VOC, g/liter: 292
  , lbs/gallon: 2.44
  Coalescent, Wt % on Polymer Solids:
    Butyl Cellosolve: 38.6
    Texanol: 4.1
  Freeze-Thaw Stability: Protect from freezing

* Using weight ratio in pounds will yield approximately 100
  gallons. With kilograms, approximately 833 liters will be
  obtained.

SOURCE: Rohm and Haas Co.: RHOPLEX WL for Acrylic Lacquers:
         Formulation WL-81-8

Lacquer Type: Flat Clear
Substrate: Wood

Materials:	Weight Ratio*	Volume Ratio*
Add the following with good agitation:		
Rhoplex WL-81	591.0	68.40
Water	77.9	9.35
Butyl Cellosolve	94.4	12.56
Texanol	9.8	1.24
BYK-301	1.2	0.15
Patcote 519	2.6	0.37
Syloid 166 Flatting Paste FP-2	65.4	7.28
14% Ammonia to pH 8.4	4.2	0.52
Flatting Paste FP-2:		
Water	749.1	89.89
Triton X-100	2.0	0.30
Deefo 806-102	9.0	1.08
Syloid 166	135.4	8.21

Formulation Constants:
    Approximate Solids, % (wt/vol): 31/28
    Syloid 166, % (wt): 15
    pH: 8.4
    Viscosity, cps: 240
    Density, lbs/gal: 8.5
    Coalescent, Wt % on Polymer Solids:
        Butyl Cellosolve: 38.5
        Texanol: 4.0
    VOC, g/liter: 299
        , lbs/gallon: 2.50
    Freeze-Thaw Stability: Protect from freezing

* Using weight ratio in pounds will yield approximetely 100
  gallons. With kilograms, approximately 833 liters will be
  obtained.

SOURCE: Rohm and Haas Co.: RHOPLEX WL for Acrylic Lacquers:
        Formulation WL-81-6

Lacquer Type: Full Gloss Topcoat
Substrate: Metal and Other Substrates

Materials:	Weight Ratio*	Volume Ratio*
Add the following with good agitation:		
Water	68.8	8.26
Rhoplex WL-93 (34 wt %)	552.8	64.20
Butyl Cellosolve	112.8	15.00
Santicizer 160	14.1	1.51
BYK-301	0.5	0.06
Dow Corning Additive #14	0.5	0.08
Water	90.6	10.87

Formulation Constants:
   Approximate Solids, % (wt/vol): 24/22
   pH: 8.2-8.6
   Viscosity, cp: 200
   Density, lbs/gal: 8.4
   VOC, g/liter: 360
     , lbs/gallon: 3.00
   Coalescent, Wt % on Polymer Solids:
     Butyl Cellsolve: 60
     Santicizer 160: 7.5
   Freeze-Thaw Stability: Protect from freezing

   * Using weight ratio in pounds will yield approximately 100
     gallons. With kilograms, approximately 833 liters will be
     obtained.

SOURCE: Rohm and Haas Co.: RHOPLEX WL for Acrylic Lacquers:
       Formulation WL-93-2

Lacquer Type: Full Gloss Topcoat
Substrate: Metal and Other Substrates

Materials:	Weight Ratio*	Volume Ratio*
Add the following with good agitation:		
Water	75.7	9.08
Rhoplex WL-93 (34 wt %)	601.9	69.91
Butyl Cellosolve	81.8	10.88
Butyl Carbitol	40.9	5.11
Santicizer 160	15.4	1.65
BYK-301	0.5	0.06
Dow Corning Additive #14	0.5	0.08
Water	27.1	3.25

Formulation Constants:
   Approximate Solids, % (wt/vol): 27/25
   pH: 8.2-8.6
   Viscosity, cps: 200
   Density, lbs/gal: 8.4
   VOC, g/liter: 363
    , lbs/gallon: 3.03
   Coalescent, Wt % on Polymer Solids:
     Butyl Cellosolve: 40
     Butyl Carbitol: 20
     Santicizer 160: 7.5
   Freeze-Thaw Stability: Protect from freezing

  * Using weight ratio in pounds will yield approximately 100
    gallons. With kilograms, 833 liters will be obtained.

SOURCE: Rohm and Haas Co.: RHOPLEX WL for Acrylic Lacquers:
      Formulation WL-93-4

Lacquer Type: Gloss Black
Substrate: Metal

Materials:	Weight Ratio*	Volume Ratio*
Add the following with good agitation:		
Kelsol 3960	92.3	10.73
Aquacat	2.0	0.24
Magnacat	3.8	0.47
28% Ammonia	2.3	0.30
Water	147.4	17.69
Patcote 519	0.9	0.13
Patcote 577	0.9	0.13
Raven 420	13.8	0.95
Pebble mill grind to 7 to 7.5 N.S. Hegman, then letdown in the following order with good agitation:		
Kelsol 3960)	154.2	17.93
28% Ammonia) Premix	3.8	0.49
Water)	182.5	21.90
Rhoplex WL-71 (41.5 wt %)	191.4	22.23
Water	32.9	3.95
Butyl Carbitol	21.9	2.74
14% Ammonia to pH 7.3-7.5	----	-----

Formulation Constants:
   Approximate Solids, % (wt/vol): 33/32
   Pigment/Binder Ratio: 5/95
   Pigment Volume Content, %: 3.2
   pH: 7.3-7.5
   Viscosity, cps: 200-400
   Density, lbs/gal: 8.5
   VOC, g/liter: 244
      , lbs/gallon: 2.04
   Freeze-Thaw Stability: Protect from freezing

Note: Reduce for spray with 87/13 water/Butyl Cellosolve.

* Using weight ratio in pounds will yield approximately 100
  gallons. With kilograms, approximately 833 liters will be
  obtained.

SOURCE: Rohm and Haas Co.: RHOPLEX WL for Acrylic Lacquers:
        Formulation WL-71-10

Lacquer Type: Gloss Black (contains Strontium Chromate)
Substrate: Ferrous Metal

Materials:	Weight Ratio*	Volume Ratio*
Add the following with good agitation:		
Kelsol 3960	89.2	10.37
Aquacat	1.9	0.23
Magnacat	3.7	0.46
28% Ammonia	2.2	0.29
Water	142.5	17.10
Patcote 519	0.9	0.13
Patcote 577	0.9	0.13
Raven 420	10.0	0.69
Strontium Chromate	10.0	0.32

Pebble mill grind to 7 to 7.5 N.S. Hegman, then letdown in the
  following order with good agitation:

Kelsol 3960)	149.0	17.33
28% Ammonia) Premix	3.7	0.48
Water)	176.4	21.17

Add in order with good agitation:

Rhoplex WL-71	185.0	21.49
Water	31.8	3.82
Butyl Carbitol	21.2	2.65
Water	21.2	2.54
14% Ammonia (to pH=7.3-7.5)	----	----

Formulation Constants:
  Approximate Solids, % (wt/vol): 33/31
  Pigment/Binder Ratio: 8/92
  Pigment Volume Content, %: 3.3
  pH: 7.3-7.5
  Viscosity, cps: 400-600
  Density, lbs/gal: 8.6
  VOC, g/liter: 244
    , lbs/gallon: 2.04
  Freeze-Thaw Stability: Protect from freezing

Note: Reduce for spray with 87/13 water/Butyl Cellosolve.

* Using weight ratio in pounds will yield approximately 100
  gallons. With kilograms, approximately 833 liters will be
  obtained.

SOURCE: Rohm and Haas Co.: RHOPLEX WL for Acrylic Lacquers:
     Formulation WL-71-9

Lacquer Type: Gloss Black Force-Dry
Substrate: Metal

Materials:	Weight Ratio*	Volume Ratio*
Add the following with good agitation:		
Kelsol 3905	89.2	10.43
Aquacat	1.9	0.23
Magnacat	3.7	0.46
28% Ammonia	2.2	0.29
Water	142.5	17.10
Patcote 519	0.9	0.13
Patcote 577	0.9	0.13
Raven 420	13.3	0.91
Pebble mill grind 7 to 7.5 N.S. Hegman, then letdown in the following order with good agitation:		
Kelsol 3905)	149.0	17.43
28% Ammonia) Premix	3.7	0.48
Water)	176.4	21.17
Note: % Neutralization=70%, pH=6.0-6.3		
Rhoplex WL-71 (41.5 wt %)	185.0	21.49
Water	53.0	6.36
Butyl Carbitol	21.2	2.66
14% Ammonia to pH 7.3-7.5	-----	-----

Formulation Constants:
   Approximate Solids, % (wt/vol): 32/31
   Pigment/Binder Ratio: 5/95
   Pigment Volume Content, %: 3.0
   pH: 7.3-7.5
   Viscosity, cps: 200-400
   Density, lbs/gal: 8.5
   VOC, g/liter: 243
      , lbs/gallon: 2.02
   Freeze-Thaw Stability: Protect from freezing

Note: Reduce for spray with 87/13 water/Butyl Cellosolve.

* Using weight ratio in pounds will yield approximately 100
   gallons. With kilograms, approximately 833 liters will be
   obtained.

SOURCE: Rohm and Haas Co.: RHOPLEX WL for Acrylic Lacquers:
        Formulation WL-71-2

<u>Lacquer Type: Gloss Green</u>
<u>Substrate: Metal</u>

Materials:	Weight Ratio*	Volume Ratio*
Add the following with good agitation:		
Chempol 10-0091	88.2	10.02
Aquacat	1.0	0.12
Magnacat	3.7	0.46
28% Ammonia	1.7	0.22
Water	142.3	17.08
Patcote 519	0.9	0.13
Patcote 577	0.9	0.13
Chrome Oxide X-1134	38.3	0.87
Imperial Brazil Yellow X-286C	6.3	0.54
Butyl Cellosolve	5.0	0.66
Butyl Carbitol	5.0	0.62

Pebble mill grind to 7 to 7.5 N.S. Hegman, then letdown in the
following order with good agitation:

Chempol 10-0091)	147.4	16.75
28% Ammonia) Premix	3.1	0.40
Water)	174.5	20.94

(Neutralization=93%, pH=6.1-6.4)
Add in order with good agitation:

Rhoplex WL-71 (41.5 wt %)	183.1	21.27
Water	52.5	6.30
Butyl Carbitol	16.0	2.00
14% Ammonia (to pH 7.3-7.5)	----	----

Formulation Constants:
    Approximate Solids, % (wt/vol): 34/30
    Pigment/Binder Ratio: 15/83
    Pigment Volume Content, %: 4.8
    pH: 7.3-7.5
    Viscosity, cps: 200-400
    Density, lbs/gal: 8.8
    VOC, g/liter: 257
      , lbs/gallon: 2.14
    Freeze-Thaw Stability: Protect from freezing

Note: Reduce for spray with 87/13 water/Butyl Cellosolve.

* Using weight ratio in pounds will yield approximately 100
  gallons. With kilograms, approximately 833 liters will be
  obtained.

SOURCE: Rohm and Haas Co.: RHOPLEX WL for Acrylic Lacquers:
    Formulation WL-71-8

Lacquer Type: Gloss Pastel Yellow
Substrate: Metal

Materials:	Weight Ratio*	Volume Ratio*
Add the following with good agitation:		
Kelsol 3909	88.2	10.20
Aquacat	1.9	0.23
Magnacat	3.7	0.46
28% Ammonia	1.7	0.22
Water	172.3	20.68
Patcote 519	0.9	0.13
Patcote 577	0.6	0.09
Hansa Yellow	40.0	3.23
Ti-Pure R-900	31.0	0.93
Butyl Cellosolve	5.0	0.66
Butyl Carbitol	5.0	0.62

Pebble mill grind to 7 to 7.5 N.S. Hegman, then letdown in the
following order with good agitation:

Kelsol 3909)	147.4	17.04
28% Ammonia) Premix	3.1	0.40
Water)	174.5	20.94
Rhoplex WL-71	183.1	21.27
Butyl Carbitol	16.0	2.00
14% Ammonia (to pH 7.3-7.5)	3.8	0.47

Formulation Constants:
    Approximate Solids, % (wt/vol): 37/34
    Pigment/Binder Ratio: 22/78
    Pigment Volume Content, %: 13.4
    pH: 7.3-7.5
    Viscosity, cps: 200-400
    Density, lbs/gal: 8.8
    VOC, g/liter: 237
      , lbs/gallon: 1.98
    Freeze-Thaw Stability: Protect from freezing

Note: Reduce for spray with 87/13 water/Butyl Cellosolve.

* Using weight ratio in pounds will yield approximately 100
  gallons. With kilograms, approximately 833 liters will be
  obtained.

SOURCE: Rohm and Haas Co.: RHOPLEX WL for Acrylic Lacquers:
        Formulation WL-71-11

Lacquer Type: Gloss Yellow
Substrate: Metal

Materials:	Weight Ratio*	Volume Ratio*
Chempol 10-0091	88.2	10.02
Aquacat	1.0	0.12
Magnacat	3.7	0.46
28% Ammonia	1.7	0.22
Water	142.3	17.08
Patcote 519	0.9	0.13
Patcote 577	0.9	0.13
Krolor Yellow KY-7810	58.3	1.77
Yellow Iron Oxide	3.8	0.11
Butyl Cellosolve	5.0	0.66
Butyl Carbitol	5.0	0.62

Pebble mill grind to 7 to 7.5 N.S. Hegman, then letdown in the following order with good agitation:

Chempol 10-0091)	147.4	16.75
28% Ammonia) Premix	3.1	0.40
Water)	174.5	20.94

Note: Neutralization=93%; pH=6.1-6.4

Rhoplex WL-71 (41.5 wt %)	183.1	21.27
Water	52.5	6.31
Butyl Carbitol	16.0	2.00
14% Ammonia (to pH 7.3-7.5)	----	----

Formulation Constants:
   Approximate Solids, % (wt/vol): 36/31
   Pigment/Binder Ratio: 20/80
   Pigment Volume Content, %: 6.2
   pH: 7.3-7.5
   Viscosity, cps: 200-400
   Density, lbs/gal: 9.0
   VOC, g/liter: 253
       , lbs/gallon: 2.11
   Freeze-Thaw Stability: Protect from freezing

Note: Reduce for spray with 87/13 water/Butyl Cellosolve.

* Using weight ratio in pounds will yield approximately 100 gallons. With kilograms, approximately 833 liters will be obtained.

SOURCE: Rohm and Haas Co.: RHOPLEX WL for Acrylic Lacquers: Formulation WL-71-3

Lacquer Type: Golden Oak Wiping Stain
Substrate: Wood

Materials:	Weight Ratio*	Volume Ratio*
Add the following with good agitation:		
Rhoplex WL-93 (34 wt %)	342.6	39.79
Butyl Cellosolve	70.0	9.31
Santicizer 160	8.8	0.94
Deefo 806-102	1.0	0.13
Water	69.5	8.34
Water	330.6	39.67
Calco Nigrosine O2P PDR	3.4	0.17
Calco CID Orange Y Ex. Conc.	6.9	0.34
Calco CID Tartazine DB1 Conc.	24.1	1.20

Formulation Constants:
    Approximate Solids, % (wt/vol): 18.6/16.2
    pH: 7.3
    Viscosity, cps: 20
    Density, lbs/gal: 8.6
    VOC, g/liter: 333
      , lbs/gallon: 2.78
    Coalescent, Wt % on Polymer Solids:
      Butyl Cellosolve: 60.1
      Santicizer 160: 7.6
    Freeze-Thaw Stability: Protect from freezing
    Heat Stability (8 days/140F): Satisfactory

  * Using weight ratio in pounds will yield approximately 100
    gallons. With kilograms, approximately 833 liters will be
    obtained.

Procedure: Wipe directly onto wood, or spray (as is) onto wood;
         immediately, wipe off excess.

SOURCE: Rohm and Haas Co.: RHOPLEX WL for Acrylic Lacquers:
       Formulation WL-93-1GO

Lacquer Type: Grey Chromate Primer
Substrate: Ferrous Metal

Materials:	Weight Ratio*	Volume Ratio*
Add the following with good agitation:		
Butyl Cellosolve	50.6	6.73
Water	148.7	17.84
Patcote 519	1.2	0.17
Igepal CTA-639	3.7	0.42
Tamol 165	16.1	1.82
Ti-Pure R-900	69.3	2.08
Strontium Chromate	10.9	0.35
Zinc Yellow	3.3	0.12
499 Talc	32.5	1.45
290 Lo-Micron Barytes	44.7	1.19

Pebble mill grind to 7 to 7.5 N.S. Hegman, then letdown in the
following order with good agitation:

Rhoplex WL-91 (41.5 wt %)	455.8	52.88
Triton X-405 (35 wt %)	21.5	2.46
Butyl Carbitol	28.4	3.55
Dibutylphthalate	19.0	2.17
Butyl Cellosolve	22.2	2.95
Colanyl Black PR-A (35%)	1.3	0.12
10% Ammonium Benzoate	22.7	2.72
14% Ammonia to pH=9.0-9.3	7.4	0.92

Adjust viscosity with water (200-600 cps)

Formulation Constants:
    Approximate Solids, % (wt/vol): 40/30
    Pigment/Binder Ratio: 35/65
    pH: 9.0-9.3
    Density, lbs/gal: 9.6
    VOC, g/liter: 279
    , lbs/gallon: 2.33
    Coalescent, Wt % on Polymer Solids:
        Butyl Cellosolve: 35
        Butyl Carbitol: 15
        Dibutylphthalate: 10
    Freeze-Thaw Stability: Protect from freezing

* Using weight ratio in pounds will yield approximately 100
  gallons. With kilograms, approximately 833 liters will be
  obtained.

SOURCE: Rohm and Haas Co.: RHOPLEX WL for Acrylic Lacquers:
        Formulation WL-91-8

Lacquer Type: Heucophos ZBZ Red Primer
Substrate: Ferrous Metal

Materials:	Weight Ratio*	Volume Ratio*
Add the following with good agitation:		
Water	120.2	14.42
Patcote 519	1.0	0.14
Igepal CTA-639	3.2	0.30
Tamol 165	15.2	1.72
Red Iron Oxide	81.0	1.89
Heucophos ZBZ	76.2	2.54
Atomite	82.2	3.65
290 Lo-Micron Barytes	115.4	3.08

Pebble mill grind to 7-7.5 N.S. Hegman, then letdown in the following order with good agitation:

Rhoplex WL-91 (41.5 wt %)	451.0	52.32
Concofoc 690	9.7	1.09
Butyl Carbitol	27.7	3.46
Dibutylphthalate	14.9	1.70
Ektasolve EEH	65.5	8.83
14% Ammonia to pH 9.0-9.3	9.7	1.21
Ammonium Benzoate (10 wt % in water)	26.5	3.18
Dow Corning Additive #14	2.7	0.41

Formulation Constants:
    Approximate Solids, % (wt/vol): 52/35
    Pigment/Binder Ratio: 65/35
    Pigment Volume Content, %: 35
    pH: 9.0-9.3
    Viscosity, cps: 200-600
    Density, lbs/gal: 11.0
    VOC, g/liter: 240
        , lbs/gallon: 2.0
    Coalescent, Wt % on Polymer Solids:
        Ektasolve EEH: 35
        Butyl Carbitol: 15
        Dibutylphthalate: 8
    Freeze-Thaw Stability: Protect from freezing

* Using weight ratio in pounds will yield approximately 100 gallons. With kilograms, approximately 833 liters will be obtained.

SOURCE: Rohm and Haas Co.: RHOPLEX WL for Acrylic Lacquers: Formulation WL-91-9

Lacquer Type: High Gloss Green
Substrate: Metal and Other Substrates

Materials:	Weight Ratio*	Volume Ratio*
Add the following with good agitation:		
Tamol 165	1.8	0.20
Triton CF-10	0.3	0.03
Patcote 519	0.3	0.04
Water	23.5	2.82
Butyl Cellosolve	16.5	2.19
Chrome Oxide X-1134	32.9	0.75
Hansa Yellow G-1230	5.4	0.44
Phthalo Blue	0.3	0.02
Pebble mill grind to 7 to 7.5 N.S. Hegman, then letdown in the following order with good agitation:		
Water	55.6	6.67
Rhoplex WL-91 (41.5 wt %)	615.1	71.36
Butyl Carbitol	25.7	3.21
Butyl Cellosolve	51.3	6.82
Dibutylphthalate	12.8	1.46
BYK-301	1.1	0.14
Patcote 519	1.1	0.16
Dow Paint Additive #14	1.1	0.17
10% HEI-SCORE-XAB	25.7	3.10
14% Ammonia to pH=8.8	3.8	0.47

Formulation Constants:
    Approximate Solids, % (wt/vol): 36/32
    Pigment/Binder Ratio: 12.6/87.4
    Pigment Volume Content, %: 3.9
    pH: 8.8
    Density, lbs/gal: 8.7
    VOC, g/liter: 259
      , lbs/gallon: 2.16
    Coalescent, Wt % on Polymer Solids:
      Butyl Cellosolve: 26.6
      Butyl Carbitol: 10.0
      Dibutylphthalate: 5.0
    Freeze-Thaw Stability: Protect from freezing
    Heat Stability (10 days/140F): Satisfactory
    Gloss (20-degree/60-degree): 70/85
    Mechanical Stability (5 minutes Waring Blender): Satisfactory

  * Using weight ratio in pounds will yield approximately 100
    gallons. With kilograms, approximately 833 liters will be
    obtained.

SOURCE: Rohm and Haas Co.: RHOPLEX WL for Acrylic Lacquers:
    Formulation: WL-91-7

Lacquer Type: Medium Gloss Topcoat
Substrate: Metal and Other Substrates

Materials:	Weight Ratio*	Volume Ratio*
Add the following with good agitation:		
Water	68.8	8.28
Rhoplex WL-93 (34 wt %)	552.8	64.22
Butyl Cellosolve	112.8	15.00
Santicizer 160	14.1	1.51
BYK-301	0.5	0.06
Dow Corning Additive #14	0.5	0.06
Water	90.6	10.87
Flatting Paste FP-2	26.9	2.99
Flatting Paste FP-2:		
Water	749.1	89.89
Triton X-100	2.0	0.30
Deefo 806-102	9.0	1.08
Syloid 166	135.4	8.21

Formulation Constants:
    Approximate Solids, % (wt/vol): 24/22
    Pigment/Binder Ratio: 2/98
    pH: 8.2-8.6
    Viscosity, cps: 200
    Density, lbs/gal: 9.0
    VOC, g/liter: 3.04
    Coalescent, Wt % on Polymer Solids:
        Butyl Cellosolve: 60
        Santicizer 160: 7.5
    Freeze-Thaw Stability: Protect from freezing

* Using weight ratio in pounds will yield approximately 100
  gallons. With kilograms, approximately 833 liters will be
  obtained.

SOURCE: Rohm and Haas Co.: RHOPLEX WL for Acrylic Lacquers:
        Formulation WL-93-2MG

Lacquer Type: Oncor F-31 Pigmented
Substrate: Ferrous Metal

Materials:	Weight Ratio*	Volume Ratio*
Add the following with good agitation:		
Epotuf 38-690	27.5	3.27
Aquacat	1.0	0.12
Magnacat	1.0	0.13
Butyl Cellosolve	13.9	1.85
Add in order:		
Triethylamine	2.4	0.40
Water	129.2	15.50
Tamol 165	2.7	0.31
Patcote 550	0.5	0.07
Raven 2000	33.8	2.32
Oncor F-31**	10.0	0.30

Pebble mill grind to 7 to 7.5 N.S. Hegman, then letdown in the
    following order with good agitation:

Epotuf 38-690	148.8	17.71
Triethylamine	12.1	2.00
Water	129.2	15.50

Mix well (recheck grind), then add the following in order with
    good agitation:

Rhoplex WL-81 (41.5 wt %)	287.4	33.26
Triton X-405 (reduced to 35 wt %)	4.8	0.55
Santicizer 160	23.7	2.53
Ammonium Benzoate (10 wt % in water)	24.0	2.88
Water	8.3	1.00

Formulation Constants:
    Approximate Solids, % (wt/vol): 37/34
    pH: 8.9
    Viscosity, cps: 300
    Epotuf 38-690/Rhoplex WL-81 (wt): 51/49
    Density, lbs/gal: 8.6
    VOC, g/liter: 223
        , lbs/gallon: 1.86
    Freeze-Thaw Stability: Protect from freezing

    * Using weight ratio in pounds will yield approximately 100
      gallons. With kilograms, approximately 833 liters will be
      obtained.
   ** Strontium Chromate can be used in place of the Oncor F-31.

SOURCE: Rohm and Haas Co.: RHOPLEX WL for Acrylic Lacquers:
        Formulation WL-81-4

Lacquer Type: Primer (contains Strontium Chromate)
Substrate: Ferrous Metal

Materials:	Weight Ratio*	Volume Ratio*
Add the following with good agitation:		
Tamol 165	15.2	1.72
Patcote 519	1.1	0.16
Igepal CTA-639	3.5	0.40
Water	117.3	14.08
Red Iron Oxide	85.3	1.99
Strontium Chromate	10.4	0.33
Zinc Yellow	3.0	0.11
499 Talc	91.8	4.08
290 Lo-Micron Barytes	128.8	3.44

	Weight Ratio*	Volume Ratio*
Cowles grind to 7 to 7.5 N.S. Hegman, then letdown in the following order with good agitation:		
Rhoplex WL-92 (42.0 wt %):	437.3	50.91
Triton X-405 (reduced to 35 wt %)	10.4	1.19
Butyl Cellosolve	27.5	3.66
Butyl Carbitol	27.5	3.44
Dowanol DPM	9.2	1.15
10% Ammonium Benzoate in Water	22.1	2.65
14% Ammonia to pH=7.9-8.1	----	----
Adjust viscosity with Acrysol ASE-60 (5% in water, pH=9.2)	79.2	9.48

Formulation Constants:
    Approximate Solids, % (wt/vol): 48/33
    Pigment/Binder Ratio: 63/37
    Pigment Volume Content, %: 31
    pH: 7.9-8.1
    Viscosity, cps: 400-600
    Density, lbs/gal: 10.8
    VOC, g/liter: 168
        , lbs/gallon: 1.40
    Coalescent, Wt % on Polymer Solids:
        Butyl Cellosolve: 15
        Butyl Carbitol: 15
        Dowanol DPM: 5
    Freeze-Thaw Stability: Protect from freezing
    Heat Stability (8 days/140F): Satisfactory

  * Using weight ratio in pounds will yield approximately 100
    gallons. With kilograms, approximately 833 liters will be
    obtained.

SOURCE: Rohm and Haas Co.: RHOPLEX WL for Acrylic Lacquers:
        Formulation WL-92-4

Lacquer Type: Primer (contains Strontium Chromate)
Substrate: Ferrous Metal

Materials:	Weight Ratio*	Volume Ratio*
Add the following with good agitation:		
Butyl Cellosolve	50.3	6.69
Water	23.8	2.86
Patcote 519	1.6	0.23
Igepal CTA-639	2.6	0.29
Tamol 165	11.1	1.26
Red Iron Oxide	118.6	2.77
Strontium Chromate	14.3	0.46
Zinc Yellow	4.2	0.15
499 Talc	42.9	1.91
290 Lo-Micron Barytes	59.3	1.58

Cowles grind to 7 to 7.5 N.S. Hegman, then letdown in the
following order with good agitation:

Rhoplex WL-92 (42.0 wt %)	596.1	69.39
Triton X-405 (reduced to 35 wt %)	14.3	1.63
Butyl Carbitol	37.6	4.70
Dibutylphthalate	12.7	1.45
10% Ammonium Benzoate in Water	33.4	4.00
14% Ammonia to pH=7.9-8.1	5.3	0.66

Adjust viscosity with Acrysol ASE-60
  (5% in water, pH=9.2)

Formulation Constants:
  Approximate Solids, % (wt/vol): 50/38
  Pigment/Binder Ratio: 49/51
  Pigment Volume Content, %: 20
  pH: 7.9-8.1
  Viscosity, cps: 200-600
  Density, lbs/gal: 10.3
  VOC, g/liter: 214
    , lbs/gallon: 1.79
  Coalescent, Wt % on Polymer Solids:
    Butyl Cellosolve: 20
    Butyl Carbitol: 15
    Dibutylphthalate: 5
  Freeze-Thaw Stability: Protect from freezing
  Heat Stability (8 days/140F): Satisfactory

  * Using weight ratio in pounds will yield approximately 100
    gallons. With kilograms, approximately 833 liters will be
    obtained.

SOURCE: Rohm and Haas: RHOPLEX WL for Acrylic Lacquers:
        Formulation WL-92-5

Lacquer Type: Strontium Chromate
Substrate: Ferrous Metal

Materials:	Weight Ratio*	Volume Ratio*
Add the following with good agitation:		
Water	45.0	5.40
Tamol 165	10.8	1.22
Triton CF-10	1.4	0.16
Patcote 519	0.4	0.06
Tiona RCL-9	152.0	4.45
Strontium Chromate	8.5	0.27

Cowles grind to 7 to 7.5 N.S. Hegman, then letdown in the following order with good agitation:

Water	33.4	4.01
Triton X-405 (25 wt %; pH=8.5-9.0)	18.0	2.01
Rhoplex WL-81 (41.5 wt %)	556.2	64.38
Butyl Cellosolve	88.6	11.78
Paraplex WP-1**	17.5	2.08
Patcote 519	0.8	0.11

Add slowly with good agitation:

Ammonia (14%) to pH 7.7-8.3	----	-----
HEI-SCORE-XAB (10 wt %)	24.7	2.98

Formulation Constants:
   Approximate Solids, % (wt/vol): 44/34
   Pigment/Binder Ratio: 40/60
   Pigment Volume Content, %: 14.7
   pH: 7.7-8.3
   Viscosity, cps: 100-400
   Density, lbs/gal: 9.7
   VOC, g/liter: 235
       , lbs/gallon: 1.96
   Coalescent, Wt % on Polymer Solids:
       Butyl Cellosolve: 38.4
       Paraplex WP-1: 7.5
   Freeze-Thaw Stability: Protect from freezing

   * Using weight ratio in pounds will yield approximately 100
     gallons. With kilograms, approximately 833 liters will be
     obtained.
  ** If Paraplex WP-1 is not added as a premix, allow 1-2 days
     for the plasticizer to be absorbed by the latex.

SOURCE: Rohm and Haas Co.: RHOPLEX WL for Acrylic Lacquers:
        Formulation WL-81-2

Lacquer Type: Strontium Chromate Primer
Substrate: Ferrous Metal

Materials:	Weight Ratio*	Volume Ratio*
Add the following with good agitation:		
Butyl Cellosolve	10.5	1.40
Water	147.1	17.65
Tamol 165	12.8	1.45
Triton CF-10	2.9	0.33
Red Iron Oxide	65.8	1.54
499 Talc	129.2	5.74
290 Lo-Micron Barytes	67.1	1.79
Strontium Chromate	15.7	0.50

Cowles grind to 6.5 N.S Hegman, then letdown in the following
order with good agitation:

		Weight Ratio*	Volume Ratio*
Kelsol 3905	)	140.7	16.46
Aquacat	)	1.3	0.16
Magnacat	)	1.2	0.15
Patcote 519	) Premix	1.0	0.14
Patcote 577	)	1.0	0.15
28% Ammonia	)	2.5	0.32
Water	)	105.0	12.60

Note: % Neutralization of the premix is only 52%, pH 5.8 to 6.0

	Weight Ratio*	Volume Ratio*
Rhoplex WL-71	253.5	29.44
Triton X-405 (reduced to 35 wt % solids)	9.0	0.98
Butyl Cellosolve	10.4	1.39
Butyl Carbitol	15.9	1.99
Dibutylphthalate	10.7	1.22
14% Ammonia to pH 7.3-7.5	3.4	0.42
87/13 Water/Butyl Cellosolve	27.8	2.73

Formulation Constants:
    Approximate Solids, % (wt/vol): 50/37
    Pigment/Binder Ratio: 56/44
    Pigment Volume Content, %: 27.5
    pH: 7.3-7.5
    Viscosity, cps: 400-600
    Density, lbs/gal: 10.5
    VOC, g/liter: 192
        , lbs/gallon: 1.60
    Freeze-Thaw Stability: Protect from freezing

Note: Reduce for spray with 87/13 water/Butyl Cellosolve.

* Using weight ratio in pounds will yield approximately 100
  gallons. With kilograms, approximately 833 liters will be
  obtained.

SOURCE: Rohm and Haas Co.: RHOPLEX WL for Acrylic Lacquers:
        Formulation WL-71-1

Lacquer Type: Strontium Chromate Primer
Substrate: Ferrous Metal

Materials:	Weight Ratio*	Volume Ratio*
Add the following with good agitation:		
Butyl Cellosolve	46.0	6.12
Water	137.0	16.44
Patcote 519	1.0	0.14
Igepal CTA-639	3.4	0.39
Tamol 165	14.8	1.68
Red Iron Oxide	82.5	1.93
Strontium Chromate	15.0	0.48
499 Talc	109.1	4.85
290 Lo-Micron Barytes	152.6	4.07

Pebble mill grind to 7 to 7.5 N.S. Hegman, then letdown in the following order with good agitation:

Rhoplex WL-91 (41.5 wt %)	420.0	48.72
Triton X-405 (reduced to 35 wt % solids)	9.9	1.13
Butyl Carbitol	26.2	3.27
Dibutylphthalate	17.4	1.99
Butyl Cellosolve	41.5	5.52
14% Ammonia to pH 9.0-9.3	6.8	0.85
Ammonium Benzoate (10 wt % in water)	20.9	2.51

Adjust viscosity with water (200-600 cps)

Formulation Constants:
   Approximate Solids, % (wt/vol): 51/34
   Pigment/Binder Ratio: 67/33
   Pigment Volume Content, %: 34.7
   pH: 9.0-9.3
   Viscosity, cps: 200-600
   Density, lbs/gal: 11.0
   VOC, g/liter: 279
     , lbs/gallon: 2.33
   Coalescent, Wt % on Polymer Solids:
      Butyl Cellosolve: 50
      Butyl Carbitol: 15
      Dibutylphthalate: 10
   Freeze-Thaw Stability: Protect from freezing

* Using weight ratio in pounds will yield approximately 100 gallons. With kilograms, approximately 833 liters will be obtained.

SOURCE: Rohm and Haas Co.: RHOPLEX WL for Acrylic Lacquers: Formulation WL-91-3

Lacquer Type: Tinsel Coating
Substrate: Plastics

Materials:	Weight Ratio*	Volume Ratio*
Add the following with good agitation:		
Rhoplex WL-51 (41.5 wt %)	612.9	71.18
Water	35.0	4.20
Butyl Cellosolve	109.3	14.54
Dibutylphthalate	12.7	1.45
Patcote 519	0.4	0.06
BYK-301	2.5	0.31

Premix the following and add to the above blend:

Stapa Hydrolac W 60 n.1	35.9	2.98
Butyl Cellosolve	17.8	2.37
Butyl Carbitol	25.3	3.16

Formulation Constants:
Approximate Solids, % (wt/vol): 34/32
pH: 6.8
Viscosity, cps: 210
       , #4 Ford Cup, seconds: 29
Density, lbs/gal: 8.5
VOC, g/liter: 359
  , lbs/gallon: 2.9
Coalescent, Wt % on Polymer Solids:
   Butyl Cellosolve: 50
   Butyl Carbitol: 10
   Dibutylpthalate: 5

* Using weight ratio in pounds will yield approximately 100 gallons. With kilograms, approximately 833 liters will be obtained.

Note: This recommended formulation contains an aluminum pigment that the manufacturer claims is stabilized against hydrogen gassing in aqueous media. The manufacturer of this pigment has suggested general guidelines to formulate aqueous one-pack coatings, and this formulation follows those guidelines. However, as conditions and methods of use of these products are beyond Rohm and Haas' control, they urge you to contact the aluminum pigment manufacturer or the distributor for further information and to adequately test this or any formulation you may develop based on this technology to ensure that it is stable for its intended use before adapting it for commercial purposes.

SOURCE: Rohm and Haas Co.: RHOPLEX WL for Acrylic Lacquers: Formulation WL-51-10

Lacquer Type: White
Substrate: Metal and Other Substrates

Materials:	Weight Ratio*	Volume Ratio*
Add the following with good agitation:		
Water	40.0	4.80
Tamol 165	10.8	1.22
Triton CF-10	1.4	0.16
Patcote 519	2.0	0.29
Tiona RCL-9	160.0	4.69

Cowles grind to 7 to 7.5 N.S. Hegman, then letdown in the following order with good agitation:

Water	54.0	6.48
Rhoplex WL-81 (41.5 wt %)	575.0	66.55
Butyl Cellosolve	91.0	12.10
Butyl Carbitol	9.5	1.19
Patcote 519	1.0	0.14
BYK-301	1.2	0.15
Dow Corning Additive #14	1.2	0.18

Add slowly with good agitation:

Ammonia (14%) to pH 7.7-8.3	---	----
HEI-SCORE-XAB (10 wt %)	24.0	2.89

Formulation Constants:
  Approximate Solids, % (wt/vol): 42/32
  Pigment/Binder Ratio: 40/60
  Pigment Volume Content, %: 15.2
  pH: 7.7-8.3
  Viscosity, cps: 300-500
  Density, lbs/gal: 9.6
  VOC, g/liter: 269
    , lbs/gallon: 2.24
  Coalescent, Wt % on Polymer Solids:
    Butyl Cellosolve: 38
    Butyl Carbitol: 4
  Freeze-Thaw Stability: Protect from freezing

  * Using weight ratio in pounds will yield approximataly 100
    gallons. With kilograms, approximately 833 liters will be
    obtained.

SOURCE: Rohm and Haas Co.: RHOPLEX WL for Acrylic Lacquers:
        Formulation: WL-81-1

Lacquer Type: White
Substrate: Metal and Other Substrates

Materials:	Weight Ratio*	Volume Ratio*
Add the following with good agitation:		
Rhoplex WL-91 (41.5 wt %)	77.4	8.98
Tamol 165	10.0	1.13
Triton CF-10	1.2	0.14
Patcote 519	1.0	0.14
Tiona RCL-9	148.7	4.36

Cowles grind to 7 to 7.5 N.S. Hegman, then letdown in the following order with good agitation:

Rhoplex WL-91 (41.5 wt %)	459.2	53.27
Water	130.5	15.66
Butyl Cellosolve	78.5	10.44
Butyl Carbitol	11.2	1.40
Dibutylphthalate	11.2	1.28

Add slowly with good agitation:

Ammonia, 14%	2.4	0.30
Ammonium Benzoate, (10 wt % in water)	22.6	2.71

Formulation Constants:
Approximate Solids, % (wt/vol): 41/31
Pigment/Binder Ratio: 39/61
Pigment Volume Content, %: 14.4
pH: 7.8-8.0
Viscosity, cps: 300-500
VOC, g/liter: 251
    , lbs/gallon: 2.09
Density, lbs/gal: 9.6
Coalescent, Wt % on Polymer Solids:
    Butyl Cellosolve: 41.2
    Butyl Carbitol: 5.9
    Dibutylphthalate: 5.9
Freeze-Thaw Stability: Protect from freezing

* Using weight ratio in pounds will yield approximately 100 gallons. With kilograms, approximately 833 liters will be obtained.

SOURCE: Rohm and Haas Co.: RHOPLEX WL for Acrylic Lacquers: Formulation WL-91-1

Lacquer Type: White
Substrate: Plastics

Materials:	Weight Ratio*	Volume Ratio*
Add the following with good agitation:		
Water	29.2	3.50
Tamol 165	9.8	1.11
Triton CF-10	1.2	0.14
Patcote 519	0.9	0.13
Tiona RCL-9	142.7	4.18

Cowles grind to 7 to 7.5 N.S. Hegman, then letdown in the following order with good agitation:

Rhoplex WL-96 (42 wt %)	509.6	59.12
Butyl Cellosolve	74.9	9.76
Butyl Carbitol	21.4	2.67
Water	162.2	19.46

Add slowly with good agitation:

Ammonia (14%) to pH approx. 8.0	-----	-----

Formulation Constants:
   Approximate Solids, % (wt/vol): 38/28
   Pigment/Binder Ratio: 40/60
   Pigment Volume Content, %: 15.0
   pH: 8.0
   Viscosity, cps: 365
   Density, lbs/gal: 9.5
   VOC, g/liter: 282
    , lbs/gallon: 2.35
   Coalescent, Wt % on Polymer Solids:
     Butyl Cellosolve: 35
     Butyl Carbitol: 10
   Freeze-Thaw Stability: Protect from freezing

  * Using weight ratio in pounds will yield approximately 100
   gallons. With kilograms, approximately 833 liters will be
   obtained.

SOURCE: Rohm and Haas Co.: RHOPLEX WL for Acrylic Lacquers:
      Formulation WL-96-3

Medium Gloss, Force Dry, Weatherborne Lacquer – White

Ingredients:	Pounds	Gallons
Chempol 20-4301	75.85	8.62
Tamol 165	10.11	1.15
Triton CF-10	1.51	0.17
Surfynol 104BC	6.07	0.81
Water	47.18	5.66
Ti-Pure R-900	131.48	3.84

Add each ingredient separately to the emulsion using good agitation. Check for signs of kick-out or seeding before adding the pigment.

Cowles to a #7 Hegman with a high speed dispersator. Adjust grind viscosity with water from below if necessary.

Let down with:		
Chempol 20-4301	510.24	57.98
2-butoxyethanol	41.06	5.47
Butyl Carbitol	26.20	3.30
Dibutyl phthalate	5.66	0.65
Nuocure CK-10	1.62	0.20
Water (hold to adjust viscosity)	101.21	12.15

Premix co-solvents, plasticizers and drier. Add this mixture to the vortex of the agitating emulsion <u>VERY SLOWLY.</u> If the mixture becomes heavy, add a small amount of water from the "adjust viscosity." Check mixture for kick-out or seeding, then add the grind to the "let down" with good agitation. Mix well and allow the paint to set overnight.

Check the pH and adjust with ammonia if necessary. Avoid pH values above 7.6, as this will cause the viscosity of the paint to increase significantly. Use the "adjust viscosity" water to bring the viscosity of the coating to spec at room temperature.

Properties:
  Total solids, % by weight: 37.11%
  Total solids, % by volume: 26.81%
  Weight/gallon: 9.58 lbs./gal.
  PVC: 14.3
  VOC: 2.01 lbs./gal.
  Viscosity: 28" #3 Zahn
  pH: 7.3
  Substrate: B-1000 CRS
  Cure: 15' @ 150F
  DFT: 1.0-1.2 mil
  Gloss, 60: 71
        20: 29
  Pencil Hardness: B-F
  Impact Resistance, reverse: Less than 20 inch
  Crosshatch adhesion: 100% retention

SOURCE: Freeman Polymers Division: CHEMPOL 20-4301-A Formulation

Pigmented Lacquer

	Parts per Hundred Weight Ratio*(Volume Basis)	
Materials:		
Add the following with good agitation:		
Water	45.0	5.6
Tamol 165	10.8	1.3
Triton CF-10	1.4	0.2
Patcote 519	0.4	0.1
TiO2	152.0	4.6
Strontium Chromate	8.5	0.3
Letdown as follows:		
Water	33.4	4.0
Surfactant XQS-20**	18.0	2.2
Rhoplex WL-81	556.2	64.7
Ethylene glycol monobutyl ether	88.6	11.8
Paraplex WP-1***	17.5	2.1
Patcote 519	0.8	0.2
Add slowly under good agitation:		
HEI-SCORE-XAB (10%)****	24.7	3.0

Formulation Constants:
   pH: 8.0+-0.3
   Viscosity, cps.: 100 to 400
   Approximate Solids, % (wt.,vol): 43, 32.5
   Density, lbs/gal.: 9.6
   PVC, %: 15
   TiO2/Binder Ratio: 40/60
   Freeze/Thaw Stability: Protect from freezing
   Mechanical Stability (5' Waring Blender): Satisfactory
   Heat Stability (10 days/140F): Satisfactory
   VOC, g/liter: 220
        lbs./gal.: 1.83
   % on Polymer Solids:
      Ethylene glycol monobutyl ether: 38.4
      Paraplex WP-1: 7.5

  * Using weight ratio in pound units will yield approximately
    100 gallons of lacquer while with kilograms, 833 liters will
    result.
 ** Reduced to 25% solids; pH adjusted to 8.5-9.0.
*** If Paraplex WP-1 is not added as a premix, allow 1 to 2 days
    for the plasticizer to be absorbed by the latex.
**** 10% solution in water.

SOURCE: Rohm and Haas Co.: Industrial Coatings: RHOPLEX WL-81:
        Formulation WL-81-2

Pigmented Lacquer--White

Materials:	Weight Ratio*	Parts Per Hundred (Volume Basis)
Add the following with good agitation:		
Water	40.0	4.8
Tamol 165	10.8	1.3
Triton CF-10	1.4	0.2
Patcote 519	2.0	0.2
TiO2	160.0	4.7

Cowles grind to 7-7 1/2 N.S. Hegman. Then, add the following in order under agitation:

Water	54.0	6.5
Rhoplex WL-81	575.0	66.9
Ethylene glycol monobutyl ether	91.0	12.1
Diethylene glycol monobutyl ether	9.5	1.2
Patcote 519	1.0	0.1
BYK-301	1.2	0.1
Dow Corning Additive #14	1.2	0.1

Finally, add slowly under agitation:

HEI-SCORE-XAB (10%)**	24.0	2.9

Formulation Constants:
    pH: 8.0+-0.3
    Viscosity, No. 4 Ford Cup, sec.: 20 to 30
    Approximate Solids, % (wt.,vol.): 41,31
    Density, lbs./gal.: 9.6
    TiO2/Binder Ratio: 40/60
    PVC, %: 15
    Freeze/Thaw Stability: Protect from freezing
    Mechanical Stability (5' Waring Blender): Satisfactory
    Heat Stability (10 days/140F): Satisfactory
    VOC, g/liter: 285 (2.4 lbs./gal.)
    % on Polymer Solids:
        Ethylene glycol monobutyl ether: 38
        Diethylene glycol monobutyl ether: 4

 * Using weight ratio in pound units will yield approximately
   100 gallons of lacquer while with kilograms, 833 liters will
   result.
** 10% solution in water.
   Adjust pH with 14% ammonia.

SOURCE: Rohm and Haas Co.: Industrial Coatings: RHOPLEX WL-81:
        Formulation WL-81-1

Waterborne Pigmented Lacquer
White, for Air Dry or Force Dry

Materials:	Weight* (Pounds)	Volume (Gallons)
Water	26.6	3.19
Tamol 165	9.3	1.05
Triton CF-10	1.1	0.12
Patcote 519	1.0	0.14
Zopaque RCL-9	140.2	4.11

Cowles grind (7 - 7 1/2 N.S.)

Letdown as follows:

Rhoplex WL-92 (42 wt. % solids)	499.2	58.11
Water	141.2	16.94
Butyl Propasol	75.3	10.29
Butyl Carbitol	21.2	2.65
10% Ammonium Benzoate in Water	21.3	2.55
14% Ammonia to pH of 7.9-8.1	----	-----

Formulation Constants:
    pH: 7.9-8.1
    Viscosity, cps: 300-500
    Approximate Solids, % (wt./vol.): 38/28
    Density, lbs./gal.: 9.4
    Pigment/Binder Ratio: 39/61
    PVC, %: 14.5
    Freeze-Thaw Stability: Protect from freezing
    Heat Stability (8 days/140F): Satisfactory
    Mechanical Stability: Satisfactory
    VOC, g/liter: 281
        lbs./gallon: 2.34
    % on Polymer Solids:
        Butyl Propasol: 35
        Butyl Carbitol: 10

   * Using weight in pound units will yield approximately 100
     gallons of lacquer. With kilograms, 833 liters will result.

SOURCE: Rohm and Haas Co.: Industrial Coatings: RHOPLEX WL-92:
        Formulation WL-92-3

Section VIII
Primers

Airdry Black Metal Primer

Ingredient:	Lbs/100 Gal
Ektasolve EP	56.01
Santicizer 160	3.93
Surfynol 104H	4.91
Diethylamine	14.24
Drew L-475	1.96
DSX 1550	9.82
Manchem APG	9.82
Water	75.67
Mix well then add:	
Pliolite 7104 (44% solid)	65.84
Mix well then add:	
Degussa Black #4	9.82
ZBZ Zinc Phosphate	235.85
Grind to 7 N.S. then add:	
Pliolite 7104 (44% solid)	458.92
Santicizer 160	20.64
Ektasolve EP	9.82
Texanol	10.81
Then add:	
10% Ammonium Benzoate	19.65

　　Air Dry: 20-30 minutes to handle
　　W.P.G.: 10.08
　　Viscosity: 40-50 seconds, No. 4 Ford Cup
　　Solids Wt%: 51.54
　　Solids Vol%: 37.40
　　VOC: 2.00 lbs/gal (235 gm/litre)
　　Pigment Binder
　　60 degree gloss: 10-15% at 1.2 mil

　　Age panel 3 to 7 days before testing

SOURCE: The Goodyear Tire & Rubber Co.: Formula #2064-24-90

Chromate-Free Anti-Corrosive Primer

Material:	Lbs.	Gals.
Water	25.0	3.00
Butyl Cellosolve	15.0	2.00
Butyl Carbitol	31.0	3.89
Colloid 226	5.0	0.50
AMP 95	5.0	0.50
Nuosept 95	1.0	0.09
Zinc Phosphate	40.0	1.50
Busan 11M-1	10.0	0.36
Tronox CR-800	6.0	0.18
#4098-D Red Oxide	13.0	0.30

High speed disperser, 15 minutes maximum.

Colloid 694	5.0	0.70
Synthemul 40-422	688.0	80.00
QR-708 (10% solution)	5.0	0.57
Natrosol 250MR (3% soln)	34.0	4.08
Water	19.4	2.32

10% QR-708 Solution:	
QR-708 (35%)	288
Propylene Glycol	424
Water	288

Analysis:
   Pigment Volume Concentration, Percent: 5.8
   Percent Solids, Weight: 45.2
   Percent Solids, Volume: 40.7
   Initial Viscosity, KU: 63-68
   Pounds/Gallon: 9.02
   pH: 8.0
   VOC (excluding water):
      Grams/Liter: 133.0
   Salt spray resistance, cold rolled steel, 1.2 mils DFT,
      64 hours:
      Blistering: Minimal (some at scribe)
      Creepage at Scribe: None (<1/8 inch)
      Overall Rust: None

SOURCE: Reichhold Chemicals, Inc.: Waterborne Handbook:
        Formulation 422-01

Force Dry Red Oxide Metal Primer

Ingredient:	Lbs/100 Gal
Ektasolve EP	91.2
Santicizer 160	6.4
DSX 1550	8.0
Surfynol 104H	6.4
Diethylamine	8.0
Water	123.2

Mix well, then add to following latex slowly, under agitation:

Pliolite 7104 (44% solid)	107.2

Then under agitation, add:

Red Oxide	32.0
Heuback ZBZ	112.0
Attagel 50	3.2
Mica 3000	32.0
Microfine Barytes	88.0

Grind to 6+ N.S. by sand mill:

Premix below, then add to grind slowly with agitation:

Pliolite 7104 (44% solid)	462.4
Santicizer 160	20.8
Ektasolve EP	20.8

Mix well then add:

10% ammonium benzoate	32.0

Dry time: 4 minutes at 200 deg F
        (after flash dry 10 minutes)
Hardness: Initial F;
        After 24 hours, F to H
60 degree gloss: 20%
VOC: 263 lb/gallon (310 gram/litre)
W.P.G.: 11.51
Viscosity: 30-35 seconds, No. 4 Ford Cup
Solids Wt%: 48.8
Solids Vol%: 38.2
Pigment/Binder ratio: 1.04:1

Age panel 3 days before testing

SOURCE: The Goodyear Tire & Rubber Co.: Formula #2064-4-90

Gray Primer

Material:	Lbs.	Gals.
Kelsol 3950-B2G-70	115.41	13.61
Butoxy Ethanol	32.52	4.33
Ammonium Hydroxide	6.49	0.87
Igepal CTA-639	4.79	0.55
Busperse-39	4.84	0.47
Titanox 2101	56.94	1.71
Atomite	55.35	2.46
106 Lo Micron Barytes	45.30	1.25
Raven 1020	3.28	0.23
Butrol 22	60.24	1.82
Water	162.61	19.52

Steel ball mill to a 6 Hegman grind.

Letdown:		
Kelsol 3950-BG-70	199.96	23.58
Butoxy Ethanol	47.16	6.28
Ammonium Hydroxide	6.65	0.89

Premix under mild agitation, then add to above:		
Cobalt Hydrocure II	6.73	0.89
Activ-8	1.12	0.14
Butoxy Ethanol	3.30	0.44
Water	175.05	21.00

Analysis:
   Pigment Volume Concentration, Percent: 22.0
   Pigment/Binder Ratio: 1/1
   Percent Solids, Weight: 49.6
   Percent Solids, Volume: 35.6
   Viscosity @ 25C, #4 Ford, Secs.: 40-60
   Pounds/Gallon: 9.87
   pH: 8.2-8.5
   VOC (excluding water):
       Grams/Liter: 365
       Pounds/Gallon: 3.03

SOURCE: Reichhold Chemicals, Inc.: Waterborne Handbook:
         Formulation 2327-2A

Gray Primer

Material:	Lbs.	Gals.
Kelsol 3962-B2G-70	124.94	14.53
Ammonium Hydroxide	8.92	1.19
Butoxy Ethanol	8.03	1.07
Secondary Butanol	8.03	1.19
Titanox 2101	94.59	2.84
Lo Micron Barytes 106	151.72	4.16
OK 412 Silica	8.03	0.44
Raven Black 1020	3.56	0.25
Strontium Chromate J-1365	22.31	0.72
Water	224.89	27.00

Steel ball mill to a 6 Hegman grind.

Letdown:		
Kelsol 3962-B2G-70	53.54	6.23
Ammonium Hydroxide	5.35	0.72
Cobalt Hydrocure II	2.49	0.33
Manganese Hydrocure II	5.18	0.61
Water	307.96	36.97
Butoxy Ethanol	7.13	0.95
Secondary Butanol	5.35	0.80

Analysis:
    Pigment Volume Concentration, Percent: 19.9
    Pigment/Binder Ratio: 1.03/1
    Percent Solids, Weight: 38.04
    Percent Solids, Volume: 22.48
    Viscosity @ 25C, #4 Ford, Secs.: 40-60
    Pounds/Gallon: 9.82
    pH: 8.2-8.5
    VOC (excluding water):
        Grams/Liter: 322.8
        Pounds/Gallon: 2.69

SOURCE: Reichhold Chemicals, Inc.: Waterborne Handbook:
        Formulation 2349-69-A

Gray Rust Inhibitive Primer

Material:	Lbs.	Gals.
Water	127.9	15.35
Tamol 731	14.1	1.53
Igepal CTA-639	2.8	0.34
Foamaster VL	1.4	0.18
Methyl Carbitol	24.0	2.89
Dowicil 75	0.5	0.04

Mix the above at slow speed, then add the following:

	Lbs.	Gals.
Titanox 2101	188.1	5.50
Zinc Yellow Y539-D	9.4	0.34
Zinc Phosphate J0852	28.1	1.06
Mica 325	30.9	1.32
Snowflake	117.5	5.22
HEC QP4400	0.8	0.08
Black Tint 888-9907	3.8	0.38

Cowles grind the above 15-20 minutes to a 4-5 Hegman.

Letdown:

	Lbs.	Gals.
Arolon 850-W-45	512.0	59.12
Butoxy Ethanol	44.7	5.95
10% Solution of Sodium Nitrite	4.0	0.48
Ammonium Hydroxide	1.8	0.24

Analysis:
  Pigment Volume Concentration, Percent: 35.2
  Pigment/Binder Ratio: 1.8/1.0
  Percent Solids, Weight: 53.0
  Percent Solids, Volume: 38.2
  Viscosity @ 25C, #4 Ford, Secs.: 18-25
  Pounds/Gallon: 11.1
  pH: 8.0-8.5
  VOC (excluding water):
    Grams/Liter: 162
    Pounds/Gallon: 1.35

SOURCE: Reichhold Chemicals, Inc.: Waterborne Handbook:
        Formulation 2154-61A

Heavy Metal Free Water-Borne Barrier Primer

	Lbs./100 gal.
Beckosol 13-402	194.0
Triethylamine	11.9
Surfynol 104 B.C.	8.5
Butyl Cellosolve	36.6
Secondary Butyl Alcohol	36.6

Mix 10 minutes

Min-U-Sil 10	50.2
Vicron 1515	51.3
YLO 3288D	149.8
Talcron MP40-27	12.0
Black Tint Base (See Below)	76.9

High Speed Disperse to a 5 N.S.

Nuocure 10% Cobalt	0.5
Nuocure 6% Manganese	1.8
Exkin #2	2.6

Add slowly

Deionized Water	392.2

Black Tint Base:	Lbs./100 gal.
Beckosol 13-402	490.8
Dimethyl Ethanolamine	27.9
Surfynol 104 B.C.	2.8
Butyl Cellosolve	245.4
Special Black #5	92.2

Ball Mill to a 7+ N.S.

    Viscosity: 66 KU at 25C
    Wt./gal.: 10.2
    N.V.W.: 44.6%
    N.V.V.: 29.6%
    P/B: 1.52/1
    PVC: 0.35
    pH: 8.4
    VOC, gm/l calculated: 380

SOURCE: U.S. Silica Co.: Formula IC-104

High Quality Stain Blocking Primer
(P-23-12 Modified to Prevent Settling)

Ingredients:	Weight Ratio*	Parts Per Hundred (Volume Basis)
Water	48.5	5.82
Exp. Dispersant QR-681M (35%)	9.3	1.02
Colloid 643	2.0	0.22
Kadox 515	12.0	0.26
Ti-Pure R-960	151.0	4.59
Atomite	53.0	2.35
Tamol 960 (40%)**	2.2	0.21
Rhoplex MV-23 (43%)	667.0	76.04
Butyl Carbitol	10.0	1.25
Ethylene Glycol	28.0	3.01
Skane M-8	2.0	0.23
Colloid 643	4.0	0.44
NH4OH (28%)	4.0	0.54
Acrysol TT-615 (30%)) Premix	2.0	0.23
Water)	10.0	1.20
Exp. Rheology Modifier QR-708		
(25% in propylene glycol/H2O)***	11.0	1.24
Water	10.3	1.24

** Must be added last since earlier addition thickens liquid
   phase.

PVC: 19.3%
Volume solids: 37.4%
Initial pH/KU/ICI: 9.9/104/.9
Equilibrated pH/KU/ICI: 9.9/105/.9

* Using weight ratio in pound units will yield approximately
  100 gallons of paint, while with kilograms, 833 liters will
  result.

***Experimental Rheology Modifier QR-708 (25% Solution):

	Weight Ratio*	Parts Per Hundred (Volume Basis)
Exp. Rheology Modifier QR-708	71.4	7.93
Propylene Glycol	17.2	2.00
Water	11.4	1.37

SOURCE: Rohm and Haas Co.: Maintenance Coatings: RHOPLEX MV-23:
        Formulation P-23-18

Latex Maintenance Primer

Materials:	Weight Ratio*	Parts Per Hundred (Volume Basis)
Water	155.23	18.63
Natrosol 250 MR (100%)	1.63	0.14
Nopco NXZ	0.84	0.11
Ammonium hydroxide (28%)	0.82	0.11
Tamol 850 (30%)	12.68	1.27
Triton CF-10	1.99	0.23
Titanium dioxide (Ti-Pure R-960)	135.84	3.98
Water ground mica, 325 mesh	22.64	0.97
Calcium carbonate (Atomite)	99.62	4.41
Barium metaborate (Busan 11-M1)	90.56	3.31
Letdown:		
Rhoplex MV-9 (45.5%)	541.84	62.31
Texanol	11.95	1.51
Ethylene glycol	21.01	2.26
Nopco NXZ	0.84	0.11
Sodium nitrite (13.8% in Water)	5.43	0.65

PVC: 32%
Volume Solids: 39.5%
Adjust pH to 9.3-9.5 with 28% ammonium hydroxide
Initial Viscosity: 65 KU

* Using weight ratio in pound units will yield approximately
  100 gallons of paint; using kilograms, about 833 liters will
  result.

SOURCE: Rohm and Haas Co.: Maintenance and Marine Coatings:
        RHOPLEX Acrylic Emulsions for Latex Maintenance Coatings:
        Formulation P-9-16

Latex Stain Blocking Formula

Raw Material:	Pounds	Gallons
Water	102.2	12.27
Nuosept 95	1.5	0.16
Nopcoside N-96	5.2	0.50
Nopcosperse 44	10.2	1.00
Igepal CO-630	2.2	0.25
Titanium Dioxide R-902	150.0	4.38
Zinc Oxide	10.0	0.22
BW-100	125.0	5.16
Nytal 300	40.0	1.68
Attagel 50	3.0	0.15
AMP 95	6.0	0.75
L-475	2.0	0.26

Grind to a minimum of NS 6. Add to the letdown after grind is achieved and the premix has been added.
Use for grind adjustment and wash:

Water	170.6	20.48

Premix next 4 items and add to the grind.

Propylene Glycol	25.8	3.00
Texanol	5.8	0.75
QR-708	4.5	0.50
Natrosol Plus	2.1	0.19

Letdown:

Aquamac 430	420.8	47.80
L-475	3.8	0.50

Properties:
    Weight Solids: 49.6%
    Volume Solids: 34.1%
    Pgmt Vol Conc: 34.0%
    Viscosity: 85-95 KU
    Gloss @ 60: 0-5
    Lbs/Gals: 10.91

Application:
    Brush: X
    Roller: X
    Spray: X

SOURCE: McWhorter: Suggested Formula Aquamac 430

Latex Wood Primer

This is a high performance, tannin stain-blocking latex wood primer based on HALOX BW-100 and Rhoplex MV-23 acrylic latex. The formula has been specifically balanced with the appropriate wetting and dispersing aids for the use of HALOX BW-100. It is recommended as a general purpose wood primer, especially over those woods which exhibit a tannin stain problem such as redwood and cedar.

Initial Grind:
Disperse the following, in order, on a high speed dispersion mill to a 5 n.s.u. grind:

	Pounds per 100 gallons	Gallons
Water	124.95	15.00
Anti-foam agent	1.70	0.23
Cellulosic Thickener	0.70	0.06
NH4OH	1.00	0.14
Ethylene Glycol	30.00	3.23
QR-681M Dispersant	18.20	2.00
Triton CF-10	2.70	0.31
Titanium Dioxide	150.00	4.38
Calcium Carbonate	200.00	8.86
Halox BW-100	50.00	2.07
Zinc Oxide	10.00	0.22

Then let down with: (at low speed)		
Rhoplex MV-23	440.23	50.00
Texanol	4.75	0.61
*Aroplaz 1271 & Driers	48.00	5.75
Defoamer	1.00	0.13
Preservative	2.00	0.23
QR-708 Thickener	5.50	0.63
Water	51.15	6.15

   * Premix the driers with the Aroplaz 1271, then slowly add this premixture to the let-down phase under agitation.

Drier combination:	Aroplaz 1271	93.4%
	Zirconium Drier 6%	5.6%
	Cobalt Drier 6%	0.5%
	Manganese Drier 6%	0.5%

   PVC: 36.51
   Weight per gallon: 11.40#
   Solids by weight: 57.83%
   Solids by volume: 42.54%
   Viscosity: 110 Ku @ 75F
   Pigment: 35.95% by weight
   Non-volatile vehicle: 21.88% by weight
   V.O.C.: 161 g/l
          133#/gal

SOURCE: Halox Pigments: Formulation A-81-08A

Lead-Free and Chromate-Free Primer

Material:	Lbs.	Gals.
Water	214.0	25.68
Ammonium Hydroxide	8.6	1.14
Kelsol 3909-BG4-75	106.9	12.41

Premix driers and cosolvent and add to batch:

	Lbs.	Gals.
Cobalt Hydrocure II	4.3	0.56
Activ-8	1.6	0.20
Butoxy Ethanol	10.7	1.42
Manganese Hydrocure	4.3	0.57
Mapico 297	53.6	1.24
106 Lo Micron Barytes	80.2	2.16
Talc 399	26.8	1.19
Halox SW-111	100.0	4.20
Atomite	48.2	2.14

Grind to a 6 Hegman on sand (bead) mill.

Letdown:	Lbs.	Gals.
Water	214.0	25.69
Ammonium Hydroxide	8.6	0.86
Kelsol 3909-BG4-75	187.3	21.74
Butoxy Ethanol	10.7	0.75

Analysis:
   Pigment Volume Concentration, Percent: 32.6
   Pigment/Binder Ratio: 1.4/1
   Percent Solids, Weight: 49.5
   Percent Solids, Volume: 34.5
   Pounds/Gallon: 10.59
   pH: 8.2-8.6
   VOC (excluding water):
      Grams/Liter: 227
      Pounds/Gallon: 1.89

SOURCE: Reichhold Chemicals, Inc.: Waterborne Handbook:
        Formulation 1855-32

Light Blue Anti-Corrosive Primer

	Pounds	Gallons
Cargill Water Reducible Self Crosslinking		
Resin 73-7390	140.2	16.46
Dimethylethanolamine	10.5	1.43
Titanium Dioxide Ti-Pure R-960	210.7	6.52
Zinc Phosphate J0852	38.7	1.45
Heliogen Blue L 6875 F	0.5	0.04
Special Black 4	0.7	0.05
Aerosil R972	4.0	0.22
Patcote 577	2.5	0.37
Deionized Water	225.0	27.01

Sand Mill to a 6 Hegman

	Pounds	Gallons
Cargill Water Reducible Self Crosslinking		
Resin 73-7390	115.2	13.52
Dimethylethanolamine	5.0	0.67
Patcote 577	2.5	0.37
Diethylene Glycol Monobutyl Ether	4.9	0.61
Ethylene Glycol Monobutyl Ether	25.8	3.43
Deionized Water	231.9	27.85

Paint Properties:
   % Nonvolatile (By Weight): 42.71
              (By Volume): 28.24
   Pigment to Binder Ratio: 1.41
   Pigment Volume Concentration: 29.33
   Weight Per Gallon: 10.18
   Theoretical VOC (Pounds/Gallon): 2.80
                 (Grams/Liter): 336
   Water:Cosolvent Weight Ratio: 78.33:21.67
   pH: 8.4-8.5
   Viscosity #4 Ford Cup (Seconds): 23

Typical Cured Film Properties:
   Cure Schedule: 10 Minutes at 325F
   Gloss (60/20): 33/0
   Pencil Hardness (Dry Film Thickness-Mils): 0.5
      Cold Rolled Steel: Pass 3H
      Zinc Phosphate Treated: Pass 6H
   % Crosshatch Adhesion:
      Cold Rolled Steel: 100
      Zinc Phosphate Treated: 100
   Impact (In. Lbs.):
      Direct: 160
      Reverse: 90

Humidity (750 Hours):	CRS	Zinc Phosphate Treated
Unexposed		
% Crosshatch Adhesion	100	100
Pencil Hardness	Pass 2H	Pass 4H

SOURCE: Cargill, Inc.: CARGILL Formulary: Formula P1887-68E

Light Duty Maintenance Primer Containing Busan 11M-1
(P-23-11A Modified to Prevent Settling)

Ingredients:	Weight Ratio*	Parts Per Hundred (Volume Basis)
Water	125.0	15.00
Methyl Carbitol	50.0	5.80
Exp. Rheology Modifier QR-708 (35%)	5.7	0.65
AMP-95	3.0	0.38
Colloid 643	2.0	0.22
Triton X-405 (70%)	2.0	0.22
Exp. Dispersant QR-681M (35%)	12.8	1.40
Kadox-515	6.0	0.13
Ti-Pure R-902	125.0	3.78
Busan 11-M1	50.0	1.82
ASP-170	117.8	5.47
Tamol 960 (40%)**	3.0	0.28
Rhoplex MV-23 (43%)	365.0	41.61
Aroplaz 1271 + Driers***	37.0	4.46
Butyl Carbitol	10.0	1.25
Colloid 643	3.0	0.40
Skane M-8	2.0	0.23
Sodium Nitrite (13.8%) aqueous	8.0	0.96
Acrysol TT-615 (30%)) Premix	2.0	0.23
Water)	10.0	1.20
Exp. Rheology Modifier QR-708		
(25% in propylene glycol/H2O)****	21.0	2.37
Water	101.2	12.14

** Must be added last since earlier addition thickens liquid
   phase.

PVC: 35.2%
Volume solids: 31.8%
Initial pH/KU/ICI: 9.4/101/1.7
Equilibrated pH/KU/ICI: 9.4/117/1.7

 * Using weight ratio in pound units will yield approximately
   100 gallons of paint, while with kilograms, 833 liters will
   result.
 *** Aroplaz 1271 + Driers:

Aroplaz 1271	93.4%
6% Cobalt Naphthenate	0.5%
6% Manganese Octoate	0.5%
6% Zirconium Octoate	5.6%

**** Experimental Rheology Modifier QR-708 (25% Solution):

	(Parts Per Hundred)	
	Weight Ratio*	(Volume Basis)
Exp. Rheology Modifier QR-708	71.4	7.93
Propylene Glycol	17.2	2.00
Water	11.4	1.37

SOURCE: Rohm and Haas Co.: Maintenance Coatings: RHOPLEX MV-23:
        Formulation P-23-17M

OEM Acrylic/Epoxy Primer

Materials:	Pounds	Gallons
Acrylic Component A:		
Grind Preparation:		

Grind the following materials using a high speed dissolver for 20 minutes:

	Pounds	Gallons
Methyl Carbitol	30.0	3.48
Water	41.1	4.93
Tamol 165	13.8	1.57
Ammonia (28%)	1.0	0.14
Triton CF-10	1.6	0.19
Drew L-405	2.0	0.29
TiPure R-900	50.0	1.46
Heucophos ZBZ	75.0	2.36
Shieldex	25.0	1.68
290 Barytes	131.1	3.50

Letdown Preparation:
Add the following in the order listed and mix thoroughly:

	Pounds	Gallons
Maincote AE-58	510.0	59.16
Water	20.0	2.40
Ammonia (28% NH3)	1.0	19.60
Grind (from above)	370.6	19.60
Ektasolve EEH	32.3	4.36
Water	18.2	2.19
Propasol B	25.0	3.38
Drew L-493	4.0	0.56
Ammonia (28%)	1.5	0.20
Sodium Nitrite (15% aqueous)	8.8	1.06
Rheology Modifier QR-708	3.0	0.34
Water	8.5	1.01

	Pounds	Gallons
Epoxy Component B:		
Genepoxy 370-H55	50.5	5.60

Primer Formulation Constants:
   Pigment Volume Content, %: 25
   Volume Solids, %: 36
   VOC, gm/l: 217
Component A Formulation Constants:
   pH: 8.5
   Stormer Viscosity, Krebs Units: 85
   ICI Viscosity, poise: 0.5

Source: Rohm and Haas Co.: Maintenance Coatings: MAINCOTE AE-58: Formulation P-58-3

Primer Formulation

Materials:	Parts by weight
Resin WB-408 (80% NVM)	255
Triethylamine	18
Amino crosslinker (1)	36
Manganese naphthenate, 6%	2
Deionized water	485
Carbon black (2)	6
Calcium carbonate (3)	94

(1) Cymel 301
(2) Raven 150
(3) Vera Blanc
Properties:
   Pigment/binder ratio: 0.42/1.00
   Melamine/resin ratio: 15/85
   Non-volatile material (NVM), wt%: 39
   Viscosity, #4 Ford cup, sec: 35-45
   Initial pH: 7.5-8.0
Compounding procedure for primer:
1. Solubilize Resin WB-408 in a mixing vessel containing a solution of triethylamine and deionized water.
2. Charge the resin solution, pigments, drier and amino crosslinker to a pebble mill. Grind overnight to a Hegman Grind of 7.
3. After milling, adjust pH to 7.5-8.0 and #4 Ford cup viscosity to 35-45 seconds.
Performance properties of thermosetting primer (1):
Substrate: B-37 (3):
   Pencil hardness: H
   Sward hardness: 18
   Crosshatch adhesion: 5B
   1/8-inch conical bend, % pass: 100
   Impact resistance, in-lb:
     direct: 80
     reverse: 50
   60 gloss, %: 46
   Mar resistance: good
   Salt spray resistance, 250 hr: pass
                       500 hr.: pass
Substrate: CRS (4):
   Pencil hardness: H
   Sward hardness: 18
   Crosshatch adhesion: 5B
   1/8-inch conical bend, % pass: 100
   Impact resistance, in-lb:
     direct: 80
     reverse: 80
   60 gloss, %: 48
   Mar resistance: good
   Salt spray resistance, 250 hr: fail
(1) Cure cycle 20 minutes at 177C (2) Overbake 30 minutes at 205C (4) Cold rolled steel panels
SOURCE: Amoco Chemical Co.: Water-Borne Bake Alkyds Based on AMOCO TMA and IPA: Tables 3 & 4

Primer Formulation
(Contains Strontium Chromate)

Materials:	Pounds	Gallons
Grind Portion:		
Butyl Cellosolve	10.5	1.40
Water	147.1	17.68
Tamol 165	12.8	1.46
Triton CF-10	2.9	0.32
Red Iron Oxide	65.8-	1.54
499 Talc	129.2	5.74
290 Lo Micron Barytes	67.1	1.88
Strontium Chromate	15.7	0.50

Cowles grind to 6 1/2 N.S. and add to the following premix:

		Pounds	Gallons
Kelsol 3905	)	140.7	16.44
Aquacat	)	1.3	0.16
Magnacat	)	1.2	0.16
Patcote 519	) Premix	1.0	0.15
Patcote 577	)	1.0	0.15
28% Ammonia	)	2.5	0.33
Water	)	105.0	12.62

(Neutralization of the premix only=52%, pH 5.8 to 6.0)
Add in order with good agitation:

	Pounds	Gallons
Rhoplex WL-71	253.5	29.48
Triton X-405 (35% non-volatiles)	9.0	1.03
Butyl Cellosolve	7.1	0.95
Butyl Carbitol	15.9	1.99
Dibutyl Phthalate	10.7	1.22
28% Ammonia	3.4	0.45
Water/Butyl Cellosolve 87/13	26.2	3.20
14% Ammonia (to pH 8.5)	1.8	0.23

Formulation Constants:
   Pounds/gallon: 10.4
   Non-volatiles (Wt.): 48.4%
           (Vol.): 34.0%
   Pigment/Binder Ratio: 56/44
   PVC: 28.6%
   Rhoplex WL-71/Kelsol 3905: 50/50
   pH: 8.5
   Viscosity: 400-600 cps.
   VOC: 1.08 lbs./gal.

SOURCE: Rohm and Haas Co.: Industrial Coatings: RHOPLEX WL-71:
      Formulation WL-71-1

Red Acrylic/Epoxy Primer

Materials:	Pounds	Gallons
Acrylic Component A:		
Grind Preparation:		
Methyl Carbitol	45.5	5.32
Tamol 165	10.0	1.13
Triton CF-10	1.0	0.11
Foamaster AP	1.0	0.14
NH4OH (28% NH3)	1.0	0.14
Water	66.0	7.92
Heucophos ZBZ Modified Zinc Phosphate	100.0	3.15
Bayferrox 120NM Red Iron Oxide	75.0	1.85
Talc	53.0	23.60

Letdown Preparation:
Add the following in the order listed and mix thoroughly:

	Pounds	Gallons
Maincote AE-58	522.7	60.75
Grind (from above)	352.5	22.12
NH4OH (28% NH3)	3.4	0.46
Dalpad-A) premix	22.2	2.42
Texanol)	11.1	1.40
Foamaster AP	3.0	0.42
Acrysol RM-1020	13.0	1.46
Water	33.1	3.97
Sodium nitrite (15% aqueous solution)	8.0	0.96

Epoxy Component B:	Pounds	Gallons
Genepoxy 370-H55	54.9	6.04

Primer Formulation Constants:
    Pigment Volume Content, %: 21.0
    Volume Solids, %: 35.0
    VOC, gm/l: 228
Component A Formulation Constants:
    pH: 8.5
    Stormer Viscosity, Kreb Units: 89
    ICI Viscosity, poise: 1.0

SOURCE: Rohm and Haas Co.: Maintenance Coatings: MAINCOTE AE-58:
        Formulation P-58-1

Red Iron Oxide Air Dry/Bake Primer

Material:	Lbs.	Gals.
Bentone 27	7.2	0.51
Isopropanol	7.2	1.10
Deionized Water	34.0	4.09
Disperse in mill 15 minutes:		
Arolon 585-W-43	185.0	20.77
Raven 1255	14.4	0.96
Oncor M50	46.1	1.38
Barytes W1430	136.5	3.75
399 Lo Micron Talc	93.6	4.18
Triton X405	6.9	0.75
6% Cobalt Naphthenate	2.0	0.26
14% Lead Naphthenate	0.5	0.05
Butoxy Ethanol	9.5	1.26
Potassium Tripolyphosphate	1.3	0.13
Drew L-475	1.3	0.23
Deionized Water	114.8	13.78
Pebble mill to 6 Hegman grind.		
Letdown:		
Arolon 585-W-43	392.0	44.04
Deionized Water	23.0	2.76

Analysis:
    Pigment Volume Concentration, Percent: 29.3
    Pigment/Binder Ratio: 1.2/1
    Percent Solids, Weight: 49.5
    Percent Solids, Volume: 35.4
    Viscosity @ 25C, #2 Zahn, Secs.: 18.25
    Pounds/Gallon: 10.75
    pH: 8.0-8.5
    VOC (excluding water):
        Grams/Liter: 69
        Pounds/Gallon: .60

SOURCE: Reichhold Chemicals, Inc.: Waterborne Handbook:
        Formulation 1914-94C

Red Iron Oxide Baking Primer

Material:	Lbs.	Gals.
Water	191.51	22.99
Ammonium Hydroxide	5.78	0.77
Kelsol 3902-BG4-75	95.81	10.95

Premix drier and solvent and add to above mixture.

Butoxy Ethanol	5.00	0.66
12% Manganese Cem-all	2.00	0.25

Then add the following, in order:

106 Lo Micron Barytes	102.43	3.09
Ti-Pure R-902	21.98	0.66
Strontium Chromate J1365	19.21	0.62
Mapico 297	28.53	0.66
Novacite L-337	57.58	2.61

Sand mill to 6 Hegman grind.

Letdown:		
Water	310.46	37.27
Ammonium Hydroxide	5.78	0.77
Kelsol 3902-BG4-75	167.65	19.16

Analysis:
    Pigment Volume Concentration, Percent: 26.3
    Pigment/Binder Ratio: 1.16/1
    Percent Solids, Weight: 42.8
    Percent Solids, Volume: 28.5
    Pounds/Gallon: 10.09
    pH: 8.2-8.5
    VOC (excluding water):
        Grams/Liter: 223
        Pounds/Gallon: 1.86

SOURCE: Reichhold Chemicals, Inc.: Waterborne Handbook:
        Formulation 1701-47

Red Iron Oxide Baking Primer

Material:	Lbs.	Gals.
Water	186.80	22.42
Ammonium Hydroxide	5.60	0.75
Kelsol 3909-BG4-75	93.40	10.86

Premix drier and solvent:		
Manganese Hydrocure	3.74	0.50
Butoxy Ethanol	4.67	0.63

106 Lo Micron Barytes	56.04	1.51
Talc 399	56.04	2.54
Novacite L-337	56.04	2.54
Mapico 297	28.02	0.64
Ti-Pure R-902	28.02	0.82
Strontium Chromate J1365	23.35	0.76

Sand mill to a 6 Hegman grind.

Letdown:		
Water	302.62	36.33
Ammonium Hydroxide	5.60	0.75
Kelsol 3909-BG4-75	163.45	19.00

Analysis:
    Pigment Volume Concentration, Percent: 29.7
    Pigment/Binder Ratio: 1.28/1
    Percent Solids, Weight: 43.4
    Percent Solids, Volume: 29.5
    Pounds/Gallon: 10.13
    pH: 8.2-8.5
    VOC (excluding water):
        Grams/Liter: 212
        Pounds/Gallon: 1.76

SOURCE: Reichhold Chemicals, Inc.: Waterborne Handbook:
        Formulation 1701-10

Red Iron Oxide Metal Primer

Ingredients:	Pounds	Gallons
Pigment Grind:		
Water	133.9	16.08
Methyl Carbitol Solvent	51.0	6.00
Dispersant (1)	17.6	2.00
Nonionic Surfactant (2)	2.2	0.25
Antifoam (3)	2.1	0.25
Urethane Associative Thickener (4)	17.2	2.00
Red Iron Oxide (5)	50.0	1.17
Calcium Carbonate (6)	225.0	9.97
Zinc Phosphate (7)	50.0	1.67
Let Down:		
Ucar Latex 430 (43.0%)	480.5	52.80
Ucar Filmer IBT	37.6	4.75
Antifoam (3)	2.1	0.25
Mix:		
Methyl Carbitol Solvent	8.5	1.00
Ucar Thickener SCT-275	8.6	1.00
Mix:		
Water	4.2	0.50
Sodium Nitrite	1.0	0.06
Ammonium Hydroxide, 28% Aqueous Solution	1.9	0.25

Suppliers:
(1) Dispersant - "Tamol" 165 or equivalent
(2) Nonionic Surfactant - "Triton" CF-10 or equivalent
(3) Antifoam - "Foamex" 1488 or equivalent
(4) Urethane Associative Thickener - "Acrysol" RM-1020 or
    equivalent
(5) Red Iron Oxide - R-2899D or equivalent
(6) Calcum Carbonate - "Gamasperse" 6352 or equivalent
(7) Zinc Phosphate - "Heucophos" ZBZ or equivalent

Paint Properties:
    Pigment Volume Concentration (PVC), %: 36.1
    Total Solids, %:
        by Volume: 35.5
        by Weight: 49.9
    Viscosity (equilibrated):
        Stormer, KU: 105
    pH, initial: 8.9
    Weight per Gallon, lb: 10.93
    VOC, g/l: 243.2

SOURCE: Ucar Emulsion Systems: UCAR Latex 430: Formulation
        Suggestion M-2206

Red Iron Oxide Primer

Material:	Lbs.	Gals.
Water	166.6	20.00
Natrosol 250	2.0	0.19
Ammonium Hydroxide	1.9	0.25
Surfynol GA	6.6	0.75
Texanol	2.0	0.25
Foamaster R	1.9	0.25
R.I.O. RO-4098	50.0	1.18
Busan 11M-1	50.0	1.82
325 Mica	25.0	1.07
Imsil A-10	50.0	2.27
Barytes #1	175.0	4.72
Cowles grind to 4-5 Hegman.		
Letdown:		
Arolon 840-W-39	491.6	56.50
Butoxy Ethanol	48.8	6.50
Water	35.4	4.25

Analysis:
  Pigment Volume Concentration, Percent: 34.8
  Pigment/Binder Ratio: 1.8/1
  Percent Solids, Weight: 49.7
  Percent Solids, Volume: 31.8
  Viscosity @ 25C, #2 Ford, Secs.: 37-42
  Pounds/Gallon: 11.1
  pH: 8.3
  VOC (excluding water):
    Grams/Liter: 159
    Pounds/Gallon: 1.3

SOURCE: Reichhold Chemicals, Inc.: Waterborne Handboook:
        Formulation 2124-94

Red Iron Oxide Primer

Material:	Lbs.	Gals.
Water	165.80	19.90
Ammonium Hydroxide	4.15	0.56
Kelsol 3960-B2G-75	82.90	9.60
Premix:		
Cobalt Hydrocure II	1.66	0.22
Manganese Hydrocure II	3.32	0.44
Butoxy Ethanol	12.44	1.66
Strontium Chromate J-1365	20.73	0.65
Mapico 297	41.45	0.96
106 Lo Micron Barytes	41.45	1.13
Talc 399	41.45	1.84
Novacite L207A	41.45	1.88
Steel ball mill to 6 Hegman grind.		
Letdown:		
Kelsol 3960-B2G-75	145.08	16.80
Ammonium Hydroxide	4.97	0.66
Water	318.34	38.22
Premix and dissolve crystals completely:		
Water	41.45	4.97
Potassium Dichromate	4.15	0.18
And then add to this premix:		
Ammonium Hydroxide	2.49	0.33
Add premix slowly on agitation.		

Analysis:
    Pigment Volume Concentration, Percent: 26.1
    Pigment/Binder Ratio: 1.1/1.0
    Percent Solids, Weight: 36.7
    Percent Solids, Volume: 24.8
    Viscosity @ 25C, #4 Ford, Secs.: 20-25
    Pounds/Gallon: 9.73
    pH: 8.2-8.6
    VOC (excluding water):
        Grams/Liter: 240
        Pounds/Gallon: 2.0

SOURCE: Reichhold Chemicals, Inc.: Waterborne Handbook:
        Formulation 2116-74-1

Red Iron Oxide Primer with Heucophos ZBZ for Brush/Roller Application

Ingredients:	Pounds	Gallons
Grind Preparation:		
Grind in Cowles Dissolver 15 minutes. Then let down at slower speed.		
Water	63.0	7.56
Methyl Carbitol	52.2	6.26
Tamol 165	12.3	1.39
Triton CF-10	3.0	0.34
Drew L-493	1.0	0.14
Bayferrox 120NM Red Iron Oxide	57.4	1.39
Atomite Calcium Carbonate	221.0	9.79
Heucophos ZBZ-Modified Zinc Phosphate	50.0	1.52
Letdown Preparation:		
Maincote HG-54	510.1	59.43
Texanol	4.0	0.50
Dibutyl Phthalate	12.9	1.48
Drew L-493	3.6	0.50
NH4OH (28% NH3)	1.7	0.20
Sodium Nitrite (15% Aqueous Solution)	8.4	1.01
Water/Acrysol RM-1020	69.5	8.10

Formulation Constants:
    Pigment Volume Content, %: 35.0
    Volume Solids, %: 36.5
    pH (pH range=8.3-8.7): 8.5
    Stormer Viscosity, Krebs Units: 87
    ICI Viscosity, poise: 1.7
    VOC, gm/liter: 150

SOURCE: Rohm and Haas Co.: Maintenance Coatings: MAINCOTE HG-54: Formulation P-54-5

Red Iron Oxide Primer with Zinc Phosphate for Airless Spray Application

Ingredients:	Pounds	Gallons
Grind Preparation:		

Grind in Cowles Dissolver for 15 minutes, then letdown at slower speed.

	Pounds	Gallons
Water	67.2	8.06
Tamol 165	11.6	1.31
Triton CF-10	1.0	0.11
Drew L-405	2.5	0.35
NH4OH (28% NH3)	1.0	0.13
Red Iron Oxide R-2899	62.4	1.46
Atomite	156.2	6.92
Sicor ZNP-M Zinc Phosphate	28.4	1.00

Letdown Preparation:

	Pounds	Gallons
Maincote HG-54	614.9	71.64
Texanol	19.1	2.41
Dibutyl Phthalate	19.1	2.19
Drew L-493	7.0	0.97
NH4OH (28% NH3)	2.8	0.39
Water	17.5	2.10
Rheology Modifier QR-708	1.0	0.12
Sodium Nitrite (15% Aqueous Solution)	7.0	0.84

Formulation Constants:
 Pigment Volume Content, %: 24.8
 Volume Solids, %: 37.9
 pH (pH range=8.3-8.7): 8.5
 Stormer Viscosity, Krebs Units: 95
 ICI Viscosity, poise: 0.4
 VOC, gm/liter: 51

SOURCE: Rohm and Haas Co.: Maintenance Coatings: MAINCOTE HG-54: Formulation P-54-6

Red Iron Oxide Primer with Zinc Phosphate for Brush/Roller Application

Ingredients:	Pounds	Gallons
Grind Preparation:		

Grind in Cowles Dissolver for 15 minutes. Then let down at slower speed.

	Pounds	Gallons
Water	31.1	3.73
Methyl Carbitol	50.0	6.00
Tamol 165	8.3	0.94
Triton CF-10	1.0	0.11
Drew L-405	2.8	0.40
Acrysol RM-1020	20.2	2.26
Red Iron Oxide R-2899	55.5	1.35
Atomite	150.0	6.65
Sicor ZNP-M Zinc Phosphate	27.3	0.96

Letdown Preparation:

	Pounds	Gallons
Maincote HG-54	591.5	68.90
Texanol	15.0	1.89
Dibutyl phthalate	15.0	1.72
Drew L-493	3.4	0.49
NH4OH (28% NH3)	4.0	0.48
Sodium nitrite (15% Aqueous Solution)	7.4	0.89
Water	26.9	3.23

Formulation Constants:
    Pigment Volume Content, %: 24.6
    Volume Solids, %: 36.4
    pH (pH range=8.3-8.7): 8.5
    Stormer Viscosity, Kreb Units: 97
    ICI Viscosity, poise: 1.9
    VOC, gm/liter: 200

SOURCE: Rohm and Haas Co.: Maintenance Coatings: MAINCOTE HG-54: Formulation P-54-4

Red Iron Oxide Rust Inhibitive Primer (Non-Lead/Non-Chromate)

Material:	Lbs.	Gals.
Water	128.58	15.43
Foamaster VL	3.04	.39
Tamol 731	23.44	2.71
Igepal CTA-639	4.09	.47
AMP-95	3.06	.39
Aroplaz 1271	51.27	6.14
Cobalt Hydrocure II	1.14	.15
Calcium Hydrocem	1.03	.13
Zirconium Hydrocem	.50	.06
Mix at low speed-then add:		
HEC QP-4400	1.96	.17
Nalzin 2	106.89	3.29
Kadox 515	13.51	.29
Atomite	190.56	8.47
Red Iron Oxide R-4098	53.93	1.26
Grind under high shear 15-20 minutes to a 6 Hegman grind.		
Letdown:		
Arolon 820-W-49	493.59	56.09
Methyl Carbitol	30.64	3.60
Butyl Carbitol	3.10	.39
Butyl Carbitol Acetate	3.10	.38

Analysis:
    Pigment Volume Concentration, Percent: 27.0
    Pigment Binder Ratio: 1.25/1
    Percent Solids, Weight: 59.20
    Percent Solids, Volume: 49.84
    Viscosity @ 25C, KU: 72-76
    Pounds/Gallon: 11.15
    pH: 8.8-9.2
    VOC (excluding water):
        Grams/Liter: 118.8
        Pounds/Gallon: .99

SOURCE: Reichhold Chemicals, Inc.: Waterborne Handbook:
        Formulation 2139-96-F1

Red Oxide Primer

	Pounds	Gallons
Cargill Water Reducible Styrenated Alkyd Copolymer 74-7422	141.3	16.72
Aqueous Ammonia 28%	8.9	1.18
Ethylene Glycol Monobutyl Ether	26.4	3.51
Byk-301	1.3	0.16
Byk-020	1.3	0.18
Atomite	68.5	3.04
Blanc Fix	72.7	1.99
Imsil A-10	15.4	0.70
Mapico Red 516 Medium	70.7	1.90
Zinc Phosphate J0852	56.5	2.11
Aerosil R972	6.6	0.36
Deionized Water	224.2	26.94

Sand Mill to a 5 Hegman

Cargill Water Reducible Styrenated Alkyd Copolymer 74-7422	70.8	8.38
Aqueous Ammonia 28%	5.2	0.69

Premix the following, then add:

Ethylene Glycol Monobutyl Ether	17.4	2.32
Intercar 6% Cobalt	1.4	0.18
Intercar 6% Zirconium	1.6	0.22
Activ-8	0.7	0.09
Exkin No. 2	0.8	0.10

Deionized Water	171.2	20.55

Deionized Water (Hold for Viscosity Adjustment)	72.3	8.68

Paint Properties:
    % Nonvolatile (By Weight): 42.60
                (By Volume): 26.47
    Pigment to Binder Ratio: 1.93
    Pigment Volume Concentration: 38.23
    Weight Per Gallon: 10.35
    Theoretical VOC (Pounds/Gallon: 2.73
                   (Grams/Liter): 327
    pH: 8.33
    Viscosity #4 Ford Cup (Seconds): 36
Typical Cured Film Properties:
    Cure Schedule: 7 Days Air Dry
    Substrate: Cold Rolled Steel
    Pencil Hardness: Pass HB
    % Crosshatch Adhesion: 80
    Dry Time (Minutes): Set to Touch: 15
                   Tack Free: 22

SOURCE: Cargill, Inc.: CARGILL Formulary: Formula P1857-208

Red Oxide Primer

Material:	Pounds	Gallons
Part A:		
Epi-Rez 510 Epoxy Resin	130.0	13.50
Red Iron Oxide	125.0	2.93
Barium Sulphate	282.2	7.70
Disperse to texture of 6-7 P.C.S. Add following at reduced speed:		
Epi-Rez 510 Epoxy Resin	32.1	3.34
Heloxy 8 Epoxy Functional Modifier	53.5	7.22
Part B:		
Epi-Cure W50-8535 Curing Agent	216.0	24.80
Glacial Acetic Acid	2.0	0.24
Water	334.0	40.27
Composite Blend:		
Part A	622.8	34.69
Part B	552.0	65.31

Typical Coating Properties:
Viscosity:
Part A: 130 Krebs Units
Part B: A (Gardner-Holdt)
Parts A + B: 119 Krebs Units
Reduced Viscosity (10:1 in water): 80-90 Krebs Units
Useable pot life: 1-2 hours
P.V.C.: 23.2%
Total Weight Solids: 61.24%
Total Volume Solids: 45.07%
Resin Composition:
Epoxy Resin: 49.96%
Curing Agent: 50.04%

Volatile Composition:	Pounds	Gallons
Water	453.84	54.58
Organic	4.16	0.50

SOURCE: Shell Chemical Co.: 24-192 Red Oxide Primer

Starting Point Formulation P-76-1 (Formerly WH-2895-C)
For Cementitious Metal Primers Based on RHOPLEX MC-76

Materials:	Weight Ratio*	Volume Ratio
Silica Flour 120	200	17.85
Portland Cement Type I**	100	7.58
RHOPLEX MC-76 (47%)	63.8	7.25
Defoamer***	0.3	0.04
Water (for silica flour)	90	10.79

Formulation Constants:
  Solids content, %: 72.7
  Lb. wt./U.S. gal.: 15.2
  % Solid polymer on cement: 30
  Aggregate/Cement ratio: 2/1

  * Using weight ratio in pound units will yield approximately
    36 gallons.
  ** White or Gray Type I Portland Cement may be used.
 *** Recommended defoamers:
     Nopco NXZ (100% active); GE Antifoam #60 (30% active)

Starting Point Formulation P-76-2 (Formerly WH-2889-J)
For Cementitious Metal Primers Based On Rhoplex MC-76

Materials:	Weight Ratio*	Volume Ratio
Water)	56.8	6.79
Thickener (2.5% Soln.)**) Premix	1.2	0.1
Ammonia (28%))	1.5	0.2
Rhoplex MC-76	619.0	70.34
Tenneco Lamp Black (Colortrend 888-9907)	2.0	0.17
Silica Flour 120	270.0	24.11
Portland Cement Type I***	540.0	40.91
Defoamer****	0.4	0.05

Formulation Constants:
  Solids content, %: 72.8
  Lb. wt./U.S. gal.: 12.8
  % Solid polymer on cement: 54
  Aggregate/Cement ratio: 1/2

   * Using weight ratio in pound units will yield approximately
     116 gallons.
  ** Acrysol RM-5 diluted 1:1 with H2O at the level of 0.5% on
     polymer solids may be substituted.
     Acrysol RM-5 is an efficient thickener that also imparts
     improved rheology characteristics.
 *** White or gray Type I Portland cement may be used.
**** Recommended defoamers:
     Nopco NXZ (100% active); GE Antifoam #60 (30% active)

SOURCE: Rohm and Haas Co.: Maintenance and Marine Coatings:
        RHOPLEX MC-76: Formulation P-76-1/Formulation P-76-2

Starting Point 45 PVC Primer Formulation

Materials:	Pounds	Gallons
Cowles Grind:		
Water	58.2	7.27
Balab 3056A	6.1	0.82
Dispersant*	38.7	4.34
Aurasperse W-7102	0.5	0.05
Ti-Pure R-960	138.6	4.22
Snowflake	254.1	11.27
Neosil A	46.2	2.13
Letdown:		
Experimental Emulsion E-2189 (40.5%)	500.6	56.48
**Surfynol 104E	2.0	0.25
Ethylene Glycol Monobutyl Ether (60% in		
H2O)	101.4	13.17

Adjust with 14% NH3 to pH 8.5. Reduce viscosity to 30 sec.
(#2 Zahn Cup) with water for DRC; to 26 sec. (#2 Zahn Cup)
for curtain coating.

**Surfynol 104E is required only for curtain coating.

Ethylene Glycol monobutyl ether at 30% based on polymer solids
is recommended to give maximum film formation under marginal
conditions.

Formulation Constants:
PVC: 45%
Weight Solids: 57%
Volume Solids: 43%
pH: 8.5
Viscosity, #2 Zahn Cup: 26 sec. for curtain coating
30 sec. for DRC

* Dispersant	Parts	
Acrysol I-62 (50%)	30	
H2O	30	
NH3 (14%)	4	(1/1 conc. NH3 (28%)/H2O)

SOURCE: Rohm and Haas Co.: Building Products Board Coatings:
Experimental Emulsion E-2189: Formulation

Tannin Block Primer/Sealer Formulation

Testing has been conducted over several staining woods, including redwood and cedar, as well as wood-base products commonly used in building materials such as ceiling tiles. In all cases the tannin block primer/sealer based on RAP 213NA performed equivalent to the industry-leading acrylic latex as well as a leading solvent-borne primer/sealer, but at lower cost.

Material:		Pounds	Gallons
Water		235.00	28.21
KTPP		1.50	0.14
Nopco NDW		1.00	0.13
Propylene Glycol	)Premix	21.50	2.48
Dowanol PPh glycol ether	)And	10.00	1.14
SCT-275	)Add	16.50	1.91
Surfynol 104E		4.00	0.48
AMP-95		4.00	0.51
Dowicil 75 Preservative		1.00	0.08
Tronox CR-821		150.00	4.49
Halox BW-100		50.00	2.07
Duramite		200.00	8.89
XX-503 ZnO		25.00	0.54
Nopco NDW		1.00	0.13
RAP 213NA		425.00	50.18

Typical Properties, As Formulated:
    Wt. per Gallon, Lbs.: 11.30
    Viscosity: 74 KU
    N.V. by Wt.: 55.7%
    N.V. by Volume: 40.0%
    PVC: 39.5%
    VOC:
        Lbs/Gal - Water: 0.90
        Gm/L - Water: 108

SOURCE: Dow Chemical USA: Formulation RAP 213NA

Tannin Stain-Blocking Exterior Primer

Material:	Pounds	Gallons
Water	166.6	20.00
Ethylene Glycol	23.2	2.50
Colloid 640	2.0	0.27
Natrosol 250 MR	1.0	0.09
Colloid 111	18.0	1.96
Igepal CO-630	3.0	0.34
Nuosept 95	2.0	0.21
Tronox CR-800	100.0	2.91
Halox BW-100	100.0	4.13
Snowflake	150.0	6.67
Nytal 300	50.0	2.11

Disperse to 3 NS - slow mill, then add:

Texanol	7.9	1.00
76 RES 4076	504.0	60.00
Colloid 643	2.0	0.28
SCT 275	1.5	0.17

Physical Properties:
   Viscosity: 77-85 KU
   pH: 8.60
   Weight per gallon: 11.02 lbs.
   % Solids:
      By weight: 58.7%
      By volume: 45.3%
   PVC: 35.0%
   VOC: 71 g/l - water

SOURCE: Unocal: Formula C-88182-B

Tannin Stain Inhibiting Latex Wood Primer

Ingredient:	Weight	Volume
Water	100.0	12.0
Cellulosic Thickener (1)	1.0	0.1
Ammonium Hydroxide	1.0	0.1
Dispersant (2)	18.2	2.0
Igepal CO-630	2.5	0.3
Surfynol 104E	2.5	0.3
Ethylene Glycol	30.0	3.2
Titanium Dioxide (5)	150.0	4.5
Halox BW-100	100.0	4.2
Snowflake	100.0	4.4
Minex 7	50.0	2.3
Styrene/Butadiene Latex (8)	485.0	57.5
Texanol	10.0	1.0
Nopco NDW	2.0	0.1
SCT 200 Thickener	10.0	1.1
Water	57.4	6.9

Raw Material Suppliers:
(1) Cellosize QP-4400 or equivalent
    SCT 200 Associative Thickener
(2) Tamol 731 Dispersant or equivalent
(5) Unitane OR-600 or equivalent
(8) Dow Latex RAP 213 NA

Formulation Constants:
    PVC: 35.4   Weight per Gallon: 11.2
    Weight Solids: 57.4   Volume Solids: 43.5
    Pigment by Weight: 35.7   NVV by Weight: 21.7
    VOC: 1.2 (lbs/gal)

SOURCE: Halox Pigments: Preliminary Formulation: AD-87-03

Tannin Stain Inhibiting Latex Wood Primer

Ingredient:	Weight	Volume
Water	100.0	12.0
Cellulosic Thickener (1)	1.0	0.1
Ammonium Hydroxide	1.0	0.1
Dispersant (2)	18.2	2.0
Igepal CO-630	2.5	0.3
Surfynol 104E	2.5	0.3
Ethylene Glycol	30.0	3.2
Titanium Dioxide (5)	150.0	4.5
Halox BW-100	150.0	6.2
Snowflake	25.0	1.1
Minex 7	25.0	1.1
Styrene/Butadiene Latex (8)	450.0	53.3
Texanol	10.0	1.0
Alkyd Modification*	24.0	2.8
Nopco NDW	2.0	0.1
SCT 200 Thickener	10.0	1.1
Water	90.0	10.8

```
    * Alkyd Modification Blend:
        Duramac 2400        94.0%
        Zr Hydrocure         5.0%
        Co Hydrocure         0.5%
        Mn Hydrocure         0.5%
```

Raw Material Suppliers:
(1) Cellosize QP-4400 or Equivalent
    SCT 200 Associative Thickener
(2) Tamol 731 Dispersant or Equivalent
(5) Unitane OR-600 or Equivalent
(8) RAP 213 S/B Latex

Formulation Constants:
    PVC: 31.6
    Weight per Gallon: 10.9
    Weight Solids: 54.2
    Volume Solids: 40.8
    Pigment by Weight: 32.1
    NVV by Weight: 22.1
    VOC: 1.2 (lbs/gal)

SOURCE: Halox Pigments: Preliminary Formulation: AD-87-04

Waterbase Acrylic Latex Primer

Grind:	Lbs/100 Gals	Grams/Liter
Water	63.00	75.60
Methyl Carbitol	52.00	62.40
Tamol 165	12.00	14.80
Triton CF-10	3.00	3.60
Drew L-493	1.00	1.20
Red Iron Oxide	57.40	68.90
Snowflake	255.00	306.00
Halox SZP-391	33.00	39.60
Letdown:		
Maincote HG-54	510.00	612.00
Texanol	4.00	4.80
Dibutyl Phthalate	12.90	15.50
Drew L-493	3.60	4.30
Ammonium Hydroxide 28%	1.70	2.00
Sodium Nitrite 10%	8.40	10.10
Water	50.00	60.00
Ucar Thickener SCT 200	7.00	8.40

Formulation Constants:
    Density lbs/gal: 10.74
    Density g/L: 1289.20
    % Weight Solids: 53.20
    % Volume Solids: 40.00
    % PVC: 35.00
    Viscosity: 95-100 KUs @ 77F (25C)
    VOC: 1.25 lbs/gal (150g/L)

The laboratory testing documented here proves that HALOX SZP-391 is an excellent choice for use in acrylic latex emulsion systems such as Maincote HG-54. HALOX SZP-391 clearly outper-formed the competition in terms of corrosion protection, package stability and cost effectiveness.

SOURCE: Halox Pigments: Starting Point Formulation

Water-Reducible Automotive Primer with Strontium Chromate Gray

Ingredients:	Pounds	Gallons
Chempol 10-0453	103.68	12.23
Byk P-104S	4.40	0.55
Surfynol 104BC	2.43	0.32
Ektasolve EP	14.60	1.92
AMP-95	4.87	0.62
Water	138.22	16.59
TiPure R-900	132.63	3.86
Raven #14 Powder	9.73	0.67
J-1365 Strontium Chromate	34.07	1.07

Premix first five ingredients and mix well. Add water slowly with agitation. Load mixture into ball mill followed by pigments.

Run in steel ball mill to a #7 Hegman. Adjust grind viscosity and/or wash mill with water from below.

Let dowm with:		
Chempol 10-0453	206.03	24.30
Ammonia, 28%	13.63	1.83
Ektasolve EP	15.49	2.04
NuoCure CK-10	8.67	1.07
NuoCure Mn	7.23	0.97
Byk 301	0.50	0.06
Water	240.74	28.90
Water (hold to adjust viscosity)	25.00	3.00

Add ammonia to the resin slowly and with good agitation; mix well. Premix cosolvent, driers, and Byk 301. Add this mixture to the neutralized resin with agitation. Mix well, then add the water slowly while still mixing. Add this mixture to the grind with continued agitation. Mix well, and allow the paint to stand overnight.

Check the pH and adjust with ammonia if necessary. Use the "adjust viscosity" water to bring the viscosity of the coating to spec at room temperature.

Note: This coating may display questionable stability properties in the form of phase separation, especially at elevated temperatures. Additional cosolvent (2-butoxyethanol) or amine (AMP-95) may be required to correct this problem; however, these additions will raise the VOC content of the coating.

Properties:
   Total solids, % by weight: 42.12
   Total solids, % by volume: 31.11
   Weight/gallon: 9.62 lbs/gal
   PVC: 18.00%
   Pigment/binder ratio: 0.77
   VOC - water: 2.84 lbs/gal
   Viscosity, #2 Zahn cup: 36"
   pH: 8.4

SOURCE: Freeman Polymers Division: CHEMPOL 10-0453-A Formulation

Water-Reducible Automotive Primer Without Strontium Chromate Gray

Ingredients:	Pounds	Gallons
Chempol 10-0453	104.76	12.35
Byk P-104S	4.45	0.56
Surfynol 104BC	2.46	0.33
AMP-95	14.75	1.96
Water	139.66	16.77
TiPure R-900	134.01	3.90
Raven #14 Powder	9.84	0.67

Premix first five ingredients and mix well. Add water slowly with agitation. Add mixture into ball mill followed by the pigments.

Run in steel ball mill to a #7 Hegman. Adjust grind viscosity and/or wash mill with water from below.

Let down with:

Chempol 10-0453	208.17	24.55
Ammonia, 28%	13.77	1.85
2-butoxyethanol	15.65	2.08
NuoCure CK-10	8.76	1.08
NuoCure Mn	7.30	0.98
Byk 301	0.50	0.06
Water	243.49	29.23
Water (hold to adjust viscosity)	25.00	3.00

Add ammonia to the resin slowly and with good agitation; mix well. Premix cosolvent, driers, and Byk 301. Add this mixture to the neutralized resin with agitation. Mix well, then add the water slowly while still mixing. Add this mixture to the grind with continued agitation. Mix well, and allow the paint to stand overnight.

Check the pH and adjust with ammonia if necessary. Use the "adjust viscosity" water to bring the viscosity of the coating to spec at room temperature.

Properties:
   Total solids, % by weight: 39.99
   Total solids, % by volume: 30.35
   Weight/gallon: 9.37 lbs/gal
   PVC: 15.07%
   Pigment/binder ratio: 0.62
   VOC - water: 2.90 lbs/gal
   Viscosity, #2 Zahn cup: 38"
   pH: 8.5
   Substrate: B-1000 CRS
   Cure: 1 week air dry
   Set-to-touch, mins.: 25
   Dust free, mins.: 25
   Tack free, mins.: 60-70
   Dry through, hours: 6-8
   DFT: 1.2-1.4 mil
   Gloss, 60: 84

SOURCE: Freeman Polymers Division: CHEMPOL 10-0453-B Formulation

Water Reducible, Gray Air Dry or Force Dry Cure Primer; Lead and Chromate Free

Formulation 10-0091-F is a water reducible primer which may be air dried or force cured on steel substrate. Applied coatings cure rapidly to produce tough, flexible and highly adherent films which have a high order of corrosion resistance. In addition, formulation 10-0091-F complies with many proposed Clean Air regulations, containing only 2.3 pounds of volatile organic compounds per gallon of paint (exclusive of water).

Ingredients:	Pounds	Gallons
Chempol 10-0091	83.88	9.53
Ammonium hydroxide	6.37	0.90
Polycompound W-953	2.49	0.30
Nalco 2309	2.49	0.30
Water	183.19	21.99
Bykumen	2.49	0.30
Zinc phosphate J0852	49.75	1.89
Ti-Pure R-900	47.75	1.39
Lampblack Superjet LB-1011	2.99	0.20
Atomite	59.70	2.69
Nytal 400	29.85	1.69
Barmite XF	59.70	1.69
EA-1075	9.95	1.49

Combine CHEMPOL 10-0091 and ammonia. Then add Polycompound W-953 and Nalco 2309. Add water, stir well and add Bykumen. Add pigments, EA-1075 and stir well.

Pebble Mill Grind 7 Hegman

	Pounds	Gallons
Chempol 10-0091	142.88	16.24
Ammonium hydroxide	10.85	1.49
Odorless mineral spirits	19.90	3.09
Hydrocure cobalt drier (5% cobalt)	1.69	0.20
Hydrocure manganese drier (5% manganese)	1.69	0.20
Activ-8	1.00	0.10
Exkin No. 2	1.99	0.30
Water	279.40	34.02

In "Let Down" combine CHEMPOL 10-0091, Ammonium Hydroxide, Odorless Mineral Spirits, driers, Activ-8 and Exkin No. 2 in order given before adding to the mill base. Final water addition should be made slowly with agitation to prevent pigment flocculation and shocking.

Properties:
    Total solids, % by weight: 42.54
    Total solids by volume: 27.34
    Viscosity, Krebs units: 96
    Reduction for spray application by volume: 4 parts paint/
                                                 1 part water
    Pigment/binder weight ratio: 1.00/0.67
    PVC, %: 35.86
    Gloss (60% Glossmeter): Flat

SOURCE: Freeman Polymers Division: CHEMPOL 10-0091-F Formulation

Water Reducible, Gray, Air Dry or Force Dry Cure Primer; Lead and Chromate Free

Ingredients:	Pounds	Gallons
Chempol 10-0095	83.88	9.64
Ammonium hydroxide (28% min)	6.72	0.95
Polycompound W-953	2.49	0.30
sec-Butanol	6.00	0.58
Nalco 2309	2.49	0.30
Water	182.84	21.92
Bykumen	2.49	0.30
Zinc Phosphate J852	49.75	1.89
Ti-Pure R-900	49.75	1.39
Lampblack Superjet LB-1011	2.99	0.20
Atomite	59.70	2.69
Nytal 400	29.85	1.69
Barmite XF	59.70	1.69
EA-1075	9.95	1.49

Combine Chempol 10-0095 and ammonium hydroxide. Then add Poly-compound W-953, sec-Butanol and Nalco 2309. Add water, mix well and add Bykumen. Add pigments and EA-1075. Mix well.
Pebble Mill Grind 7 Hegman
Let down with:

Chempol 10-0095	142.88	16.42
Ammonium hydroxide (28% min)	11.45	1.57
sec-Butanol	10.20	0.98
Odorless mineral spirits	19.90	3.09
Hydrocure cobalt drier (5% cobalt)	1.69	0.20
Hydrocure manganese drier (5% manganese)	1.69	0.20
Activ-8	1.00	0.10
Exkin No. 2	1.99	0.30
Water	268.55	32.20

In "Let Down" combine Chempol 10-0095, ammonium hydroxide, sec-Butanol, odorless mineral spirits, driers, Activ-8 and Exkin No. 2 in order given before adding to the mill base. Final water addition should be made slowly with agitation to prevent pigment flocculation and shocking.

Properties:
   Total solids, % by weight: 42.3
   Total solids, % by volume: 28.4
   Reduction for spray application: 4 parts paint/
      (by volume)   1 part water
   Viscosity spray application, seconds, Zahn #4: 20-23 seconds
   Pigment/binder weight ratio: 1.5/1.0

SOURCE: Freeman Polymers Division: CHEMPOL 10-0095-A Formula

White Acrylic/Epoxy Primer

Materials:	Pounds	Gallons
Acrylic Component A:		
Grind Preparation:		

Grind the following materials using a high speed dissolver for 20 minutes:

	Pounds	Gallons
Methyl Carbitol	50.0	6.00
Tamol 165	13.8	1.57
NH4OH (28% NH3)	1.0	0.12
Triton CF-10	1.6	0.19
Drew L-405	2.0	0.28
TiPure R-900	50.0	1.46
Heucophos ZBZ	75.0	2.38
Shieldex	25.0	1.68

Letdown Preparation:
Add the following in the order listed and mix thoroughly:

	Pounds	Gallons
Maincote AE-58	492.8	57.25
Water	58.2	7.07
Ammonia (28% NH3)	1.0	0.12
Grind (from above)	218.4	13.68
Ektasolve EEH	31.5	4.25
Drew L-493	3.0	0.42
Acrysol RM-1020	14.0	1.58
Water	33.2	3.99
Chroma-Chem 896-9901 (lampblack)	0.3	0.03
Sodium Nitrite (15% aqueous solution)	8.8	1.09

Epoxy Component B:	Pounds	Gallons
Genepoxy 370-H55 (55%)	94.8	10.52

Primer Formulation Constants:
    Pigment Volume Content, %: 16.5
    Volume Solids, %: 34.7
    VOC, gm/l: 216
Component A Formulation Constants:
    pH: 8.5
    Stormer Viscosity, Krebs Units: 82
    ICI Viscosity, poise: 1.2

SOURCE: Rohm and Haas Co.: Maintenance Coatings: MAINCOTE AE-58: Formulation P-58-2

White Alkyd Wood Primer

Ingredient:	Weight	Volume
Halox BW-100	25.0	1.0
Titanium Dioxide	150.0	4.4
Zinc Oxide	25.0	0.5
Beaverwhite 325 Talc	325.0	14.1
Kraygel 36	6.0	0.4
Cargill 5150	362.5	48.3
Raw Linseed Oil	27.7	3.6
Nuosperse 657	3.0	0.4
Mineral Spirits	169.0	26.4
Zr Drier (12%)	3.6	0.5
Co Drier (12%)	1.5	0.2
Mn Drier (6%)	0.9	0.1
Methanol: Water (95:5)	2.0	0.3
Skino #1	1.0	0.1

Formulation Constants:
PVC: 45.3%   Solids by Volume: 45.0%
Weight per gallon: 11.0   Pigment by Weight: 48.2%
Solids by Weight: 66.9%   NVV by Weight: 19.2%
Preliminary Formulation: A-84-03

Clear Latex Primer/Sealer
(Water/Smoke Damage)

Ingredient:	Weight	Volume
Water	100.0	12.0
Cellulosic Thickener	1.0	0.1
Ammonium Hydroxide	1.0	0.1
Dispersant	18.2	2.0
Igepal CO-630	2.5	0.3
Surfynol 104E	2.5	0.3
Ethylene Glycol	30.0	3.2
Halox BW-100	100.0	4.2
Snowflake	100.0	4.4
Minex 7	100.0	4.6
Styrene/Butadiene Latex	465.0	55.0
Texanol	10.0	1.0
Nopco NDW	2.0	0.1
SCT 200	10.0	1.1
Water	96.5	11.6

Formulation Constants:
PVC: 33.7   Pigment by Weight: 28.9
Weight per Gallon: 10.4   NVV by Weight: 21.6
Weight Solids: 50.5   VOC: 1.3 (lbs/gal)
Volume Solids: 39.2

SOURCE: Halox Pigments: Preliminary Formulation: AD-87-05

White House Paint Primer

Material:	Lbs.	Gals.
Water	116.5	14.00
Triethylamine	3.5	0.57
Methyl Carbitol	10.4	1.23
Patcote 550	0.9	0.12
Kelsol 3922-G-80	86.3	10.00
Ti-Pure R-902	237.3	7.12
Daper Novacite	172.6	7.82
Cowles to a 5 Hegman grind.		

Letdown:		
Kelsol 3922-G-80	153.6	17.80
Surfynol 104E	1.3	0.16
Aroplaz 1272	22.4	2.68
Ammonium Hydroxide	10.4	1.38
Butyl Carbitol	12.1	1.52
Water	280.5	33.68

Premix:		
Activ-8	0.4	0.05
Cobalt Hydrocure II	3.0	0.40
Calcium Hydrocem	3.0	0.38
Zirconium Hydrocem	2.6	0.31

Premix:		
Water	3.5	0.20
Acrysol ASE-60	3.5	0.40

Analysis:
    Pigment Volume Concentration, Percent: 38.0
    Pigment/Binder Ratio: 1.92/1
    Percent Solids, Weight: 55.9
    Percent Solids, Volume: 39.6
    Viscosity @ 25C, KU: 85-90
    pH: 8.2-8.6
    Pounds/Gallon: 11.24
    VOC (excluding water):
      Grams/Liter: 187
      Pounds/Gallon: 1.55

SOURCE: Reichhold Chemicals, Inc.: Waterborne Handbook:
        Formulation 2121-53-1

Woodstain-Resistant/Corrosion-Resistant Primer Containing Acrysol RM-825

Materials:	Weight/Pounds	Volume/Gallons
Water	125.0	15.00
Methyl Carbitol	50.0	5.80
Acrysol RM-5 (25%)	8.0	0.92
AMP-95	3.0	0.38
Colloid 643	2.0	0.22
Triton X-405 (70%)	2.0	0.22
Experimental Dispersant QR-681M (35%)	12.8	1.40
Kadox 515	6.0	0.13
Ti-Pure R-902	125.0	3.78
Busan 11M-1	50.0	1.82
ASP-170	117.8	5.47
Tamol 960 (40%)*	3.0	0.28
Rhoplex MV-23 (43%)	365.0	41.61
Aroplaz 1271 + Driers**	37.0	4.46
Butyl Carbitol	4.5	0.56
Colloid 643	3.0	0.40
Super-Ad-It	4.0	0.50
Sodium Nitrite (13.8%) aqueous	8.0	0.96
Acrysol TT-615 (30%))Premix	2.0	0.23
Water)	10.0	1.20
Acrysol RM-825 (25%)	21.0	2.41
Water	102.0	12.25

* Must be added last since earlier addition thickens liquid phase.

Formulation Constants:
    PVC: 35.2%
    Volume Solids: 31.8%
    pH: 9.4
    Initial Viscosity:
        KU: 101
        ICI: 1.7
    Equilibrated Viscosity:
        KU: 101
        ICI: 1.7

** Aroplaz 1271 + Driers:
    Aroplaz 1271   93.4%
    6% Cobalt Naphthenate   0.5%
    6% Manganese Octoate   0.5%
    6% Zirconium Octoate   5.6%

SOURCE: Rohm and Haas Co.: Trade Sales Coatings: Acrysol RM-825: Formulation P-23-22A

20 PVC Waterborne Pigmented Primer
(Contains Strontium Chromate)

Materials:	Weight* (Pounds)	Volume (Gallons)
Butyl Cellosolve	50.3	6.69
Water	23.8	2.86
Patcote 519	1.6	0.23
Igepal CTA-639	2.6	0.29
Tamol 165	11.1	1.26
Red Iron Oxide	118.6	2.77
Strontium Chromate	14.3	0.46
Zinc Yellow	4.2	0.15
499 Talc	42.9	1.91
290 Lo Micron Barytes	59.3	1.58

Cowles grind (7-7 1/2 N.S.)

Letdown with:

Rhoplex WL-92 (42 wt. % solids)	596.1	69.39
Triton X-405 (reduced to 35 wt. % solids)	14.3	1.63
Butyl Carbitol	37.6	4.70
Dibutyl Phthalate	12.7	1.45
10% Ammonium Benzoate in Water	33.4	4.00
14% Ammonia to pH of 7.9-8.1	5.3	0.66

Adjust viscosity with ASE-60 (5% in water,
    pH of 9.2), if required

Formulation Constants:
    pH: 7.9-8.1
    Viscosity, cps: 200-600
    Approximate Solids, % (wt./vol.): 50/38
    Density, lbs./gal.: 10.3
    Pigment/Binder Ratio: 49/51
    PVC, %: 20
    Freeze-Thaw Stability: Protect from freezing
    Heat Stability (8 days/140F): Satisfactory
    VOC, g/liter: 214
        lbs./gallon: 1.79
    Weight % on Polymer Solids:
        Butyl Cellosolve: 20
        Butyl Carbitol: 15
        Dibutyl Phthalate: 5

* Using weight in pound units will yield approximately 100
    gallons of lacquer. With kilograms, 833 liters will result.

SOURCE: Rohm and Haas Co.: Industrial Coatings: RHOPLEX WL-92:
        Formulation WL-92-5

31 PVC Waterborne Pigmented Primer
(Contains Strontium Chromate)

Materials:	Weight* (Pounds)	Volume (Gallons)
Tamol 165	15.2	1.72
Patcote 519	1.1	0.16
Igepal CTA-639	3.5	0.40
Water	117.3	14.08
Red Iron Oxide	85.3	1.99
Strontium Chromate	10.4	0.33
Zinc Yellow	3.0	0.11
499 Talc	91.8	4.08
290 Lo Micron Barytes	128.8	3.44

Cowles grind (7-7 1/2 N.S.)

Letdown with:

Rhoplex WL-92 (42 wt. % solids)	437.3	50.91
Triton X-405 (reduced to 35 wt % solids)	10.4	1.19
Butyl Cellosolve	27.5	3.66
Butyl Carbitol	27.5	3.44
Dowanol DPM	9.2	1.15
10% Ammonium Benzoate in Water	22.1	2.65
14% Ammonia to pH of 7.9-8.1	----	----
Adjust viscosity with ASE-60 (5% in water, pH of 9.2)	79.2	9.48

Formulation Constants:
  pH: 7.9-8.1
  Viscosity, cps: 400-600
  Approximate Solids, % (wt./vol.): 48/33
  Density, lbs./gal.: 10.8
  Pigment/Binder Ratio: 63/37
  PVC, %: 31
  Freeze-Thaw Stability: Protect from freezing
  Heat Stability (8 days/140F): Satisfactory
  VOC, g/liter: 168
      lbs./gallon: 1.40
  Weight % on Polymer Solids:
    Butyl Cellosolve: 15
    Butyl Carbitol: 15
    Dowanol DPM: 5

  * Using weight in pound units will yield approximately 100
    gallons of lacquer. With kilograms, 833 liters will
    result.

SOURCE: Rohm and Haas Co.: Industrial Coatings: RHOPLEX WL-92:
        Formulation WL-92-4

45 PVC Exterior Hardboard Primer

Grind:	Pounds	Gallons
Water	90.2	10.82
Acrysol I-62 (25%, pH 9.5 with DMAE)	39.2	4.51
Balab 3056A	3.6	0.49
Aurasperse W-7012	0.8	0.08
Ti-Pure R-960 (TiO2)	78.5	2.37
Snowflake (CaCO3)	207.8	9.18
Minex 4 (Silicate)	207.8	9.58
Bentone LT	0.7	0.05
Letdown:		
Resimene 745	25.3	2.53
Rhoplex AC-1822 (46.5%)	488.4	55.48
Butyl Cellosolve (60%)	41.6	5.33
Dimethylaminoethanol (50%)	1.3	0.16
Cellosize HEC, QP-4400 (2.5%)	8.6	1.04

Constants:
    PVC: 45%
    pH: 9.0
    Viscosity: 25 sec., #4 Ford Cup
    Acrylic Binder/Melamine (weight ratio): 90/10
    Weight Solids: 64%
    Volume Solids: 47%

Note: Weight in pounds will yield about 100 gallons of paint;
      using kilograms, the yield will be about 833 liters.

SOURCE: Rohm and Haas Co.: Building Products Board Coatings:
        RHOPLEX AC-1822: Formulation Table IV

45 PVC Exterior Hardboard Primer

Grind:	Pounds	Gallons
Water	90.2	10.82
Acrysol I-62 (25%, pH 9.5 with DMAE)	39.2	4.51
Balab 3056A	3.6	0.49
Aurasperse W-7012	0.8	0.08
Ti-Pure R-960 (TiO2)	78.5	2.37
Snowflake (CaCO3)	207.8	9.18
Minex 4	207.8	9.58
Bentone LT	0.7	0.05

Letdown:		
Resimene 745	25.3	2.53
Rhoplex AC-1822 (46.5%)	488.4	55.48
Butyl Cellosolve (60%)	41.6	5.33
Dimethylaminoethanol (50%)	1.3	0.16
Cellosize HEC, QP-4400 (2.5%)	8.6	1.04

Constants:
    PVC=45%
    pH=9.0
    Viscosity=25 sec., #4 Ford Cup
    Acrylic Binder/Melamine (weight ratio)=90/10
    Weight Solids: 64%
    Volume Solids: 47%

Note: Weight in pounds will yield about 100 gallons of paint;
          using kilograms, the yield will be about 833 liters.

SOURCE: Rohm and Haas Co.: Building Products Board Coatings:
          RHOPLEX AC-1822: Formulation Table V

55 PVC Primer

Materials:	Pounds	Gallons
Cowles Grind:		
Acrysol I-62 (22.7%)*	49.4	5.54
Balab 3056A	4.4	0.60
Aurasperse W-7012	1.0	0.10
TiPure R-960	98.7	3.00
Snowflake CaCO3	261.5	11.59
Minex 4	261.5	12.06
Bentone LT	1.0	0.07
Water	125.3	15.04
Letdown:		
Rhoplex AC-2045 (45.0%)	400.4	45.48
Cymel 373 (85%)	25.1	2.26
Ethylene glycol monobutyl ether (60% in water)	33.4	4.27

Formulation Constants:
   pH: adjust to 8.3-8.8 with DMAE (50% in H2O)
   Viscosity (#4 Ford Cup), sec.: 35
   % PVC: 55
   % Weight Solids: 66
   % Volume Solids: 49
   Acrylic/Melamine (solids ratio): 90/10

   * Acrysol I-62 Neutralization:

Acrysol I-62 (50%)	40.0
Water	40.0
DMAE (50%)	8.0

SOURCE: Rohm and Haas Co.: Building Products Hardboard Coatings:
     RHOPLEX AC-2045: Formulation #P-2045-1

55 PVC Primer

Materials:	Pounds	Gallons
Cowles Grind:		
Acrysol I-62 (Neutralized at 23.7%)*	54.4	6.42
Balab 3056A	4.6	0.58
TiPure R-960	102.2	3.11
Snowflake	270.9	12.03
Neosil A	270.8	12.27
Aurasperse W-7012	1.0	0.10
Bentone LT	1.0	0.06
Water	96.9	11.63

Letdown:		
Experimental Emulsion E-2573 (47.5%)) 2nd	392.6	45.54
Cymel 373 (85%)) 1st) Pre-	24.3	2.31
Triethylamine (TEA) Premix) mix	5.5	0.92
Ethylene Glycol Monobutyl Ether (60% in		
water)	38.9	5.05

Formulation Constants:
    Adjust pH with TEA to: 9.0
    Viscosity (#4 Ford Cup), sec.: 16
    % Pigment Volume Content: 55
    % Weight Solids: 67.5
    % Volume Solids: 50
    Binder/Melamine (solids ratio): 90/10

    * Acrysol I-62 Neutralization:
      Acrysol I-62 (50%)   40.0
      Water   40.0
      Triethylamine   4.0

SOURCE: Rohm and Haas Co.: Building Products Hardboard Coatings:
        Experimental Emulsion E-2573: Formulation #P-2573-1

55 PVC Primer Formulation

Materials:	Pounds	Gallons
Acrysol I-62 (22.7% T.S.)*	50.0	5.75
Balab 3056A	4.5	0.61
Aurasperse W-7012	1.0	0.10
Rutile TiO2, Ti-Pure R-960	100.0	3.02
Snowflake	265.0	11.77
Minex 4	265.0	12.22
Bentone LT	1.0	0.07
Water	86.4	10.37

Grind at high speed and let down as follows:

Rhoplex AC-22 (46%)	417.0	47.40
Cymel 373 (85%)	25.1	2.40
Butyl Cellosolve (60%)	35.5	4.55

Adjust pH to 9.0 with DMAE 50%.

Primer Constants:
  PVC: 55.0%
  Viscosity (#4 Ford Cup): 24"
  Weight Solids: 68%
  Volume Solids: 50%

  * Neutralization of Acrysol I-62:

Acrysol I-62 (50%)	40.0
Water	40.0
DMAE (50%)	8.0

SOURCE: Rohm and Haas Co.: Building Products Board Coatings:
        RHOPLEX AC-1822: Formulation Table V

Section IX
Sealers

Clear Concrete Sealer

Material:	Lbs.	Gals.
Arolon 840-W-39	415.4	47.75
Water	129.1	15.50
Premix and add:		
Spensol L44	163.7	18.75
Water	70.8	8.50
Premix and add:		
Ektasolve DE	33.0	4.00
Water	36.4	4.37
Add in following order with agitation:		
Tributoxy Ethyl Phosphate	6.4	0.75
Lodyne S107B	1.7	0.18
SWS 211	0.2	0.03
Ammonium Hydroxide	1.3	0.17

Analysis:
   Ratio of AROLON 840 to SPENSOL L44: 75/25
   Percent Solids, Weight: 25.2
   Percent Solids, Volume: 22.8
   Viscosity @ 25C, #2 Zahn, Secs.: 16
   pH: 8.0-8.8
   VOC (excluding water):
      Grams/Liter: 202
      Pounds/Gallon: 1.7

Formulation 2076-70

Clear Urethane/Acrylic Wood Finish

Material:	Lbs.	Gals.
Arolon 840-W-39	378.1	43.46
Ammonium Hydroxide	2.0	0.26
Byk 301	3.0	0.37
Michem Emulsion 43040	24.6	2.96
Rheolate 255	2.5	0.29
Spensol L44	460.8	52.66

Analysis:
   Percent Solids, Weight: 35.2
   Percent Solids, Volume: 32.2
   Viscosity @ 25C, #2 Zahn, Secs.: 25
   Pounds/Gallon: 8.7
   pH: 8.2-8.8
   VOC (excluding water):
      Grams/Liter: 159
      Pounds/Gallon: 1.3

SOURCE: Reichhold Chemicals, Inc.: Waterborne Handbook:
         Formulation 2384-65B

Clear Sealer
A clear sealer formulation for concrete, chalk, etc.
Ideal starting formulation for clear lacquer.

Ingredients:	Pounds	Gallons
Water	158.9	19.08
Fluorosurfactant (1)	0.5	0.06
Antifoam (2)	0.5	0.07
Butyl Carbitol Solvent	47.5	6.00
Propylene Glycol	25.9	3.00
Plasticizer (3)	23.1	3.00
Ammonium Hydroxide, 28% Aqueous Solution	1.9	0.25
Ucar Latex 430 (43.0%)	596.3	68.54

Suppliers:
(1) Fluorosurfactant - FC-120 or equivalent
(2) Antifoam - SWS-211 or equivalent
(3) Plasticizer - Tributoxyethyl Phosphate KP-140 or equivalent

Clear Wet Properties:
Total Solids, %:
by Volume: 28.2
by Weight: 30.1
Viscosity, #4 Ford Cup, sec: 21
pH, initial: 9.4
Weight per gallon, lb: 8.55
Freeze-thaw, 3 cycles: Pass
Heat Stability, 2 wks at 120F (49C): Pass
VOC, g/l: 216.8

Film Properties:
Gloss, 60: 86
Whitening, water, 24-hr day: None
Low-Temperature Coalescence: Excellent
Chalk Adhesion (1): Excellent
(1) ASTM #2 Chalk

SOURCE: Ucar Emulsions Systems: UCAR Latex 430: Formulation Suggestion V-2207

Concrete Sealer

	Pounds	Gallons
Cargill Water Reducible Epoxy Ester 73-7331	278.0	32.90
Aqueous Ammonia 28%	16.7	2.22

Premix the following, then add:

	Pounds	Gallons
Ethylene Glycol Monobutyl Ether	15.0	1.99
Activ-8	1.0	0.13
Intercar 6% Cobalt	2.1	0.27
Intercar 6% Manganese	2.9	0.38
Byk-301	4.0	0.49

	Pounds	Gallons
Deionized Water	463.4	55.63
Deionized Water (Hold for Viscosity Adjustment)	49.9	5.99

Paint Properties:
    % Nonvolatile (By Weight): 23.99
                 (By Volume): 22.40
    Weight Per Gallon: 8.33
    Theoretical VOC (Pounds/Gallon): 2.92
                 (Grams/Liter): 350
    Water:Cosolvent Weight Ratio: 84.20:15.80
    pH: 8.48
    Viscosity #4 Ford Cup (Seconds): 30

It should be noted that this coating will tend to yellow more than conventional solvent-based products.

SOURCE: Cargill, Inc.: CARGILL Formulary: Formula P1877-19G

Low-Solids Sanding Waterborne Sealer Based on Rhoplex CL-104
Emulsion for Air-Dry or Low-Force-Dry Application

	Weight	Volume
Materials:	Pounds	Gallons
Rhoplex CL-104 emulsion	514.1	59.6
Dee Fo 3000	1.0	0.1

Premix and add under agitation:

Butyl Carbitol	19.8	2.5
Butyl Cellosolve	49.5	6.6
Water	162.6	19.5

Add:

Isopropanol	62.7	9.6
Liquazinc AQ-90	11.9	1.4
QR-708 Rheology Modifier	5.6	0.7

Formulation Constants:
   Approximate solids, % (wt/vol): 25.0/23.1
   pH: 7.0-7.5
   Viscosity, Brookfield LV Spindle #1, cP @ 30 rpm: 25-30
   Viscosity, #2 Zahn cup, seconds: 16-22
   VOC, g/liter: 382
      lb/gallon: 3.18
   Coalescent, wt % on polymer solids:
      Butyl Cellosolve: 25
      Butyl Carbitol: 10
   Rheology modifier, wt % on polymer solids: 1.0
   Freeze/thaw stability: Protect from freezing
   Heat/age stability (140F/10 days): Passes

SOURCE: Rohm and Haas Co.: RHOPLEX CL-104 Acrylic Emulsion:
      Formulation WR-74-5LS

Sanding Sealer Based on Experimental Emulsion E-2955

	Weight	Volume
Materials:	Pounds	Gallons
Experimental emulsion E-2955	660.0	76.9
Water	84.0	10.1

Mix above then add slowly with good agitation:

14% ammonia (to adjust pH to 8.5)

Premix and add under agitation:

Butyl Cellosolve	31.7	4.2
Ektasolve EEH	4.9	0.7
Isopropynol (15%)	36.6	5.6

Add:

Water	21.0	2.5

   (adjust viscosity with water to a range of 20 to 26 seconds
    on a #2 Zahn cup)

Formulation Constants:
   Approximate solids, % (wt/vol): 29.1/26.9
   pH: 7.5-8.0
   Flash point, closed cup, F: 120
   Viscosity, #2 Zahn cup, seconds: 20 to 26
   VOC, g/liter: 235
      lb/gallon: 1.95
   Coalescent, wt % on polymer solids:
     Butyl Cellosolve: 13
     Ektasolve EEH: 2
   Freeze/thaw stability: Protect from freezing
   Heat/age stability (140F/10 days): Passes

SOURCE: Rohm and Haas Co.: Experimental Emulsion E-2955:
     Formulation WR-55-2

Sanding Waterborne Sealer Based on Rhoplex CL-104 Emulsion
For Air-Dry or Low-Force-Dry Application

	Weight Pounds	Volume Gallons
Materials:		
Rhoplex CL-104 emulsion	695.1	81.3
Dee Fo 3000	1.6	0.2

Premix and add under agitation:

Butyl Carbitol	26.7	3.3
Butyl Cellosolve	67.0	8.9
Water	15.9	1.9

Add:

Liquazinc AQ-90	16.0	1.9

14% ammonia
   (adjust viscosity with water to a range of 20 to 26 seconds
   on a #2 Zahn cup)

Water	19.4	2.3

Formulation Constants:
   Approximate solids, % (wt/vol): 32.5/30.8
   pH: 7.5-8.0
   Viscosity, #2 Zahn cup, seconds: 20 to 26
   VOC, g/liter: 259
          lbs/gallon: 2.15
   Coalescent, wt % on polymer solids:
      Butyl Cellosolve: 25
      Butyl Carbitol: 10
   Freeze/thaw stability: Protect from freezing
   Heat/age stability (140F/10 days): Passes

SOURCE: Rohm and Haas Co.: RHOPLEX CL-104 Acrylic Emulsion:
        Formulation WR-74-5

Sealer

Materials:	Weight Pounds	Volume Gallons
1. E-2774 (38.5% nv)	694.6	80.6
2. Dehydran 1620 (100% active)	2.5	0.3

Premix 3-5 and add to 1 and 2 with agitation

3. Butyl Carbitol	26.7	3.4
4. Butyl Cellosolve	66.9	8.9
5. Water	20.1	2.4
6. Liquazinc AQ-90 (50%)	16.0	1.3
7. 14% Ammonia	1.4	0.2
8. Water	24.5	2.9

Formulation Constants:
   Approximate Solids, % (wt/vol): 32.3/30.0
   pH: 7.5
   Viscosity, #2 Zahn, seconds: 18
   VOC,* g/liter: 264
       lbs/gallon: 2.20
   Coalescent, wt% Polymer Solids:
     Butyl Cellosolve: 25
     Butyl Carbitol: 10
   Freeze/Thaw Stability: Protect from freezing
   Heat Age Stability: Satisfactory

SOURCE: Rohm and Haas Co.: Formulation WR-74-9

Waterproofing Sealer: White Masonry Paint

	lb/100 gal
Propyl Cellosolve	40
Methyl Propasol	19
Kodaflex TXIB	18
Triethylamine	11
Nopco NDW	5

Mix well; then add:

Pliolite 7104	400

Mix well; then add:

Titanox 2101	90
Atomite	225
Minusil 30	179
Mica C-3000	113
Nopcocide N-96	3

High speed disperse to 3 NS. Then add the following in order
   shown:

Water	57
28% NH4OH Solution	10
Nopco NDW	4
4% Cellosize QP-4400 Solution	45

    VOC: 1.22 lb/gal=147 g/l (-H20)
    Density: 12.25 lb/gal
    PVC: 54%
    NVW: 66%
    NVV: 48%
    KU Viscosity: 100 to 110 KU
    Freeze Thaw Stability: pass 5 cycles
    Oven Stability: pass 4 weeks 120 degrees F
    Dry Time @ 2 mil DFT: 1 hour
    Recoat Time: 4 hours
    Method of Application: brush or roll
    Hydrostatic Water Pressure Test: pass 4 psi no leaks or
                                     blisters

SOURCE: The Goodyear Tire & Rubber Co.: Formula 2074-41-1

Section X
Stains

Brown Pigmented Spray Stain Based on Acrysol I-62

Solubilize the Acrysol I-62 by combining materials 1, 2 and 3. Disperse pigments 4 through 7 with the Butyl Carbitol (8) in a media mill to a 7 Hegman Grind. Letdown with 9 through 11. Some settling may occur during storage. Agitate and check the Hegman grind before application. This formulation is similar to the pigmented wiping stain but is let down with water to achieve the proper color concentrations after spraying.

	Weight	Volume
Materials:	Pounds	Gallons
1. Acrysol I-62 (50% nv)	11.6	1.4
2. Water	9.4	1.1
3. 14% Ammonia	0.9	0.1
4. Bentone LT	0.5	0.0
5. Van Dyke Brown	7.4	0.3
6. Burnt Umber	2.3	0.1
7. Raw Umber	8.2	0.3
8. Butyl Carbitol	5.8	0.7
9. Dipropylene glycol	19.7	2.5
10. Butyl Carbitol	48.1	6.1
11. Water	63.7	7.6
12. Water	664.4	79.8

Formulation Constants:
    Approximate Solids, % (wt/vol): 2.9/1.4
    pH: 6.7
    Viscosity, #2 Zahn, seconds: 15
    VOC,* g/liter: 88
        lbs/gallon: .74
    Freeze/Thaw Stability: Protect from Freezing
    Heat Age Stability: Satisfactory

    * California's South Coast Air Management District allows
      VOC's for stains with less than one pound solids per gallon
      to be calculated on the basis of pounds of VOC per gallon
      plus water. Accordingly, that convention is used here.

SOURCE: Rohm and Haas Co.: Formulation PST-1

Brown Pigmented Wiping Stain

Solubilize the Acrysol I-62 by combining materials 1, 2 and 3. Disperse pigments 4 through 7 with the Butyl Carbitol (8) in a media mill to 7 Hegman grind. Let down with 9 through 11. Some settling may occur during storage. Agitate and check the Hegman grind before application.

Materials:	Weight Pounds	Volume Gallons
1. Acrysol I-62 (50% nv)	32.4	3.8
2. Water	26.4	3.2
3. 14% Ammonia	2.6	0.3
4. Bentone LT	1.4	0.1
5. Van Dyke Brown	20.8	0.9
6. Burnt Umber	6.4	0.2
7. Raw Umber	23.0	0.9
8. Butyl Carbitol	16.2	2.0
9. Di propylene glycol	55.4	6.9
10. Butyl Carbitol	135.0	17.0
11. Water	178.5	21.4
12. Water	360.1	43.2

Formulation Constants:
    Approximate solids, % (wt, vol): 7.9/4.0
    pH: 7.0
    Viscosity, #2 Zahn, seconds: 15
    VOC*, g/liter: 248
        lbs/gallon: 2.07
    Freeze/Thaw Stability: Protect from Freezing
    Heat/Age Stability: Satisfactory

  * California's South Coast Air Management District allows VOC's for stains with less than one pound solids per gallon to be calculated on the basis of pounds of VOC per gallon plus water. Accordingly, that convention is used here.

SOURCE: Rohm and Haas Co.: Formulation PWS-1

Redwood Stain--Opaque

Material:	Lbs.	Gals.
Kelsol 3931-WG4-45	166.0	20.00
Butoxy Ethanol	15.2	2.00
Mapico 347A	50.0	1.16
Tan 10A	25.0	0.57
Syloid 234	25.0	1.50

Grind the above to a 6 Hegman, then add the following and disperse 10-15 minutes:

Michemlube 743	28.7	3.50

Letdown:

Kelsol 3931-WG4-45	207.5	25.00

Premix the next two materials, then add under good agitation:

Manganese Hydrocure II	6.4	0.75
Butoxy Ethanol	7.5	1.00

Premix the next four materials, then add under good agitation:

Aerosol OT-75	2.3	0.25
Intercide T-O	4.9	0.50
Polyphase AF-1	7.3	0.75
Butoxy Ethanol	7.5	1.00

Add the following under good agitation:

Ammonium Hydroxide	1.9	0.25
Water	347.9	41.77

Analysis:
   Pigment Volume Concentration, Percent: 12.7
   Pigment/Binder Ratio: 0.6/1.0
   Percent Solids, Weight: 32.2
   Percent Solids, Volume: 25.5
   Viscosity @ 25C, #2 Zahn, Secs.: 23
   Pounds/Gallon: 9.0
   pH: 8.3
   VOC (excluding water):
      Grams/Liter: 264
      Pounds/Gallon: 2.2

SOURCE: Reichhold Chemicals, Inc.: Waterborne Handbook:
         Formulation 2262-12A

Redwood Stain--Semi-Transparent

Material:	Lbs.	Gals.
Kelsol 3931-WG4-45	166.0	20.00
Butoxy Ethanol	15.2	2.00
Mapico 347A	15.0	0.35
Syloid 234	25.0	1.50

Grind the above to a 6 Hegman, then add the following and disperse
10-15 minutes:

Michemlube 743	28.7	3.50

Letdown:

Kelsol 3931-WG4-45	207.5	25.00

Premix the next two materials, then add under good agitation:

Manganese Hydrocure II	6.4	0.75
Butoxy Ethanol	7.5	1.00

Premix the next four materials, then add under good agitation:

Aerosol OT-75	2.3	0.25
Intercide T-O	4.9	0.50
Polyphase AF-1	7.3	0.75
Butoxy Ethanol	7.5	1.00

Add the following under good agitation:

Ammonium Hydroxide	1.2	0.25
Water	359.4	43.15

Analysis:
   Pigment Volume Concentration, Percent: 7.7
   Pigment/Binder Ratio: 0.2/1.0
   Percent Solids, Weight: 26.8
   Percent Solids, Volume: 24.1
   Viscosity @ 25C, #2 Zahn, Secs.: 25
   Pounds/Gallon: 8.2-8.6
   pH: 8.5
   VOC (excluding water):
      Grams/Liter: 264
      Pounds/Gallon: 2.2

SOURCE: Reichhold Chemicals, Inc.: Waterborne Handbook:
        Formulation 2262-12B

Redwood Stain--Semi-Transparent

Material:	Lbs.	Gals.
Kelsol 3937-WG4-45	166.0	19.80
Butoxy Ethanol	12.2	1.62
Mapico 347A	15.0	0.35
Syloid 234	25.0	1.50

Grind the above to a 6 Hegman, then add the following and disperse 15 minutes:

Michemlube 743	28.7	3.50

Letdown:

Kelsol 3937-WG4-45	207.5	24.76

Premix the next two materials, then add under agitation:

Manganese Hydrocure II	6.4	0.75
Butoxy Ethanol	7.5	1.00

Premix the next four materials, then add under agitation:

Aerosol OT-75	2.3	0.25
Intercide T-O	4.9	0.50
Polyphase AF-1	7.3	0.75
Butoxy Ethanol	7.5	1.00
Ammonium Hydroxide	1.2	0.25
Water	366.3	43.97

Analysis:
    Pigment Volume Concentration, Percent: 9.0
    Pigment/Binder Ratio: .24/1
    Percent Solids, Weight: 24.2
    Percent Solids, Volume: 21.3
    Viscosity @ 25C, #2 Zahn, Secs.: 25-35
    Pounds/Gallon: 8.58
    pH: 8.2-8.6
    VOC (excluding water):
       Grams/Liter: 244
       Pounds/Gallon: 2.04

SOURCE: Reichhold Chemicals, Inc.: Waterborne Handbook:
       Formulation 2402-54A

Semi-Transparent Cherry Stain

Material:	Lbs.	Gals.
Water	284.1	34.10
2% Natrosol 250MR Solution	100.0	11.93
Propylene Glycol	45.0	5.21
Butyl Cellosolve	34.5	4.59
AMP 95	0.6	0.06
Proxel GXL	1.0	0.10
Synthemul 40-423	370.0	42.05
Brown Iron Oxide Disp.	27.6	1.95

Analysis:
   Pigment Volume Concentration, Percent: 3.6
   Percent Solids, Weight: 21.5
   Percent Solids, Volume: 18.4
   Initial Viscosity, KU: 50-55
   pH: 8.3-8.8
   Pounds/Gallon: 8.63
   VOC (excluding water):
      Grams/Liter: 338.0
   60 Gloss* (comm. alkyd=4-6): 9-10
   20 Gloss* (comm. alkyd=0.5-1.0): 1.0-1.5
   85 Sheen* (comm. alkyd=14-18): 18-22

   * Two Coats

SOURCE: Reichhold Chemicals, Inc.: Waterborne Handbook:
        Formulation 423-01

Shingle Stain Tint Base

Material:	Lbs.	Gals.
Water	549.00	65.89
Linaqua	168.00	20.70
Kelecin 1081	8.20	0.95
Cobalt Hydrocure II	3.00	0.39
Manganese Hydrocure	3.00	0.40
Activ-8	0.74	0.09
Bentone 38	5.10	0.36
Titanox 2101	76.00	2.23
Zinc Oxide XX601	102.00	2.19
Talc 399	153.00	6.80

Analysis:
  Pigment Volume Concentration, Percent: 38.2
  Percent Solids, Weight: 46.0
  Percent Solids, Volume: 29.3
  Pounds/Gallon: 10.68
  VOC (excluding water):
      Grams/Liter: 98.4
      Pounds/Gallon: 0.82

SOURCE: Reichhold Chemicals, Inc.: Waterborne Handbook:
        Formulation 1936-54

Sierra Redwood Stain--Semi-Transparent

Material:	Lbs.	Gals.
Kelsol 3939-WG4-45	166.0	19.80
Butoxy Ethanol	12.2	1.62
Mapico 297	10.0	0.23
Mapico 1075	4.0	0.12
Mapico Black	1.0	0.02
Syloid 234	25.0	1.50

Grind the above to a 6 Hegman, then add the following and disperse 15 minutes:

Michemlube 743	28.7	3.50

Letdown:

Kelsol 3939-WG4-45	207.5	24.75

Premix the next two materials, then add under good agitation:

Manganese Hydrocure II	6.4	0.75
Butoxy Ethanol	7.5	1.00

Premix the next four materials, then add under good agitation:

Aerosol OT-75	2.3	0.25
Intercide T-O	4.9	0.50
Polyphase AF-1	7.3	0.75
Butoxy Ethanol	7.5	1.00
Ammonium Hydroxide	1.2	0.25
Water	366.3	43.96

Analysis:
    Pigment Volume Concentration, Percent: 9.0
    Pigment/Binder Ratio: .24/1
    Percent Solids, Weight: 24.2
    Percent Solids, Volume: 21.3
    Viscosity @ 25C, #2 Zahn, Secs.: 25-35
    Pounds/Gallon: 8.58
    pH: 8.2-8.6
    VOC (excluding water):
        Grams/Liter: 244
        Pounds/Gallon: 2.04

SOURCE: Reichhold Chemicals, Inc.: Waterborne Handbook:
        Formulation 2402-34A

Stain Formulation

Materials:	Parts by Weight
WB-151 Resin (at 75% NVM)	205.0
Deionized Water	450.0
Ammonium Hydroxide	11.0
Ethylene glycol monobutyl ether	15.0
Secondary butanol	23.0
Cobalt drier (1)	6.0
Opaque stain (2)	----
Red Iron Oxide Pigment (3)	11.0
Flow Agent (4)	1.0
Foam Control Agent (5)	2.0
Fungicide (6)	2.0

Stain Properties:
    Non-volatile material, (NVM), wt %: 23.0
    Pigment/binder Ratio: 0.07
    Cosolvent/Water Ratio, wt%: 17/83
    Viscosity, #2 Zahn Cup, sec: 28.0
    pH: 8.02
    Dry times at 20C, 50% RH:
        set to touch, min: 13.0
        tack free, min: 20.0
        dry hard, min: 20.0
    Volatile Organic Compound, (VOC), lb/gal: 2.90

1. Cobalt Hydrocure
2. WA-8
3. RO-3097
4. BYK-301
5. Drewplus L-418
6. Nopcocide N-40-D

Compounding Procedure:
1. Disperse pigment into resin using a high speed mixer.
2. Neutralize the resin/pigment dispersion with ammonium hydroxide and deionized water; mix thoroughly.
3. Add cosolvent, drier, flow agent, foam control agent and fungicide; mix thoroughly.

SOURCE: Amoco Chemical Co.; Water-borne Alkyd Stain/Sealer for Wood Based on Amoco IPA and TMA: Formula WB-1

Stain Formulation

Materials:	Parts by Weight
WB-151 Resin (at 75% NVM)	205.0
Deionized Water	500.0
Ammonium Hydroxide	11.0
Ethylene glycol monobutyl ether	15.0
Secondary butanol	23.0
Cobalt drier (1)	6.0
Opaque stain (2)	11.0
Flow Agent (4)	1.0
Foam Control Agent (5)	2.0
Fungicide (6)	5.0

Stain Properties:
    Non-volatile material, (NVM), wt%: 22.0
    Pigment/binder Ratio: 0.07
    Cosolvent/Water Ratio, wt%: 15/85
    Viscosity, #2 Zahn Cup, sec: 21.0
    pH: 8.33
    Dry times at 20C, 50% RH:
       set to touch, min: 14.0
       tack free, min: 30.0
       dry hard, min: 32.0
    Volatile Organic Compound, (VOC), lb/gal: 2.90

1. Cobalt Hydrocure
2. WA-8
4. BYK-301
5. Drewplus L-418
6. Nopcocide N-40-D

Compounding Procedure:
1. Disperse pigment into resin using a high speed mixer.
2. Neutralize the resin/pigment dispersion with ammonium hydroxide and deionized water; mix thoroughly.
3. Add cosolvent, drier, flow agent, foam control agent and
   fungicide; mix thoroughly.

SOURCE: Amoco Chemical Co.: Water-borne Alkyd Stain/Sealer for
        Wood Based on AMOCO IPA and TMA: Formula WB-2

Waterborne Redwood Solid Color Stain

	Pounds	Gallons
Water Reducible Modified Polyolefin 73-7358	70.0	9.03
Synthetic Red Iron Oxide RO-3097	50.0	1.23
Aerosil R972	2.5	0.14
Patcote 577	2.5	0.36

High Speed to a 6 Hegman

Add under agitation:

	Pounds	Gallons
Water Reducible Modified Polyolefin 73-7358	90.0	11.61
OK 412	25.0	1.58
Ethylene Glycol Monobutyl Ether	80.0	10.65
Manganese Hydrocure II)	9.5	1.10
Ethylene Glycol Monobutyl Ether)Premix	24.0	3.19
Polyphase AF-1)	8.0	0.85
Aqueous Ammonia 28%	11.5	1.53
Deionized Water	468.4	56.23
Michem Lube 511	20.0	2.50

Paint Properties:
    % Nonvolatile (By Weight): 29.54
                (By Volume): 25.41
    Pigment to Binder Ratio: 0.44
    Pigment Volume Concentration: 11.92
    Weight Per Gallon: 8.61
    Theoretical VOC (Pounds/Gallon): 2.84
                (Grams/Liter): 341
    Water:Cosolvent Weight Ratio: 80.68:19.32
    pH: 8.8
    Viscosity #2 Zahn Cup (Seconds): 33

Typical Cured Film Properties:
    Dry Time (Hours):
        200g Zapon: 5
        500g Zapon: 6 1/4

SOURCE: Cargill, Inc.: CARGILL Formulary: Formulation 1396-70

Waterborne Semi-Transparent Russet Brown Stain

	Pounds	Gallons
Water Reducible Modified Polyolefin 73-7358	80.0	10.32
Titanium Dioxide Ti-Pure R-900	3.0	0.09
Special Black 4	1.0	0.10
Synthetic Red Iron Oxide RO-3097	8.0	0.20
Aerosil R972	4.0	0.22

High Speed to a 6 Hegman

Add under agitation:

	Pounds	Gallons
Water Reducible Modified Polyolefin 73-7358	100.0	12.91
OK 412	28.0	1.77
Ethylene Glycol Monobutyl Ether	90.0	11.98
Manganese Hydro Cure II)	9.5	1.10
Ethylene Glycol Monobutyl Ether) Premix	24.0	3.20
Polyphase AF-1)	8.0	0.85
Aqueous Ammonia 28%	14.7	1.96
Deionized Water	440.1	52.80
Michem Lube 511	20.0	2.50

Paint Properties:
    % Nonvolatile (By Weight): 28.94
                 (By Volume): 27.31
    Pigment to Binder Ratio: 0.23
    Pigment Volume Concentration: 8.89
    Weight Per Gallon: 8.30
    Theoretical VOC (Pounds/Gallon): 2.85
                 (Grams/Liter): 342
    Water:Cosolvent Weight Ratio: 78.58:21.42
    pH: 8.0-8.5
    Viscosity #2 Zahn Cup (Seconds): 20

Typical Cured Film Properties:
    Dry Time (Hours):
        200g Zapon: 5
        500g Zapon: 6 1/4

SOURCE: Cargill, Inc.: CARGILL Formulary: Formula 1396-71

Section XI
Texture Paints

Interior Texture Paint

A low cost texture paint formulated with PYRAX WA to eliminate need for asbestos or mica to prevent mudcracking. It is easy to apply with brush or roller, yet films retain whatever texture profile developed during application.

Material:	Pounds	Gallons
Water	460	55.2
Vancide TH	1	0.1
Darvan 7	8	0.8
Wetting Agent (1)	2	0.2
Coalescent (2)	6	0.8
Defoamer (3)	1	0.1
Calcined Clay (4)	150	6.8
Wollastonite (5)	75	3.1
Pyrax WA	425	18.2
Defoamer	1	0.1
Latex (6)	100	11.1
Rhodopol 23) Combined	10	0.8
Ethylene Glycol)	25	2.7

1) Triton CF-10 or equivalent
2) Texanol or equivalent
3) Nopco NDW or equivalent
4) Iceberg or equivalent
5) Nyad G or equivalent
6) Everflex GT or equivalent

Formulation Constants:
    Solids by weight, %: 57
    Solids by volume, %: 35
    Pigment volume concentration, %: 83
    Pigment/binder weight ratio: 12/1

Paint Preparation:
    Materials are added in the order listed in the formula using suitable dispersion equipment. RHODOPOL 23 is wetted with the ethylene glycol and added all at once under vigorous agitation to avoid the large viscosity build-up until the end of paint preparation.

SOURCE: R.T. Vanderbilt Co., Inc.: Development Formulation No. TX-305A

Interior Texture Paint

Material:	Pounds	Gallons
Water	363	43.6
Van-Gel	6	0.3
Thickener (1)	6	0.5
Vancide TH	1	0.1
Dispersant (2)	15	1.5
Wetting agent (3)	1	0.1
Coalescent (4)	4	0.5
Ethylene glycol	25	2.7
Defoamer (5)	1	0.1
Titanium dioxide (6)	25	0.8
Vansil W-10	500	20.7
Pyrax WA	325	13.9
Defoamer	1	0.1
Vinyl latex (7)	136	15.1

1) Natrosol 250HR or equivalent
2) Tamol 731 or equivalent
3) Triton CF-10 or equivalent
4) Texanol or equivalent
5) Nopco NDW or equivalent
6) Ti-Pure R-931 or equivalent
7) Everflex GT or equivalent

Formulation Constants:
    Solids by weight, %: 67
    Solids by volume, %: 44
    Pigment volume concentration, %: 82
    Pigment/binder weight ratio: 11.3/1

This formulation exhibits a smooth texture finish with sharp peaking.

Paint Preparation:
    Materials are added in the order listed in the formula using suitable dispersion equipment. VAN-GEL is added to water using slow agitation followed by high-shear mixing for five minutes. Thickener and remaining ingredients are added in the usual manner. A portion of PYRAX WA, if necessary, may be added after the latex to maintain a desirable working viscosity.

SOURCE: R.T. Vanderbilt Co., Inc.: Development Formulation No. TX-302

Interior Texture Paint

Material:	Pounds	Gallons
Water	277	33.3
Vancide TH	1	0.1
Dispersant (1)	15	1.5
Ethylene glycol	25	2.7
Defoamer (2)	1	0.1
Titanium dioxide (3)	50	1.6
Vansil W-10	600	24.8
Pyrax WA	300	12.9
Vinyl-acrylic latex (4)	200	22.2
Defoamer	1	0.1
Hydroxyethyl cellulose (5)) Premix	2	0.2
Coalescent)	4	0.5

1) Tamol 731 or equivalent
2) Nopco NDW or equivalent
3) Ti-Pure R-931 or equivalent
4) Everflex GT or equivalent
5) Cellosize QP-4400 or equivalent
6) Texanol or equivalent

Formulation Constants:
    Solids by weight, %: 72
    Solids by volume, %: 51
    Pigment volume concentration, %: 78
    Pigment/binder weight ratio: 8.6/1

This formulation exhibits a smooth texture finish. Sharper peaking and coarser appearance may be obtained by adding 10 pounds MiniFibers Grade 13 038 to 100 gallons paint.

Paint Preparation:
    Materials are added in the order listed in the formula using suitable dispersion equipment. PYRAX WA disperses easily and, if necessary, a portion (e.g., 100 pounds) may be added after the latex to maintain a desirable working viscosity. Agitation should be continued for 10 minutes after addition of the premixed hydroxyethyl cellulose/coalescent slurry.

SOURCE: R.T. Vanderbilt Co., Inc.: Development Formulation No. TX-301

White Textured Coating

Material:	Lbs.	Gals.
Water	50.00	6.00
Ammonium Hydroxide	1.90	0.25
Bentone EW	5.00	0.24

Cowles grind the above three materials 15-20 minutes, then add
    the following under good agitation:

Arolon 850-W-45	476.30	55.00
Surfynol GA	4.40	0.50
Foamaster O	1.90	0.25
Titanox 2101	150.00	4.29

Pebble mill or sand mill grind the above materials to a 7 Hegman.

Letdown:		
Arolon 850-W-45	114.70	13.25
Butoxy Ethanol	80.70	10.75
Water	50.00	6.00
Texanol	2.00	0.25
Surfynol 104H	4.00	0.50
Michemlube 155	12.50	1.50
Cab-O-Sil M5	10.00	0.55
Acrysol ASE-60	4.20	0.50
Ammonium Hydroxide	1.90	0.25

Optional: A polyfunctional aziridine crosslinking agent, such as
          XAMA-7, can be added to this formula at a level of
          9 to 10 lbs/100 gallons to maximize all physical prop-
          erties of the enamel.

Analysis:
    Pigment Volume Concentration, Percent: 11.9
    Pigment/Binder Ratio: 0.6/1.0
    Percent Solids, Weight: 45.7
    Percent Solids, Volume: 35.9
    Viscosity @ 25C, KU: 100-105
    Pounds/Gallon: 9.7
    pH: 8.0-8.5
    VOC (excluding water):
        Grams/Liter: 208
        Pounds/Gallon: 1.7

SOURCE: Reichhold Chemicals, Inc.: Waterborne Handbook:
        Formulation 2154-23

Section XII
Miscellaneous

Ball Mill Paste-I

Mapico Yellow 1000	58%
Polywet ND-2 (25%)	8%
Nopco 1907	0.05%
Water (variable to 100%)	33.95%

Ball Mill Paste-II

Mapico Yellow 3100	45%
Polywet ND-2 (25%)	8%
Nopco KFS	0.05%
Methocel HG-400	0.45%
Water (variable to 100%)	46.50%

Ball Mill Paste-III

Mapico Red 297	65%
Polywet ND-2 (25%)	6%
Nopco KFS	0.05%
Water (Variable to 100%)	28.95%

Ball Mill Paste-IV

Mapico Brown 422	55%
Polywet ND-2 (25%)	8%
Nopco KFS	0.05%
Water (variable to 100%)	36.95%

Formulae I and II are used to develop optimum pigment loading for yellow iron oxides; Formula III is used to develop maximum pigment loading for red, roasted copperas iron oxides; Formula IV for brown and black iron oxies.

SOURCE: Uniroyal Chemical Co., Inc.: POLYWET ND-2 for Color Concentrates: Formulas

Clear Coating and Concrete Curing Membrane

Ingredient:	lbs/100 Gal
Santicizer 160	23.80
Butyl Carbitol	29.80
AMP 95	9.00
Surfynol 104	5.80
Fluorad FC-431	0.13
Fluorad FC-129	0.13
Dehydran 1293	3.00

Mix the above ingredients and add to the following with mixing:

Water	50.00
Pliolite 7104 (44% Solids)	325.80

After mixing well, add the following ingredients in order:

Water	385.50
Ethylene Glycol	9.50

    Solids Weight Percent: 21.2
    Solids Volume Percent: 20.2
    Pounds Per Gallon: 8.4
    VOC (gm/l): 226.5
    pH: 10.6
    Freeze/Thaw, 5 Cycles: Pass
    Viscosity Stability, 30 days, 120F: Pass

    Age panel per ASTM C-309
    Meets MVTR Requirements of ASTM C-309

    Formula #1910-102-6

Concrete Sealer/CCM (ASTM C 309 Compliant)

Propyl Cellosolve	4.8 lbs
Surfynol 104H	2.5
FC-129	0.1
Water	478
Butyl Dipropasol	11.6
Santicizer 160	22
28% Ammonium Hydroxide	3
Mix well, then add under agitation:	
Pliolite 7104	319

    1. Pass ASTM C 309
    2. Pass oven stability 120 deg F, 4 weeks
    3. NVV 18%
    4. Protect from Freezing
    Note: If 5 cycles freeze thaw stability is needed, substitute
          38 lbs of water with 35 lbs of Methyl Propasol.

SOURCE: The Goodyear Tire & Rubber Co.: Formula 2074-40-5

Corrosion Inhibitor Comparison in an Air Dry Waterborne Alkyd System

Grind (Pebble Mill):	% by Weight
Cargill 7474	12.67
Ti-Pure R902	20.00
Sec-Butanol	0.50
N-Butanol	0.54
2-Butoxyethanol	1.20
28 Ammonia	0.28
Water	4.30

Letdown:	
Cargill 7474	14.10
28 Ammonia	0.90
n-Butanol	1.10
2-Butoxyethanol	0.90
Cobalt Hydrocure	0.58
Activ 8	0.10
Water	42.83

Corrosion Inhibitor   1.0% active on total weight

Characteristics:
    Total Solids: 40.0%
    Pigment/Binder: 1/1
    % Pigment: 20.0
    % Resin Solids: 20.0
    pH: 8.5

Film Properties:
    Substrate: Bonderite 1000 Iron Phosphated CRS
    Cure: 7 Days at Room Temperature
    Film Thickness: 0.8-1.0 mils

NACORR 1151:
    Pencil Hardness: HB   Tukon Hardness: 7.46
    MEK Rubs (2X): 9   Gloss: 90.9
NACORR 1351:
    Pencil Hardness: HB   Tukon Hardness: 8.88
    MEK Rubs (2X): 13   Gloss: 91.8
NACORR 1651:
    Pencil Hardness: H   Tukon Hardness: 9.28
    MEK Rubs (2X): 15   Gloss: 92.6
NA-SUL 412:
    Pencil Hardness: B   Tukon Hardness: 5.22
    MEK Rubs (2X): 7   Gloss: 92.9
NA-SUL MPD:
    Pencil Hardness: 2B   Tukon Hardness: 5.85
    MEK Rubs (2X): 10   Gloss 60: 93.4

SOURCE: King Industries: Formulation WR-22

Corrosion Inhibitor Comparison in a Waterborne Polyester/ Melamine Bake System

In this study, the NACORR line of liquid corrosion inhibitors were compared to several commercially available competitive products. The coating used was a pigmented waterborne polyester/ melamine enamel. All of the inhibitors were post added at a rate of 1% active ingredient on the total weight of the coating. Salt Spray testing was carried out over both iron phosphated panels (B1000) and zinc phosphated panels (B37).

Grind (Pebble Mill):	% by Weight
Cargill 7203	18.00
Ti-Pure R960	24.00
2-Butoxyethanol	4.00
DMEA	0.69
Water	1.31

Letdown:	
Cargill 7203	6.00
DMEA	1.00
Cymel 325	7.50
10% Silwet 7605 in 2-Butoxyethanol	1.00
Water	36.50

Characteristics:
    Total Solids: 48.0%
    Pigment/Binder: 1/1
    % Pigment: 24.0
    % Resin Solids: 24.0
    Polyester/Melamine: 75/25
    pH: 8.55

Film Properties:
    Substrate: Bonderite 1000 Iron Phosphated CRS
    Cure: 15 minutes at 300F (150C)
    Film Thickness: 0.9-1.1 mils

NACORR 1151:
    Pencil Hardness: 2H-3H   Tukon Hardness: 25.6
    MEK Rubs (2X): 100   Gloss 60: 94.0
NACORR 1351:
    Pencil Hardness: 2H-3H   Tukon Hardness: 25.9
    MEK Rubs (2X): 100   Gloss 60: 94.7
NACORR 1552:
    Pencil Hardness: 2H-3H   Tukon Hardness: 24.7
    MEK Rubs (2X): 100   Gloss 60: 92.9
NA-SUL SRC-X:
    Pencil Hardness: 2H-3H   Tukon Hardness: 15.6
    MEK Rubs (2X): 100   Gloss 60: 87.9

SOURCE: King Industries: Formulation WR-23

Gloss White Tile-Like Formulation XG-105-2
Based on MAINCOTE TL-5 Polymer...
For Brush and Roller Applications

The choice and amount of each ingredient in Formulation XG-105-2 are significant to the performance of the coating. Substitutions should be evaluated carefully.

Materials:	Pounds	Gallons
Acrylic Component A:		
Methyl Carbitol	45.0	5.26
Tamol 731	10.8	1.18
Triton CF-10	2.0	0.22
Patcote 519	0.5	0.07
TiPure R-900 HG	225.0	6.48

Grind the above using a high speed Cowles dissolver for 25 minutes, then letdown at a lower speed as follows:

Water		42.0	5.04
Ethylene Glycol		18.0	1.94
Maincote TL-5 (41.5%)		517.9	60.25
Dalpad A		21.6	2.35
Texanol		10.8	1.37
Colloid 643		1.8	0.23
Sodium Nitrite (15% in water)		6.6	0.79
NH4OH (14% NH3)		4.5	0.59
Water	)	30.4	3.65
NH4OH (28% NH3)	) Premix	2.0	0.27
Acrysol RM-5 (30%)	)	21.6	2.46
Water		16.3	1.96

Epoxy Component B:		
Genepoxy 370 H-55	53.1	5.89

Formulation Constants:
    PVC, %: 19.4
    Volume Solids, %: 33.3
    Weight Solids, %: 46.2
    CARB (grams VOM/liter-water): 247
    pH: 8.3-8.5

Typical Viscosities, KU/ICI:
Component A:
    Initial: 93/1.1
    Equilibrated: 100/1.1
Components A + B:
    Initial: 95/1.1
    6 hours: 95/1.0

SOURCE: Rohm and Haas Co.: Maintenance and Marine Coatings: MAINCOTE TL-5: Formulation XG-105-2

Gloss White Tile-Like Formulation XG-105-3
Based on MAINCOTE TL-5 Polymer...
For Spray Applications

The choice and amount of each ingredient in this formulation are significant to the performance of the coating. Substitutions should be evaluated carefully.

Materials:	Pounds	Gallons
Acrylic Component A:		
Methyl Carbitol	46.7	5.45
Tamol 731	11.2	1.22
Triton CF-10	2.0	0.22
Patcote 519	0.5	0.07
Ti-Pure R-900 HG	233.4	6.72

Grind the above using a high speed Cowles dissolver for 25 minutes, then letdown at a lower speed as follows:

Water	84.0	10.08
Maincote TL-5 (41.5%)	537.2	62.48
Dalpad A	22.4	2.44
Texanol	11.2	1.41
Experimental Rheology Modifier QR-708 (35%)	1.9	0.21
Sodium Nitrite (15% in water)	6.6	0.79
NH4OH (14% NH3) to adjust to pH 8.3-8.8	5.0	0.64
Water	18.0	2.16

Epoxy Component B:		
Genepoxy 370 H-55	55.1	6.11

Formulation Constants:
    PVC, %: 19.4
    Volume Solids, %: 34.6
    Weight solids, %: 47.0
    CARB (grams VOM/liter-water): 214

Spray Information:
    Pressure Pot Spray Gun (DeVilbiss MBC):
    Fluid tip No.: AV-15FX
    Nozzle cap No.: 704
    Pot pressure: 28 psi
    Recommended film thickness: 5 to 7 wet mils
                                (1.7 to 2.4 dry mil)

Typical Viscosities, KU/ICI:
Component A:
    Initial 82/0.4
    Equilibrated 94/0.4
Component A + B:
    Initial: 93/0.4
    6 hours: 93/0.4

SOURCE: Rohm and Haas Co.: Maintenance and Marine Coatings:
        MAINCOTE TL-5: Formulation XG-105-3

Gloss White Tile-Like Formulation XG-105-4 Based on
Maincote TL-5 Polymer
For Brush and Roller Applications

Materials:	Pounds	Gallons
Acrylic Component A:		
Methyl Carbitol	44.0	5.14
Tamol 731	10.6	1.15
Triton CF-10	2.0	0.22
Patcote 519	0.5	0.07
Ti-Pure R-900 HG	220.0	6.34

Grind the above using a high speed Cowles dissolver for 25
minutes, then letdown at a lower speed as follows:

	Pounds	Gallons
Water	84.9	10.19
Experimental Rheology Modifier QR-1001 (20.0%)	23.5	2.64
Maincote TL-5 (41.5%)	506.4	58.91
Ethylene Glycol	18.0	1.94
Dalpad A	21.1	2.30
Texanol	10.6	1.34
Colloid 643	1.8	0.23
Sodium Nitrite (15% in water)	6.6	0.79
NH4OH (14% NH3) (to adjust to pH 8.3-8.8)	4.5	0.59
Water	20.0	2.40

	Pounds	Gallons
Epoxy Component B:		
Genepoxy 370 H-55	51.9	5.75

Formulation Constants:
    PVC, %: 19.4
    Volume Solids, %: 32.6
    Weight Solids, %: 44.7
    CARB (grams VOM/liter-water): 254

Typical Viscosities (KU/ICI):
Component A:
    Initial 90/1.7
    Equilibrated 97/1.8
Components A&B:
    Initial 96/1.7
    6 Hours: 96/1.7

SOURCE: Rohm and Haas Co.: Maintenance and Marine Coatings:
        MAINCOTE TL-5: Formulation XG-105-4

Gray EMI Shielding Coating

Material:	Lbs.	Gals.
Water	168.7	20.25
Ammonium Hydroxide	1.9	0.25
Bentone EW	5.0	0.24
Cellosize QP-4400	6.2	0.54

Cowles grind the above four materials for 20-25 minutes, then add the following in order under good agitation.

Arolon 850-W-45	510.9	59.00
Foamaster O	1.9	0.25
Butoxy Ethanol	58.2	7.75
Dibutyl Phthalate	11.1	1.25
Nickel HCT	700.0	9.80
Surfynol 104H	4.0	0.50
Ammonium Hydroxide	5.6	0.75

Analysis:
    Pigment Volume Concentration, Percent: 26.3
    Pigment/Binder Ratio: 3/1
    Percent Solids, Weight: 64.8
    Percent Solids, Volume: 37.2
    Viscosity @ 25C, KU: 80-90
    Pounds/Gallon: 14.65
    pH: 8.2-8.8
    VOC (excluding water):
        Grams/Liter: 133
        Pounds/Gallon: 1.1

SOURCE: Reichhold Chemicals, Inc.: Waterborne Handbook:
        Formulation 2154-13 Revised

Green, High pH Formulation

This formulation represents an all-purpose formulation for farm implement or general maintenance type coatings.

	Parts Per Hundred	
Materials:	Weight Ratio*(Volume Basis)	
Pebble Mill Grind: to 7-1/2 N.S.		
Chrome Oxide Green X-1134	37.3	0.87
Imperial Brazil Yellow X-2866	4.9	0.42
Phthalo Blue BT-417D	0.7	0.05
Acryloid WR-748 (60%)	115.1	13.83
Water	310.7	37.27
NH4OH** (28%)	6.5	0.78

Adjust pH to 8.8+-0.1 with ammonim hydroxide, if necessary.

Letdown:		
Above Grind	475.2	53.22
Acryloid WR-748 (60%)	99.3	11.93
Butyl Cellosolve	34.2	4.56
Water	240.9	28.89
NH4OH** (28%)	7.6	0.90

Mix well, then add:		
Cobalt Naphthenate	2.1	0.26
Manganese Naphthenate	0.4	0.05
Activ 8	1.5	0.19

Adjust pH to 8.8+-0.1 with 28% ammonium hydroxide**; if necessary
Reduce to spray viscosity of 55+-5" #4 Ford Cup with 80/20 by
  volume water/cosolvent.

Formulation Constants:
  Pigment/Binder: 25/75
  Water/cosolvent ratio (by volume): 80/20
  Drier level: 0.10% Co/0.02% Mn/1.2% Activ 8
  % Solids: 19-21
  VOC: lb./gal. less water: 3.83
      gms./liter less water: 459

  ** DMAE can be substituted for ammonia if a baking finish is
     desired. Solids in such a system increase to 20-22%.
   * Using weight ratio in pound units will yield approximately
     100 gallons of paint, while with kilograms, 833 liters will
     result.

SOURCE: Rohm and Haas Co.: Industrial Coatings: ACRYLOID WR-748:
        Formulation WR-748-6

Green, Low pH, High Solids Formulation

This formulation is a low pH, higher solids version of previous formulation. It should be used by those who can exercise relatively tight pH control.

Materials:	Weight Ratio*	Parts Per Hundred (Volume Basis)
Pebble Mill Grind to 7-1/2 N.S.:		
Chrome Oxide Green X-1134	52.3	1.22
Imperial Brazil Yellow X-2866	6.9	0.59
Phthalo Blue BT-417D	0.9	0.07
Acryloid WR-748 (60%)	161.3	19.39
Water	435.4	52.22
NH4OH (28%)**	9.2	1.10

Adjust pH to 8.8+-0.1 with ammonium hydoxide, if necessary.

Letdown:		
Above Grind	666.0	74.59
Acryloid WR-748 (60%)	139.2	16.72
Water	66.6	7.99

Mix well, then add:		
Cobalt Naphthenate	3.0	0.37
Manganese Naphthenate	0.7	0.07
Activ 8	2.2	0.27

Adjust pH to 8.0+-1.0 with 28% ammonium hydroxide**, if
   necessary. Reduce to spray viscosity with water to 45+-5"
   #4 Ford Cup.

Formulation Constants:
   Pigment/Binder: 25/75
   Water/cosolvent ratio (by volume): 80/20
   Drier level: 0.1%
   % Solids: 27-28
   VOC: lb./gal. less water: 3.24
        gms./liter less water: 388

   ** DMAE can be substituted for ammonia if a baking finish
      is desired. Solids in such a system increase to 28-30%.
    * Using weight ratio in pound units will yield approximately
      100 gallons of paint, while with kilograms, 833 liters will
      result.

SOURCE: Rohm and Haas Co.: Industrial Coatings: ACRYLOID WR-748:
        Formulation WR-748-7

High Performance Traffic Paint

	Lb/100 gallons
Isopropyl Alcohol	73
Methyl Propasol	21
Colloid 226	5
Triethylamine	6
Santicizer 160	23
Butyl Dipropasol	24
Water	29
Nopco NDW	4

Mix well; then add:

Pliolite 7103	312

Mix well; then add:

Titanox 2160	130
Mica 325	21
Atomite	260
Minusil 30	208

HSD to 3 N.S. Then add:

Pliolite 7103	104

Lb/gallon: 12.20
VOC: 1.83 lb/gal=220 g/l
PVC: 52%
NVV: 50%
NVM: 68%
No Pick Up: Pass 5 minutes @ 75F and 58% R.H.
            Pass 5 minutes @ 36F and 75% R.H.
Cold Temperature Coalescence: Pass 36F and 75% R.H.
Oven Stability: Pass 1 week 120F
Freeze Thaw Stability: Pass 5 cycles.

SOURCE: The Goodyear Tire & Rubber Co.: Formula 2074-55-1

Red Semi-Gloss Traffic Paint

	Pounds	Gallons
Cargill Water Reducible Short TOFA Alkyd		
74-7461	137.7	16.20
M-P-A 1075	4.4	0.64
Aqueous Ammonia 28%	11.6	1.55
Byk-301	1.2	0.15
Byk-020	1.3	0.18
Toluidine Red RD-0027	30.4	2.44
X-2806 Red Orange	10.1	0.22
Mapico Red 516 Medium	4.5	0.12
Atomite	252.4	11.18
Deionized Water	153.5	18.42

Sand Mill to a 6 Hegman

Cargill Water Reducible Chain Stopped		
Alkyd 74-7487	120.5	13.60
Aqueous Ammonia 28%	6.1	0.81
Patcote 577	1.3	0.19
Butyl Carbitol	3.6	0.45

Premix the following, then add:

Ethylene Glycol Monopropyl Ether	19.4	2.56
12% Zirconium Hydro Chem	5.9	0.71
Activ-8	1.3	0.16
Deionized Water	250.6	30.08
Exkin No. 2	2.6	0.34

Paint Properties:
    % Nonvolatile (By Weight): 48.83
                (By Volume): 35.21
    Pigment to Binder Ratio: 1.50
    Pigment Volume Concentration: 39.89
    Weight Per Gallon: 10.18
    Theoretical VOC (Pounds/Gallon): 2.09
                (Grams/Liter): 250
    Water:Cosolvent Weight Ratio: 80.00:20.00
    pH: 8.72
    Viscosity Krebs-Stormer (KU): 75

Typical Cured Film Properties:
    Cure Schedule: 7 Days Air Dry
    Substrate: Cold Rolled Steel
    Gloss (60/20): 51/11
    Pencil Hardness: Pass 7B
    % Crosshatch Adhesion: 100
    Impact (In. Lbs.): Direct: 150
                Reverse: 50

SOURCE: Cargill, Inc.: CARGILL Formulary: Formula P1887-36E

Rhoplex WL-81/Epotuf 38-690 (Epoxy Ester) Blend System

	Parts per Hundred	
Materials:	Weight Ratio*	Volume Basis
Epotuf 38-690	26.7	3.18
Aquacat	1.0	0.12
Magnacat	1.0	0.12
Ethylene glycol monobutyl ether	13.5	1.80
Triethylamine	2.3	0.39
Water	125.4	15.05
Tamol 165	2.6	0.30
Patcote 550	0.5	0.07
Raven 2000	32.8	2.25

Ball mill to 7 1/2+ N.S., add the following in order:

Epotuf 38-690	144.5	17.21
Triethylamine	11.7	1.93
Water	125.4	15.05

Mix well (recheck grind) and add the following in order with
  good agitation:

Rhoplex WL-81	279.1	32.84
Triton X-405 (35% NV)	4.7	0.56
Santicizer 160	23.0	2.48
Water	31.4	3.78
HEI-SCORE-XAB (10%)	23.3	2.91

Formulation Constants:
  Pounds/Gallon: 8.5
  pH: 8.9
  Viscosity, cps.: 330
  Grind: 7 1/2+
  Solids % (wt.): 34.5
         (vol.): 31.3
  P/B: 11/89
  Epotuf 38-690/Rhoplex WL-81: 51/49
  VOC (lbs./gal.): 1.52
  % on Polymer Solids:
    Ethylene glycol monobutyl ether: 27.5
    Santicizer 160: 9.7

* Using weight ratio in pound units will yield approximately
  100 gallons of lacquer while with kilograms, 833 liters
  will result.

SOURCE: Rohm and Haas Co.: Industrial Coatings: RHOPLEX WL-81:
        Formulation WL-81-5

Rhoplex WL-81/Epotuf 38-690 (Epoxy Ester) Blend System
(Chromate Containing)

	Parts per Hundred	
Materials:	Weight Ratio*(Volume Basis)	
Epotuf 38-690	26.6	3.17
Aquacat	1.0	0.12
Magnacat	1.0	0.12
Ethylene glycol monobutyl ether	13.5	1.79

Mix well, then add in order:

Triethylamine	2.3	0.39
Water	125.1	15.01
Tamol 165	2.6	0.30
Patcote 550	0.5	0.07
Raven 2000	32.7	2.25
Oncor F-31	9.7	0.29

Ball mill to 7 1/2 N.S., add the following in order:
Epotuf 38-690	144.1	17.16
Triethylamine	11.7	1.93
Water	125.1	15.01

Mix well (recheck grind) and add the following in order with
   good agitation:
Rhoplex WL-81	278.2	32.74
Triton X-405 (35% NV)	4.6	0.56
Santicizer 160	22.9	2.47
Water	31.3	3.77
HEI-SCORE-XAB (10%)	23.2	2.90

Formulation Constants:
   Pounds/Gallon: 8.6
   pH: 8.9
   Viscosity, cps.: 290
   Grind: 7 1/2
   Solids % (wt.): 35.3
         (vol.): 31.5
   Epotuf 38-690/Rhoplex WL-81: 51/49
   VOC (lbs./gal.): 1.52
   % on Polymer Solids:
      Ethylene glycol monobutyl ether: 27.5
      Santicizer 160: 9.7

   * Using weight ratio in pound units will yield approximately
     100 gallons of lacquer while with kilograms, 833 liters will
     result.
   ** Strontium Chromate can be used in place of the Oncor F-31.

SOURCE: Rohm and Haas Co.: Industrial Coatings: RHOPLEX WL-81:
        Formulation WL-81-4

Traffic Marking Paint

Material:	Lbs.	Gals.
Synthemul 97-799	405.42	46.60
Nopcosperse 44	7.0	0.70
Colloid 643	1.0	0.14

Premix the following, then add:

Methyl Carbitol	22.9	2.69
Natrosol 250MR	1.0	0.09
Titanox 2020	200.0	5.81
Duramite	401.4	17.84
Silica #19	137.9	6.25

High speed disperser, 15 minutes maximum.

Water	0.6	0.08
28% Ammonia	1.0	0.12
Igepal CO-830	2.0	0.26
Colloid 643	3.0	0.40
Synthemul 97-799	147.9	17.00
Methanol	10.0	1.51
Polyphase AF-1	5.0	0.51

Analysis:
    Pigment Volume Concentration, Percent: 49.5
    Percent Solids, Weight: 75.5
    Percent Solids, Volume: 60.3
    Initial Viscosity, KU: 80.0
    Pounds/Gallon: 13.46
    pH: 8.9
    VOC (excluding water):
        Grams/Liter: 59
    Dry Time, 3 mils Wet, Min.: <15
        (77F Dry)
    Blister Resistance: Exc.
    Wet Adhesion (Alkyd): Exc.
    Early Glass (Wet) Adhesion: Exc.
    Contrast Ratio (3 Mils Wet): 0.990

SOURCE: Reichhold Chemicals, Inc.: Waterborne Handbook:
        Formulation 799-02

Wash Coat Based on Emulsion E-2774

This low solids wash coat is used as a partial sealer between stains to hold in colors and provide a more uniform surface for subsequent stains to be applied to. Premix 1 through 4 and add to Emulsion E-2774 (5) with agitation. Add Liquazinc AQ-90 (6) with agitation.

Materials:	Weight Pounds	Volume Gallons
1. Deefo 3000	0.4	0.1
2. Butyl Carbitol	6.5	0.8
3. Butyl Cellosolve	16.4	2.2
4. Water	10.9	1.3
5. Emulsion E-2774	170.1	19.7
6. Liquazinc AQ-90 (50%)	0.5	0.0
7. 14% Ammonia	6.0	0.7
8. Water	625.9	75.1

Formulation Constants:
  Approximate solids, % (wt/vol): 7.9/7.2
  pH: 7.0
  Viscosity, #2 Zahn, seconds: 15 seconds
  VOC*, g/liter: 27
        lbs/gallon: .23
  Coalescent, wt% Polymer Solids:
    Butyl Cellosolve: 25
    Butyl Carbitol: 10
  Freeze/Thaw Stability: Protect from Freezing
  Heat Age Stability: Satisfactory

  * California's South Coast Air Management District allows
    VOC's for wash coats with less than one pound of solids
    per gallon to be calculated on the basis of pounds of VOC
    per gallon plus water. Accordingly, that convention is
    used here.

SOURCE: Rohm and Haas Co.: Formulation WR-74-3

Water-Borne Air Dry Primer and Topcoat System
Primer Formulation

Materials:	Parts by weight
Resin WB-408 (65% NVM)	292.7
Ammonium hydroxide, 28%	15.7
Deionized water	356.0
Cobalt naphthenate, 6% (1)	2.9
Anti-skin agent (2)	0.9
Red iron oxide (3)	97.6
Calcium carbonate (4)	146.4
China clay (5)	29.3
Talc (6)	58.5

Properties:
    Non-volatile material (NVM), wt%: 52.2
    Viscosity, #4 Ford Cup, sec: 30-35
    pH: 8.0
    Dry times at 20C, 50% RH:
        set to touch, min: 20-30
        tack-free, min: 60
        dry hard, min: 120

Compounding procedure:
1. Neutralize resin with ammonium hydroxide and add 260 parts water.
2. Add drier and anti-skin agent.
3. Add fillers and pigment and stir into resin solution.
4. Charge mixture to a pebble mill and grind overnight.
5. Adjust to application viscosity with remaining water.

(1) Product of the Huls America Inc.
(2) Exkin #1
(3) R-4098
(4) Duramite
(5) ASP 400
(6) Nytal 300

SOURCE: Amoco Chemical Co.: Water-Borne Air Drying Alkyds
        Based on Amoco IPA and TMA: Formula Table 4

Water-Borne Air Dry Primer and Topcoat System
Topcoat Formulation

Materials:	Parts by weight
Resin WB-408 (65% NVM)	320.6
Ammonium hydroxide, 28%	27.4
Deionized water	348.9
Cobalt naphthenate, 6% (1)	5.4
Manganese naphthenate, 6% (1)	2.7
Drier accelerator (2)	1.4
Anti-skin agent (3)	1.4
Flow additive (4)	0.7
Titanium dioxide (5)	291.5

Properties:
    Non-volatile material (NVM), wt%: 50.0
    Viscosity, #4 Ford Cup, sec: 44
    pH: 9.3
    Dry times at 20C, 50% RH:
        set to touch, min: 30-45
        tack free, hr: 6.5-7.0
        dry hard: overnight

Compounding Procedure:
1. Neutralize resin with ammonium hydroxide and add 320 parts water.
2. Add driers, anti-skin agent and flow additive.
3. Add pigment and stir into resin solution.
4. Charge mixture to a pebble mill. Grind overnight to a Hegman Grind of 7+.
5. Adjust to application viscosity with remaining water.

(1) Product of Huls America Inc.
(2) Activ-8
(3) Exkin #1
(4) FC-430
(5) Ti-Pure R-900

SOURCE: Amoco Chemical Co.: Water-Borne Air Drying Alkyds Based on AMOCO IPA and TMA: Formula Table 5

Water Reducible Alkyd

Ingredient:	Lbs/100 Gals	Grams/Liter
Chempol 10-0091	150.0	180.0
Butyl Cellosolve	25.0	30.0
Water	200.0	240.0
TEA	4.5	5.4
Ammonium Hydroxide	4.0	4.8
Byk 301	1.0	1.2
Yellow Iron Oxide	75.0	90.0
Halox SZP-391	33.0	40.0
Snowflake	75.0	90.0
Beaverwhite 325	75.0	90.0
Korthix	1.0	1.2
Chempol 10-0091	100.0	120.0
Butyl Cellosolve	30.0	36.0
Water	150.0	180.0
TEA	3.5	4.2
Ammonium Hydroxide	4.0	4.8
Cobalt Hydrocure Drier	1.5	1.8
Calcium Hydrocure Drier	3.5	4.2
Zirconium Hydrocure Drier	2.0	2.4
Activ 8	1.0	1.2
N-Butanol	25.0	30.0
Water	32.8	39.4

PVC: 40.0%
Density: 9.4 lbs/gal (1193 g/L)
Weight Solids: 45.9%
Volume Solids: 32.0%
VOC: 2.94 lbs/gal (350 g/L)

SOURCE: Halox Pigments: HALOX SZP-391 in a Water Reducible Alkyd

Water-Reducible Automotive Under-Hood Coating Black

Ingredients:	Pounds	Gallons
Chempol 10-0453	71.79	8.47
Byk P-104S	1.58	0.20
Surfynol 104BC	2.51	0.33
Ammonia, 28%	5.02	0.67
Water	100.49	12.06
Raven #14 Powder	15.78	2.02

Premix first three ingredients and mix well. Add ammonia slowly and with good agitation. Mix until well incorporated. Add the water in the same manner and mix well. Load mixture into the ball mill followed by the black pigment.

Run in steel ball mill to a #7 Hegman. Adjust grind viscosity and/or wash mill with water from below.

Let down with:		
Chempol 10-0453	262.87	31.00
Ammonia, 28%	15.07	2.02
2-butoxyethanol	20.10	2.68
NuoCure CK-10	9.36	1.15
NuoCure Mn	7.80	1.05
Byk 301	293.97	35.29
Water	293.97	35.29
Water (hold to adjust viscosity)	25.00	3.00

Add ammonia to the resin slowly and with good agitation; mix well. Premix cosolvents, driers, and Byk 301. Add this mixture to the neutralized resin with agitation. Mix well, then add the water slowly with continued agitation. Add this mixture to the grind with agitation and mix well. Allow the paint to stand overnight.

Check the pH and adjust with ammonia if necessary. Use the "adjust viscosity" water to bring the viscosity of the coating to spec at room temperature.

Properties:
    Total solids, % by weight: 31.11
    Total solids, % by volume: 26.46
    Weight/gallon: 8.40 lbs/gal
    PVC: 3.80%
    Pigment/binder ratio: 0.06
    VOC - water: 2.91 lbs/gal
    Viscosity, #2 Zahn cup: 38"
    pH: 8.5
    Substrate: B-1000 CRS
    Cure: 1 week air dry
    Set-to-touch, mins.: 25
    Dust free, mins.: 30
    Tack free, mins.: 60-65
    Dry through, hours: 6-8 hours
    DFT: 1.1-1.3 mil

SOURCE: Freeman Polymers Division: CHEMPOL 10-0453-C Formulation

Water Reducible Epoxy: Polyol Modified-B

Grind:
Epoxy Ester P-2017-S-11   97.7
Triethylamine   7.8
K-FLEX 188 (75% in IPA)   14.1
Water   75.0
Titanox 2020   99.5
Byk 301   3.0

Letdown:
Cymel 303   31.8
Water   169.0

Enamel Characteristics:
   Epoxy/Polyol/Melamine: 60/10/30
   Enamel solids, wt %: 43.2
   TRS, Wt %: 21.2
   Pigment/Binder: 0.94/1
   VOC, Calc. (lbs/gal): 0.95

Film Properties:
   Substrate: Bonderite 1000 Iron Phosphated CRS
   Cure: 15' at 300F.
   Film Thickness: 0.7-0.8 mil

   Epoxy/Polyol/Melamine: 60/10/30
   Pencil Hardness: 2-3H
   MEK Rubs: 200+
   Impact Resistance, (in-lbs):
      Direct: 60-70
      Reverse: 5-10
   60 Gloss: 87
   Salt Spray, 150 hrs. mm Creep/Blister: 0-3 Scat/NA
   50C Water Soak, 500 hrs. 60 Gloss/Blister: 83/VLT 9
   QUV*, 60 Gloss:
     600 hrs: 55
    1000 hrs: 42

   * 8 hrs. UV 50C, 4 hrs. Humidity, 40C

Enamel Stability, 50C:
   No. 4 Ford Cup, Seconds at 25C:
   60/10/30:
      Initial: 26"
     1 Week: 236"
     2 Weeks: Soft Paste, Too Visc. to test

SOURCE: King Industries: Formulation WR-8B

Water Reducible Acrylic Enamel: Polyol Modified

WR-7 demonstrates the advantages of formulating with K-FLEX 188 polyol as a modifier in a water reducible acrylic coating. The addition of a suitable cosolvent to the polyol such as isopropanol will render it water reducible at typical modification levels of 10-20% on total resin solids. K-FLEX 188 improves solvent, impact, QUV and salt spray corrosion resistance properties.

Grind:
Titanox 2020	480
Acrysol WS-68	126.3
Water	149.4
Butyl Cellosolve	21.4
n-Butanol	20.0
DMEA	2.9

Pebble Mill Grind Overnight

Letdown:
Grind	326.4
Acrysol WS-68	412.3
Water	114.7
n-Butanol	6.9
Dow Corning No. 57 (30% in Butyl Cellsolve)	1.9
DMEA (10% in Water)	5.1
Cymel 303	88.14
K-Flex 188 (50% in IPA)	58.76

Enamel Characteristics:
    Total Solids, %: 48.3
    Pigment/Binder: 40/60
    Acrylic/Polyol/Melamine: 60/10/30
    Total Resin Solids, %: 29.0
    VOC, Calc. (lbs/gal): 1.38
    pH: 7.8
    Viscosity, Sec. (No. 4 Ford Cup): 19

Film Properties:
    Substrate: Bonderite 1000 Iron Phosphated CRS
    Cure: 30' at 275F.
    Film Thickness: 0.6-0.7 mil

Acrylic/Polyol/Melamine: 60/10/30
I. No Catalyst:
    Pencil Hardness: 2-3H
    MEK Rubs: 200+

II. Catalyst: 0.8% Nacure X49-110 on TRS:
    Pencil Hardness: 2-3H
    MEK Rubs: 200+

SOURCE: King Industries: Formulation WR-7B

White, High pH, General All-Purpose Ambient Cure System with an Optimum Balance for Key Property Development

Materials:	Parts per Hundred Weight Ratio*(Volume Basis)	
Roller Mill Grind: to >7-1/2 N.S.		
Titanium dioxide (Ti-Pure R-902)	101.9	2.98
Acryloid WR-748 (60%)	98.0	11.79
Butyl Cellosolve**	3.8	0.51
Letdown:		
Above Grind	203.7	15.28
Acryloid WR-748 (60%)	109.2	13.12
Butyl Cellosolve**	37.9	5.05
Water	207.8	24.91
NH4OH (28%)	16.3	1.96
Mix well then add***:		
Cobalt Naphthenate	2.1	0.26
Manganese Naphthenate	0.4	0.05
Activ 8 (as supplied)	1.5	0.19
Mix well then add:		
Water	326.5	39.15

   Adjust pH to 8.8+-0.1 with NH4OH if necessary. Reduce to spray viscosity of 55+-5" #4 Ford Cup with 80/20 by volume water/Butyl Cellosolve; add 0.1% Byk-301 on total enamel weight, if desired.

Formulation Constants:
   Pigment/Binder: 45/55
   Water/cosolvent ratio (by volume): 80/20
   Drier level: 0.10% Co/0.02% Mn/1.2% Activ 8
   % Solids: 25+-1
   VOC: lb./gal. less Water: 3.76
       gms./liter less water: 451

  ** Propasol P can be substituted for Butyl Cellosolve, and will
     result in a system with a slightly reduced tack-free time
     but with somewhat reduced gloss.
 *** For a system with better color and better projected exterior
     durability, a drier system of 0.15% Cobalt/0.15% Calcium
     (Naphthenate) is recommended. On the above formulation, in
     lieu of listed driers, add 1.5g Cobalt Naphthenate and 2.3g
     Calcium Naphthenate.
   * Using weight ratio in pound units will yield approximately
     100 gallons of paint, while with kilograms, 833 liters will
     result.

Source: Rohm and Haas Co.: Industrial Coatings: ACRYLOID WR-748:
        Formulation WR-748-1

White, Low pH, High Solids Formulation
Based on Acryloid WR-748 Resin (1)

Materials:	Weight Ratio*	Parts Per Hundred (Volume Basis)
Roller Mill Grind: 7-1/2 N.S.:		
Titanium dioxide (Ti-Pure R-902)	148.8	4.35
Acryloid WR-748 (60%)	143.3	17.22
Butyl Cellosolve**	5.6	0.75
Letdown:		
Above Grind	297.7	22.32
Acryloid WR-748 (60%)	159.6	19.17
Water	469.8	56.33
NH4OH (28%)	11.9	1.43
Mix well, then add***:		
Cobalt Naphthenate	3.0	0.39
Manganese Naphthenate	0.6	0.07
Activ 8	2.2	0.28

Adjust pH to 8.0+-0.1 NH4OH, if necessary. Reduce to spray viscosity of 45+-5" #4 Ford Cup with water if necessary; add 0.1% Byk-301 on total enamel weight, if desired.

Formulation Constants:
Pigment/Binder: 45/55
Water/cosolvent ratio:
    by volume: 67/33
Drier level: 0.10% Co/0.02% Mn/1.2% Activ 8
% Solids: 35
VOC: lb./gal. less water: 3.07
    gms./liter less water: 368

(1) This formulation represents Rohm and Haas' best recommend-ation for maximum solids from Acryloid WR-748. This form-ulation also offers fastest drying potential but should be used by those who can exercise relatively tight pH control.
** Propasol P can be substituted for Butyl Cellosolve, and will result in a system with a slightly reduced tack-free time but with somewhat reduced gloss.
*** For a system with better color and better projected exter-ior durability, a drier system of 0.15% Cobalt/1.5% Calcium (Naphthenate) is recommended. On the above formulation, in lieu of listed driers, add 1.5g Cobalt Naphthenate and 2.3g Calcium Naphthenate.
* Using weight ratio in pound units will yield approximately 100 gallons of paint, while with kilograms, 833 liters will result.

Source: Rohm and Haas Co.: Industrial Coatings: ACRYLOID WR-748: Formulation WR-748-2

White, Pebble Mill Formulation

Formulation provides same paint as formulation WR-748-1 but is designed for those who cannot use a roller mill.

	Parts Per Hundred	
Materials:	Weight Ratio*	(Volume Basis)
Pebble Mill Grind: to 7-1/2 N.S.:		
Titanium dioxide (Ti-Pure R-902)	101.7	2.96
Acryloid WR-748 (60%)	45.3	5.52
Water	119.4	14.26
NH4OH (28%)	4.9	0.59
Letdown:		
Above Grind	271.3	23.33
Acyloid WR-748 (60%)	162.0	19.48
Water	415.5	49.82
NH4OH (28%)	10.9	1.30
Butyl Cellosolve	41.6	5.54
Mix well, then add:		
Cobalt Naphthenate	2.1	0.27
Manganese Naphthenate	0.4	0.05
Activ 8	1.5	0.19

Adjust pH to 8.8+-0.1 with ammonium hydroxide, if necessary. Reduce to spray viscosity of 55+-5" #4 Ford Cup with 80/20 by volume water/Butyl Cellosolve; add 0.1% Byk-301 on total enamel weight, if desired.

Formulation Constants:
    Pigment/Binder: 45/55
    Water/cosolvent ratio (by volume): 80/20
    Drier level: 0.10% Co/0.02% Mn/1.2% Activ 8
    % Solids: 25+-1
    VOC: lb./gal. less water: 3.76
        gms./liter less water: 451

* Using weight ratio in pound units will yield approximately 100 gallons of paint, while with kilograms, 833 liters will result.

SOURCE: Rohm and Haas Co.: Industrial Coatings: ACRYLOID WR-748: Formulation WR-748-5

White, Water Grind Formulation, High pH

Formulations designed for those who need to maximize the volume of paint resulting from a unit grind. Some attrition in humidity-salt spray resistance can be expected.

Materials:	Weight Ratio*	Parts Per Hundred (Volume Basis)
Mix Thoroughly:		
Water	22.3	2.67
Tamol 165	6.9	0.75
Triton CF-10	0.9	0.10
While stirring on a Cowles, add in gradually:		
Titanium dioxide (RCL-9)	101.8	2.99
Grind co 7 1/2 N.S.		
Letdown:		
Above Grind	131.9	6.51
Acryloid WR-748 (60%)	207.6	24.95
Water	459.0	59.35
NH4OH (28%)	16.3	1.95
Butyl Cellosolve	50.5	6.72
Mix well, then add:		
Cobalt Naphthenate	2.1	0.26
Manganese Naphthenate	0.4	0.05
Activ 8 (as supplied)	1.5	0.19

Adjust pH to 8.8+-0.1 with NH4OH, reduce to spray viscosity of 55+-5", #4 Ford Cup with 80/20 water/cosolvent.

Formulation Constants:
   Pigment/Binder: 45/55
   Water/cosolvent ratio by volume: 80/20
   % Solids: 26.0
   VOC: lb./gal. less water: 3.57
        gms./liter less water: 428

* Using weight ratio in pound units will yield approximately 100 gallons of paint, while with kilograms, 833 liters will result.

SOURCE: Rohm and Haas Co.: Industrial Coatings: ACRYLOID WR-748: Formulation WR-748-3

White, Water Grind Formulation, Low pH

Formulations designed for those who need to maximize the volume of paint resulting from a unit grind. Some attrition in humidity-salt spray resistance can be expected.

Materials:	Parts Per Hundred Weight Ratio*(Volume Basis)	
Mix Thoroughly:		
Water	32.7	3.92
Tamol 165	10.1	1.09
Triton CF-10	1.3	0.15
While striiring on a Cowles, add in gradually:		
Titanium dioxide (RCL-9)	149.2	4.38
Grind to 7-1/2 N.S.		
Above Grind	193.3	9.54
Acryloid WR-748	304.2	36.56
Water	434.5	52.10
NH4OH (28%)	8.9	1.07
Mix well, then add:		
Cobalt Naphthenate	3.0	0.39
Manganese Naphthenate	0.6	0.06
Activ 8 (as supplied)	2.2	0.27

Adjust pH to 8.0+-0.1 with NH4OH, reduce to spray viscosity of 45+-5" #4 Ford Cup with water.

Formulation Constants:
   Pigment/Binder: 45/55
   Water/cosolvent ratio by volume: 71/29
   % Solids: 35.0
   VOC: lb./gal. less water: 3.19
        gms./liter less water: 382

   * Using weight ratio in pound units will yield approximately 100 gallons of paint, while with kilograms, 833 liters will result.

SOURCE: Rohm and Haas Co.: Industrial Coatings: ACRYLOID WR-748: Formulation WR-748-4

White, Water Reducible Acrylic/Melamine-A

K-FLEX UD-320W modification of this white water reducible acrylic/melamine system demonstrated several advantages including: improved flow/leveling; higher gloss; improved gloss retention on QUV exposure; higher solids (lower VOC) and lower cost per pound on a solids basis.

Grind:	Weight%
Joncryl 540	41.8
AMP-95	1.0
Surfynol 104	1.0
Ti-Pure R-960	48.6
Water	7.6

Grind to 6+ Hegman on the ball mill

Letdown:	
Grind	100.0
Joncryl 540	58.3
Cymel 303LF	19.9
Butyl Cellosolve	11.4
K-Flex UD-320W	----
Nacorr 1552	1.3
Water	43.0

Enamel Characteristics:
  Acryl/K-Flex/Mel: 70/0/30
  % Solids, 1 Hr./110C: 47.7
  Cost/pound (solids): $1.62
  Viscosity, cps (25C): 205
  pH: 8.70
  VOC, lbs./gal.: 1.604
  Pigment:Binder: 0.75:1.0
    * Excluding water, but including amines.

Application & Cure:
  Substrate: Bonderite 1000
  Cure Schedule: 15'/200F
  DFT (mils): 0.9+-0.01

Film Properties:
  Pencil Hardness: 4H-5H
  Knoop Hardness(25g): 17.2
  MEK Resistance (2X): 100
  Crosshatch Adhesion: 100%
  Impact Resistance:
    Direct (in-lbs): 100-110
    Reverse (in-lbs): 20-30
  Gloss, 60 Degree: 83.6
         20 Degree: 47.1

SOURCE: King Industries: Formulation UDW-12A

White, Water Reducible Acrylic/Melamine-B

Grind:	Weight%
Joncryl 540	41.8
AMP-95	1.0
Surfynol 104	1.0
Ti-Pure R-960	48.6
Water	7.6

Grind to 6+ Hegman on the ball mill

Letdown:	
Grind	100.0
Joncryl 540	50.9
Cymel 303LF	19.9
Butyl Cellosolve	----
K-Flex UD-320W	3.6
Nacorr 1552	1.3
Water	23.0

Enamel Characteristics:
    Acryl/K-Flex/Mel: 65/5/30
    % Solids, 1 Hr./110C: 56.6
    Cost/pound (solids): $1.56
    Viscosity, cps (25C): 200
    pH: 8.75
    VOC, lbs./gal.*: 0.551
    Pigment:Binder: 0.76:1.0
        * Excluding water, but including amines.

Application & Cure:
    Substrate: Bonderite 1000
    Cure Schedule: 15'/300F
    DFT (mils): 0.9+-0.01

Film Properties:
    Pencil Hardness: 4H-5H
    Knoop Hardness(25g): 17.0
    MEK Resistance (2X): 100
    Crosshatch Adhesion: 100%
    Impact Resistance:
        Direct (in-lbs): 100-110
        Reverse (in-lbs): 20-30
    Gloss, 60 Degree: 91.1
          20 Degree: 65.6
    Cleveland Humidity Resistance:
        500 hours: 9/M
        1200 hours: 3/MD

SOURCE: King Industries: Formulation UDW-12-B

White, Water Reducible Acrylic/Melamine-C

Grind:	Weight%
Joncryl 540	41.8
AMP-95	1.0
Surfynol 104	1.0
Ti-Pure R-960	48.6
Water	7.6

Grind to 6+ Hegman on the ball mill

Letdown:	
Grind	100.0
Joncryl 540	39.1
Cymel 303LF	29.1
K-Flex UD-320W	7.2
Nacorr 1552	1.3
Water	25.0

Enamel Characteristics:
    Acryl/K-Flex/Mel: 57/10/33
    % Solids, 1 Hr./110C: 57.9
    Cost/pound (solids): $1.57
    Viscosity, cps (25C): 200
    pH: 8.74
    VOC, lbs./gal.*: 0.546
    Pigment:Binder: 0.76:1.0
        * Excluding water, but including amines.

Application & Cure:
    Substrate: Bonderite 1000
    Cure Schedule: 15'/300F
    DFT (mils): 0.9+-0.01

Film Properties:
    Pencil Hardness: 4H-5H
    Knoop Hardness(25g): 18.4
    MEK Resistance (2X): 100
    Crosshatch Adhesion: 100%
    Impact Resistance:
        Direct (in-lbs): 100-110
        Reverse (in-lbs): 20-30
    Gloss, 60 Degree: 91.5
            20 Degree: 67.0
    Cleveland Humidity Resistance:
        500 hours: 9/MD
        1200 hours: 8/MD

SOURCE: King Industries: Formulation UDW-12C

White, Water Reducible Acrylic/Melamine-D

Grind:	Weight%
Joncryl 540	41.8
AMP-95	1.0
Surfynol 104	1.0
Ti-Pure R-960	48.6
Water	7.6

Grind to 6+ Hegman on the ball mill.

Letdown:	
Grind	100.0
Joncryl 540	28.7
Cymel 303LF	23.3
K-Flex UD-320W	10.9
Nacorr 1552	1.3
Water	21.0

Enamel Characteristics:
    Acryl/K-Flex-Mel: 50/15/35
    % Solids, 1 Hr./110C: 59.1
    Cost/pound (solids): $1.59
    Viscosity, cps (25C): 197
    pH: 8.70
    VOC, lbs./gal.*: 0.429
    Pigment:Binder: 0.74:1.0
        * Excluding water, but including amines.

Application & Cure:
    Substrate: Bonderite 1000
    Cure Schedule: 15'/300F
    DFT (mils): 0.9+-0.01

Film Properties:
    Pencil Hardness: 4H-5H
    Knoop Hardness (25g): 18.9
    MEK Resistance (2X): 100
    Crosshatch Adhesion: 100%
    Impact Resistance:
        Direct (in-lbs): 80-90
        Reverse (in-lbs): 5-10
    Gloss, 60 Degree: 92.6
           20 Degree: 73.0
    Cleveland Humidity Resistance:
        500 hours: 8/M
       1200 hours: 6/D

SOURCE: King Industries: Formulation UDW-12-D

Wiping Stain for Media Mills

Materials:	Weight Pounds	Volume Gallons
Premix and mill:		
Water	428.2	51.4
Triton CF-10	10.1	1.2
Dehydran 1620	0.6	0.1
Van Dyke Brown	101.0	8.1
Bentone LT	1.3	0.1

Add:

Aqueous Ammonia (28%) to adjust pH>8	20.2	2.6

Add to above uniform mixture:

Experimental emulsion E-2955	227.2	26.1

  (grind all of above with shot mill for at least 30 minutes
   or until pigment is sufficiently dispersed)

Premix and add slowly under agitation:		
Ethylene glycol	69.4	7.5
Butyl Carbitol	12.6	1.6
Butyl Cellosolve	6.3	0.8
Ektasolve EEH	3.8	0.5

Add 7% ammonia (adjust pH between 7.5 and 8.0)

Formulation Constants:
   Approximate solids, % (wt/vol): 22.3/18.3
   pH: 7.5-8.0
   VOC, g/liter: 373
      lb/gallon: 3.2
   Coalescent, wt % on polymer solids:
      Butyl Carbitol: 15
      Butyl Cellosolve: 7.5
      Ektasolve EEH: 4.5
   Edge control solvent, wt % on polymer solids ethylene
      glycol: 82
   Freeze/thaw stability: Protect from freezing
   Heat/age stability (140F/10 days): Passes

SOURCE: Rohm and Haas Co.: Experimental Emulsion E-2955:
        Formulation WPS 55-1

Yellow Implement Enamel

	Pounds	Gallons
Cargill Water Reducible Chain Stopped		
Alkyd 74-7474	80.0	9.10
Ethylene Glycol Monobutyl Ether	18.0	2.40
Aqueous Ammonia 28%	4.4	0.58
Titanium Dioxide Ti-Pure R-902	31.0	0.91
Synthetic Yellow Iron Oxide YLO-3288-D	57.1	1.70
Synthetic Red Iron Oxide RO-3097	2.1	0.05
Aerosil R972	1.7	0.09
Surfynol 104 E	1.5	0.19
Byk-020	1.5	0.21
Deionized Water	162.6	19.52

Sand Mill to a 7 Hegman

Cargill Water Reducible Chain Stopped		
Alkyd 74-7474	160.9	18.29
Ethylene Glycol Monobutyl Ether	10.3	1.38
Aqueous Ammonia 28%	11.8	1.57

Premix the following, then add:

Ethylene Glycol Monobutyl Ether	7.8	1.04
5% Cobalt Hydro Cure II	1.7	0.22
Manganese Hydro Cure II	1.3	0.15
Activ-8	0.9	0.11
sec-Butanol	7.3	1.08
Exkin No. 2	1.5	0.20
Deionized Water	343.4	41.21

Paint Properties:
    % Nonvolatile (By Weight): 30.40
                (By Volume): 22.00
    Pigment to Binder Ratio: 0.50
    Pigment Volume Concentration: 12.65
    Weight Per Gallon: 9.07
    Theoretical VOC (Pounds/Gallon): 3.00
                (Grams/Liter): 360
    Water:Cosolvent Weight Ratio: 82.00:18.00
    Viscosity #4 Ford Cup (Seconds): 47

Typical Cured Film Properties:
    Cure Schedule: 7 Days Air Dry
    Substrate: Cold Rolled Steel
    Gloss (60/20): 91/85
    Pencil Hardness: Pass 2B
    Impact (In. lbs.):
        Direct: 30
        Reverse: <5

SOURCE: Cargill, Inc.: CARGILL Formulary: Formula P1848-135C

Yellow Iron Oxide Paint

	Weight Ratio*	Parts Per Hundred (Volume Basis)
Roller Mill:		
Yellow Iron Oxide YLO-1888D	52.43	1.56
Acryloid WR-748 (60%)	55.85	6.93
Butyl Cellosolve	5.70	0.76
Letdown:		
Above Grind	113.98	9.25
Acryloid WR-748 (60%)	148.05	18.37
Water	539.14	64.65
28% NH4OH	11.40	1.37
Butyl Cellosolve	42.63	5.68
Byk 301	1.60	0.20
Cobalt Naphthenate	2.04	0.25
Manganese Naphthenate	0.41	0.05
Activ 8	1.47	0.19

Formulation Constants:
    Pigment/Binder: 30/70
    Water/cosolvent ratio (by volume): 80/20
    Drier level: 0.10% Co/0.02% Mn/1.2% Activ 8
    Solids %: 22+-1
    Spray viscosity (#4 Ford Cup): 55+-5"
    pH: 8.8+-0.1
    VOC: lbs./gal. less water: 3.87
         gms./lit. less water: 465

    * Using weight ratio in pound units will yield approximately
      100 gallons of paint, while with kilograms, 833 liters will
      result.

SOURCE: Rohm and Haas Co.: Industrial Coatings: ACRYLOID WR-748:
        Formula MT-748-3Y

Section XIII
Trade Named Raw Materials

The information given in this section is based on data supplied by the manufacturer, as available. Where information is incomplete, data was not available at time of publication.

RAW MATERIAL	CHEMICAL DESCRIPTION	SOURCE
A-187	Silane ester adhesion promoter	Union Carbide
A-4434	Tacoma blue	Ciba-Geigy
AC-1024	Acrylic emulsion	Rohm & Haas
Acryloid WR-97	Acrylic resin	Rohm & Haas
Acrysol ASE-60	Water-soluble acrylic resin	Rohm & Haas
Acrysol I-62	Water-soluble acrylic resin	Rohm & Haas
Acrysol I-98	Water-soluble acrylic resin	Rohm & Haas
Acrysol QR708	Water-soluble acrylic resin	Rohm & Haas
Acrysol RM-5	Water-soluble acrylic resin	Rohm & Haas
Acrysol RM-825	Water-soluble acrylic resin	Rohm & Haas
Acrysol RM-1020	Water-soluble acrylic resin	Rohm & Haas
Acrysol SCT-275	Water-soluble acrylic resin	Rohm & Haas
Acrysol TT-615	Water-soluble acrylic resin	Rohm & Haas
Acrysol TT-935	Water-soluble acrylic resin	Rohm & Haas
Acrysol WS-68	Water-soluble acrylic resin	Rohm & Haas
Activ-8	Drier, accelerator, stabilizer	Vanderbilt

RAW MATERIAL	CHEMICAL DESCRIPTION	SOURCE
Additol XW 395	Additive	Hoechst
Aerosil R-972	Fumed silica	Degussa
Aerosol OT	Surface active agent	Am Cyan
Aerosol OT-B	Surface active agent	Am Cyan
Aerosol OT-75	Surface active agent	Am Cyan
Aerosol TR-70	Surface active agent	Am Cyan
Airflex 738	Vinyl acetate-ethylene copolymer emulsion	Air Prod.
Alcolec 439-C	Water dispersible lecithin	Am Lec
Alcophor 827	Paint additive	Henkel
AMP-95	Dispersant	Angus
Anti-Sag WR300		Byk
Aquacat	Drier	Ultra
Aquamac 430		McWhort
Arcosolv PTB	Propylene glycol monotertiary butyl ether	Arco
Arcosolv DPM	Dipropylene glycol monoethyl ether/Arco	
Arcosolv PM	Propylene glycol monomethyl ether Arco	
Arcosolv PP	Coalescent	Arco
Arolon 465	Polyester. % Solids: 80	Reich-
Arolon 557	Acrylic polymer. % Solids: 70	Reich-
Arolon 559	Acrylic polymer. % Solids: 70	Reich-
Arolon 580	Oxidizing alkyd. % Solids: 42	Reich-
Arolon 585	Oxidizing alkyd. % Solids: 43	Reich-
Arolon 820	Acrylic polymer. % Solids: 49	Reich-
Arolon 840	Acrylic polymer. % Solids: 39	Reich-

RAW MATERIAL	CHEMICAL DESCRIPTION	SOURCE
Arolon 845	Acrylic polymer. % Solids: 45	Reich-
Arolon 850	Acrylic polymer. % Solids: 45	Reich-
Arolon 860	Acrylic polymer. % Solids: 45	Reich-
Arolon 921	Oxidizing alkyd. % Solids: 70	Reich-
Arolon 970	Oxidizing alkyd. % Solids: 70	Reich-
Aroplaz 1271	Long-oil linseed alkyd resin	NLChem
Aroplaz 1272	100% solids version of Aroplaz 1271	NLChem
ASA 35.3	Wax emulsion	Chem. Corp.
Asbestine 3X	Talc	IntTalc
ASP-NC2 Clay	Aluminum silicate extender	Engel-
ASP-170	Aluminum silicate extender	Engel-
ASP-400	Aluminum silicate extender	Engel-
Atomite	Calcium carbonate	ECC
Attagel 50	Attapulgite thickening agent	Engel-
Azo Rubine Ex Conc.		
Aurasperse W-7012	Carbon black	Harshaw
Balab 3046A	Defoamer	Witco
Balab 3056A	Defoamer	Witco
Barimite XF	Natural barites	Thomp-
Barytes W1430	Barytes	Pfizer
Barytes #1	Barytes	WC&D
Bayferrox 120NM	Synthetic iron oxide pigment	Miles-

RAW MATERIAL	CHEMICAL DESCRIPTION	SOURCE
Beaverwhite 325	Talc	Cyprus
Beckosol 13-402	Synthetic resin	Reich-
Beetle 65	Urea-formaldehyde resin	Am Cyan
Bentone EW	Rheological additive	Rheox
Bentone LT	Rheological additive	Rheox
Bentone 27	Rheological additive	Rheox
Bentone 38	Rheological additive	Rheox
Benzotriazole	Inhibitor	Sandoz
Black Tint 888-9907	Black tint	Huls
Blanc Fix		Sachtl-
Borchigen DFN	Nonionic surfactent	Borch-
Borchigen ND	Dispersant	Borch-
Brown Iron Oxide Disp.	Pigment dispersion	Huls/
Brown Iron Oxide 444	Brown	Columb-
Burgess No. 98	Hydrous clay	Burgess
Busan 11M-1	Microbiocide	Buckman
Busperse-39	Dispersing agent	Buckman
Butrol 22	Corrosion and scale inhibitor	Buckman
Butyl Carbitol	Diethylene glycol monobutyl ether	Union Carbide
Butyl Carbitol Acetate	Diethylene glycol monobutyl ether acetate	Union Carbide
Butyl Cellosolve	Ethylene glycol monobutyl ether acetate	Union Carbide
Butyl Dipropasol	Coalescent	Union Carbide

RAW MATERIAL	CHEMICAL DESCRIPTION	SOURCE
Butyl Propasol	Coalescent	Union Carbide
n-Butyl Propionate	Coalescent	Union Carbide
Byk 020	Defoamer	Byk-
Byk 035	Defoamer	Byk-
Byk 080	Additive	Byk-
Byk 104S	Additive	Byk-
Byk 156	Additive	Byk-
Byk 301	Additive	Byk-
Byk 306	Additive	Byk-
Byk 344	Mar aid	Byk-
Byk-Catalyst 451	Additive	Byk-
Byk P-104S	Additive	Byk-
Bykumen	Wetting/dispersing additive	Byk-
Byk VP020	Additive	Byk-
Byk VP-155	Additive	Byk-
Byk VP-321	Mar proofing and slip aid	Byk-
Cab-O-Sil M5	Fumed silica	Cabot
Calcium Hydro-Cem	6% Calcium drier catalyst	OMGroup
Calcium Hydrocure	Catalyst	OMGroup
Calco Stains	Stains	Am Cyan
Cal/Ink 8800 Series	Colorants	Huls
Camel-Carb	Calcium carbonate	Genstar

RAW MATERIAL	CHEMICAL DESCRIPTION	SOURCE
Carbitol	Solvent	Union Carbide
Carbitol Acetate	Solvent	Union Carbide
Cargill 17-7240	Water reducible acrylic	Cargill
Cargill 23-2317	High solids polymeric methylated melamine	Cargill
Cargill 23-2347	High solids monomeric methylated melamine	Cargill
Cargill 72-7203	Water reducible oil free poly-ester	Cargill
Cargill 72-7230	Waterborne polyester dispersion	Cargill
Cargill 72-7231	Waterborne polyester dispersion	Cargill
Cargill 72-7232	Waterborne polyester dispersion	Cargill
Cargill 72-7289	Water reducible oil-free poly-ester	Cargill
Cargill 73-7331	Water reducible epoxy ester	Cargill
Cargill 73-7390	Water reducible self crosslinking resin	Cargill
Cargill 74-7412	Water reducible forest products resin	Cargill
Cargill 74-7422	Water reducible styrenated alkyd copolymer	Cargill
Cargill 74-7425	Water reducible styrenated co-polymer alkyd	Cargill
Cargill 74-7432	Water reducible acrylic modified alkyd	Cargill
Cargill 74-7435	Water reducible silicone modified alkyd	Cargill
Cargill 74-7450	Water reducible short soya alkyd	Cargill

RAW MATERIALS	CHEMICAL DESCRIPTION	SOURCE
Cargill 74-7451	Water reducible short tofa alkyd	Cargill
Cargill 74-7455	Water reducible cioconut oil alkyd	Cargill
Cargill 74-7461	Water reducible short tofa alkyd	Cargill
Cargill 74-7470	Water reducible chain stopped alkyd	Cargill
Cargill 74-7472	Water reducible fatty acid alkyd	Cargill
Cargill 74-7474	Water reducible chain stopped alkyd	Cargill
Cargill 74-7476	Water reducible fatty acid alkyd	Cargill
Cargill 74-7478	Water reducible phenolic modified alkyd	Cargill
Cargill 74-7487	Water reducible chain stopped alkyd	Cargill
Cargill 74-7495	Water reducible chain stopped alkyd	Cargill
Cargill 108-1775	Waterborne polyester dispersion	Cargill
Cargill 5150		Cargill
Cargill 7203	Oil free polyester	Cargill
Cargill 7474	Waterborne alkyd resin	Cargill
Catalyst 600	Organic acid catalyst	Am Cyan
Celite 281	Diatamaceous silica	Celite
Celite 499	Diatomaceous silica	Celite
Cellosize ER-15M	Hydroxyethyl cellulose	Un Carb
Cellosize ER-30M	Hydroxyethyl cellulose	Un Carb
Cellosize ER-4400	Hydroxyethyl cellulose	Un Carb
Cellosize QP-40	Hydroxyethyl cellulose	Un Carb
Cellosize QP-4400	Hydroxyethyl cellulose	Un Carb

RAW MATERIAL	CHEMICAL DESCRIPTION	SOURCE
Cellosize QP-15,000	Hydroxyethyl cellulose	Un Carb
Cellosize QP-15,000H	Hydroxyethyl cellulose	Un Carb
Chemcor AS-35-3	Wax emulsion	ChemCor
Chempol 10-0091	Water reducible short oxidizing oil alkyd	Freeman
Chempol 10-0094	Water reducible short oxidizing oil alkyd	Freeman
Chempol 10-0095	Water reducible short oil alkyd	Freeman
Chempol 10-0097	Water reducible short oil alkyd	Freeman
Chempol 10-0173	Water reducible chain stopped alkyd	Freeman
Chempol 10-0453	Water reducible styrenated epoxy ester	Freeman
Chempol 10-0501	High performance polyester	Freeman
Chempol 10-0503	Saturated polyester	Freeman
Chempol 10-0509	Water reducible hydroxy functional thermosetting acrylic	Freeman
Chempol 10-1105	Water reducible coconut alkyd	Freeman
Chempol 10-1210	Water reducible short oil alkyd	Freeman
Chempol 10-1300	Water reducible acrylic alkyd copolymer	Freeman
Chempol 10-1313	Water reducible acrylic alkyd copolymer	Freeman
Chempol 10-1706	Water reducible acrylic copolymer	Freeman
Chempol 20-4301	Water-based acrylic dispersion	Freeman
Chroma-Chem 896-9901	Industrial colorant	Huls
Chrome Oxide X-1134	Pigment	Ciba-

RAW MATERIAL	CHEMICAL DESCRIPTION	SOURCE
Chrome Yellow X3356	Yellow	Ciba-
CMD J60-8290	Curing agent	Rhone-
CMD WJ60-8537	Curing agent	Rhone-
CMD 9012	Acrylic dispersion	Rhone-
CMD 9790	Acrylic dispersion	Rhone-
Cobalt Cyclodex-6%	Drier	Huls
Cobalt Hydrocure-5%	Drier, catalyst	OMGroup
Cobalt Hydrocure II	Drier, catalyst	OMGroup
Colanyl Colorants	Predispersed pigments	Hoechst
Colloid X-0137-MC-33	Antifoam	Rhone-
Colloid 111	Dispersant	Rhone-
Colloid 226	Dispersant	Rhone-
Colloid 226/35	Dispersant	Rhone-
Colloid 261	Dispersant	Rhone-
Colloid 270	Dispersant	Rhone-
Colloid 286	Dispersant	Rhone-
Colloid 600	Defoamer	Rhone-
Colloid 640	Defoamer	Rhone-
Colloid 643	Defoamer	Rhone-
Colloid 650	Antifoam	Rhone-
Colloid 653	Antifoam	Rhone-
Colloid 677	Antifoam	Rhone-
Colloid 679	Antifoam	Rhone-
Colloid 681-F	Antifoam	Rhone-

RAW MATERIAL	CHEMICAL DESCRIPTION	SOURCE
Colloid 694	Defoamer	Rhone-
Colloid 697	Defoamer	Rhone-
Color-Sperse 188-A	Nonionic surfactant	Henkel
Colortrend 888-9907	Lamp black	Huls
Composition T	Dispersant	Calgon
Concofoc 690	Surfactant	Contin-
Copperas Red R-7098	Red	Columb-
Coroc A-2678-M	Additive	Freeman
Cosan 145	Preservative	Cosan
CR-800		
CR-813		
CT-136		AirProd
Cycat 4040	Catalyst	AmCyan
Cyclodex Cobalt (6%)	Drier	Huls
Cyclodex Manganese (6%)	Drier	Huls
Cyclodex Zirconium (12%)	Drier	Huls
Cymel 301	Melamine-formaldehyde resin	Am Cyan
Cymel 303	Melamine-formaldehyde resin	Am Cyan
Cymel 303LF	Melamine-formaldehyde resin	Am Cyan
Cymel 325	Melamine-formaldehyde resin	Am Cyan
Cymel 327	Melamine-formaldehyde resin	Am Cyan
Cymel 350	Melamine-formaldehyde resin	Am Cyan
Cymel 370	Melamine-formaldehyde resin	Am Cyan
Cymel 373	Melamine-formaldehyde resin	Am Cyan

RAW MATERIAL	CHEMICAL DESCRIPTION	SOURCE
Cymel 385	Cross-linking agent	Am Cyan
Cyzac 4040	Catalyst	Am Cyan
Dalamar Yellow YT820D		Harmon
Dalpad-A	Coalescent	Dow
Daper Novacite		Malvern
Dapro 1181	Foam suppressor	Daniel
Darvan No. 7	Dispersing agent	Vander-
DC-29	Silicone	DowCorn
Deefo 806-102	Defoamer	Ultra
Deefo 916	Defoamer	Ultra
DeeFo 3000	Defoamer	Ultra
Degussa TS-100		Degussa
Dehydran 1293	Defoamer	Henkel
Dehydran 1620	Defoamer	Henkel
Dextrol OC-50	Wetting agent	Dexter
Dimethylethanolamine		Pennwal
Dispersant QR-681M		Rohm&
Disperbyk	Wetting/dispersing additive	Byk
Disperse-Ayd W22		Daniel
DMAMP-80	2-Dimethylamino-2-methyl-1- 20% water	IMC
DMEA	Neutralizing amine	Un Carb
Dow 65	Defoamer	Dow

RAW MATERIAL	CHEMICAL DESCRIPTION	SOURCE
Dow 762-W	Latex polymer solution	Dow
Dowanol DB	Glycol ether	Dow
Dowanol DPM	Glycol ether	Dow
Dowanol EB	Glycol ether	Dow
Dowanol PNB	Glycol ether	Dow
Dowanol PPh	Glycol ether	Dow
Dow Corning 14	Additive	DowCorn
Dow Corning #57	Additive	DowCorn
Dowicil 75	Preservative	Dow
Dow Latex RAP 213NA	Styrene/butadiene latex	Dow
Drewplus L-405	Foam control agent	Drew
Drewplus L-418	Foam control agent	Drew
Drewplus L-464	Foam control agent	Drew
Drewplus L-475	Foam control agent	Drew
Drewplus L-484	Foam control agent	Drew
Drewplus L-493	Foam control agent	Drew
Drewplus Y-250	Foam control agent	Drew
DSX 1514	Rheology modifier	Henkel
DSX 1550	Rheology modifier	Henkel
Duramac 2400	Soya alkyd resin	McWhort
Duramite	Calcium carbonate	ECC

RAW MATERIAL	CHEMICAL DESCRIPTION	SOURCE
EA-1075	Anti-settling aid	NL Chem
Ektasolve DE	Solvent	Eastman
Ektasolve EB	Solvent	Eastman
Ektasolve EEH	Solvent	Eastman
Ektasolve EP	Ethylene glycol monopropyl ether	Eastman
Emulsion E-2774		Rohm&
Enco Aromatic 150	High flash aromatic hydrocarbon	Exxon
Epi-Cure W50-8535	Epoxy curing agent	Rhone-
Epi-Rez WJ-3520	Epoxy resin	Rhone-
Epi-rez WJ-5522	Epoxy resin	Rhone-
Epi-rez 510	Epoxy resin	Rhone-
Epotuf 38-690	Epoxy resin	Reichh-
Epoxy Ester P-2017-S-11		McClosk
Everflex E	Vinyl-acrylic latex	Grace
Everflex GT	Vinyl copolymer emulsion	Grace
Exkin #1	Anti-skinning agent	Huls
Exkin #2	Anti-skinning agent	Huls
Emulsion E-2189		Rohm&
Emulsion E-2573		Rohm&
Emulsion E-2955		Rohm&
Exxate 600	Oxo-acetate	Exxon
Exxate 1300	Oxo-acetate	Exxon

RAW MATERIAL	CHEMICAL DESCRIPTION	SOURCE
FC-120	Fluorosurfactant	3M
FC-129	Fluorosurfactant	3M
FC-430	Fluorosurfactant	3M
Flatting Agent TS-100	Flatting agent	Degussa
Flexbond 325	Vinyl-acrylic emulsion	AirProd
Flexbond 380 DEV	Vinyl-acrylic emulsion	AirProd
Flexbond 471	Acrylic emulsion	AirProd
Fluorad FC 129/431	Fluorosurfactant	3M
Foamaster AP	Defoamer	Henkel
Foamaster DS	Defoamer	Henkel
Foamaster G	Defoamer	Henkel
Foamaster O	Defoamer	Henkel
Foamaster R	Defoamer	Henkel
Foamaster TCX	Defoamer	Henkel
Foamaster VF	Defoamer	Henkel
Foamaster VL	Defoamer	Henkel
Foamburst 363		
Foamex 1488	Antifoam	Goldsc-
Foamkill FBF	Foam control agent	Crucib-
Foamkill 639Q	Foam control agent	Crucib-

RAW MATERIAL	CHEMICAL DESCRIPTION	SOURCE
Gamasperse 6532	Calcium carbonate	Georgia
Garnet Toner X-2433	Toner	Ciba-
GE60	Defoamer	GE
Genepoxy 370-H55	Epoxy emulsion	Daubert
Glomax LL	Calcined aluminum silicate	DryBran
Gold Bond R	Silica	GoldBon
GT-674 Phthalo Green	Green	duPont
Halox BW-100	Calcium barium phosphosilicate	Halox
Halox SW-111 Halox	Calcium strontium phosphosilicate	Halox
Halox SZP-391	Non-lead inhibitive pigment	Halox
Hansa Yellow G-1230	Pigment	Harshaw
HEI-SCORE-XAB	Corrosion inhibitor	Whittak
Heliogen Blue L6875F	Phthalo blue pigment	BASF
Heliogen Blue L6975F	Phthalo blue pigment	BASF
Heliogen Blue L7071F	Phthalo blue pigment	BASF
Heliogen Green L8690	Green phthalocyanine pigment	BASF
Heloxy 8	Epoxy modifier	Rhone-
Hercules SGL	Defoamer	Aqualon
Hercules 7H3SF	Sodium CMC	Aqualon
Hercules 501	Defoamer	Aqualon
Heubach Y-539-D	Zinc yellow	Heubach
Heucophos ZPO		Heubach
Heucophos ZBZ	Zinc yellow	Heubach

RAW MATERIAL	CHEMICAL DESCRIPTION	SOURCE
Heucophthal Blue RF BT-627-D	Blue	Heubach
Heucophthal Blue RF BT-698-D	Blue	Heubach
Hexyl Cellosolve	Coalescent	UnCarb
Hi-Sol 15	High flash aromatic hydrocarbon	Ashland
Hoover Brown 7148	Brown	Hoover
Horsehead RF-30	Zinc oxide	NJ Zinc
Horsehead XX-503	Zinc oxide	NJ Zinc
Huber 70C		Huber
Hubercarb Q-6	Calcium carbonate	Huber
Hubercarb Q-325	Calcium carbonate	Huber
Hydro-Cem Calcium	Catalyst	OMGroup
Hydro-Cen Zirconium	Catalyst	OMGroup
Hydro-Cure Cobalt	Catalyst	OMGroup
Hydro-Cure Manganese	Catalyst	OMGroup
Hydro-Paste 8726		Silber-
Hydro-Paste 82255		Silber-
Hypermer PS2	Polymeric dispersant, stabilizer	ICI
Iceberg	Anhydrous aluminum silicate	Burgess
Igepal CO-610	Non-ionic surfactant	GAF
Igepal CO-630	Non-ionic surfactant	GAF
Igepal CTA-639	Surfactant	GAF

RAW MATERIAL	CHEMICAL DESCRIPTION	SOURCE
Imperial Brazil Yellow X-2860	Pigment	Ciba–
Imperial Brazil Yellow X-2866	Pigment	Ciba–
Imsil A-8	Micronized microcrystalline silica	Unimin
Imsil A-10	Micronized microcrystalline silica	Unimin
Imsil A-15	Micronized microcrystalline silica	Unimin
Imsil A-25	Micronized microcrystalline silica	Unimin
Intercar 4% Calcium	Drier	Interst
Intercar 6% Cobalt	Drier	Inter–
Intercar 6% Manganese	Drier	Inter–
Intercar 6% Zirconium	Drier	Inter–
Intercide T-O	Microbiostatic agent	Inter–
IT-325	Talc	Vander–
Joncryl SCX-1520	Acrylic polymer	Johnson
Joncryl 56	Acrylic polymer	Johnson
Joncryl 540	Acrylic polymer	Johnson
Kadox-515	Zinc oxide	NJ Zinc
Kadox 911	Zinc oxide	NJ Zinc
Kaopaque 105	Aluminum silicate	DryBran

RAW MATERIAL	CHEMICAL DESCRIPTION	SOURCE
Kathon LX	Fungicide	Rohm&
K-Cure 1040	Catalyst	King
Kelecin 1081	Soybean oil	Reich-
Kelsol 860	Oil copolymer	Reich-
Kelsol 3900	Oxidizing alkyd. % solids: 75	Reich-
Kelsol 3902	Oxidizing alkyd. % solids: 75	Reich-
Kelsol 3904	Oxidizing alkyd. % solids: 75	Reich-
Kelsol 3905	Water-reducible alkyd	Reich-
Kelsol 3906	Oxidizing alkyd. % solids: 75	Reich-
Kelsol 3907	Oxidizing alkyd. % solids: 75	Reich-
Kelsol 3909	Oxidizing alkyd. % solids: 75	Reich-
Kelsol 3910	Oxidizing alkyd. % solids: 75	Reich-
Kelsol 3918	Oxidizing alkyd. % solids: 70	Reich-
Kelsol 3919	Oxidizing alkyd. % solids: 70	Reich-
Kelsol 3922	Oxidizing alkyd. % solids: 80	Reich-
Kelsol 3931	Modified oil. % solids: 45	Reich-
Kelsol 3937	Modified oil. % solids: 45	Reich-
Kelsol 3939	Modified oil. % solids: 45	Reich-
Kelsol 3950	Oxidizing alkyd. % solids: 70	Reich-
Kelsol 3960	Oxidizing alkyd. % solids: 75	Reich-
Kelsol 3961	Oxidizing alkyd. % solids: 75	Reich-
Kelsol 3962	Oxidizing alkyd. % solids: 70	Reich-
Kelsol 3963	Oxidizing alkyd. % solids: 70	Reich-
Kelsol 3970	Oxidizing alkyd. % solids: 75	Reich-

RAW MATERIAL	CHEMICAL DESCRIPTION	SOURCE
Kelsol 3980	Non-oxidizing alkyd. % solids: 75	Reich-
Kelsol 3990	Oxidizing alkyd. % solids: 70	Reich-
Kelsol 4069	Acrylic polyester. % solids: 55	Reich-
Kelsol 4097	Acrylic polyester. % solids: 55	Reich-
Kelsol 5293	Polyester. % solids: 75	Reich-
K-Flex UD-320W	Polyol	King
K-Flex 188	Polyol	King
Kodaflex TXIB	Plasticizer	Eastman
Korthix	Thixotropic agent	Kaopol-
KP-140	Plasticizer	FMC
Kraygel 136		
Krolor Yellow KY-7810	Pigment	duPont
L-5310	Flow control agent	UnCarb
L-7605	Flow control agent	UnCarb
Lactimon WS	Wetting/dispersing additive	Byk-
Linaqua	Modified oil. % solids: 85	Reich-
Liquazinc AQ-90	Mar/sanding aid	Witco
Lodyne S-107		Ciba-
Lodyne S-107B		Ciba-
Lubrizol 2063		Lubriz-

RAW MATERIAL	CHEMICAL DESCRIPTION	SOURCE
Magnacat	Drier	Ultra
Maincote AE-58	Acrylic latex	Rohm&
Maincote HG-54	Acrylic latex	Rohm&
Maincote TL-5	Acrylic latex	Rohm&
Makon 10	Surfactant. Nonoxynol-10	Stepan
Manchem APG	Adhesion promoter	Manchem
Manganese Cem-All 12%	Synthetic drier	OMGroup
Manganese Cyclodex(6%)	Drier	Huls
Manganese Hydrocure(5%)	Catalyst	OMGroup
Manganese Hydrocure II	Catalyst	OMGroup
Mapico 1075	Iron oxide	Columb-
Mapico Brown 422	Iron oxide	Columb-
Mapico Black	Iron oxide	Columb-
Mapico Red 297	Iron oxide	Columb-
Mapico Red #347	Iron oxide	Columb-
Mapico Red 516 Medium	Iron oxide	Columb-
Mapico 347A	Iron oxide	Columb-
Mapico Yellow 1000	Iron oxide	Columb-
Mapico Yellow 1050	Iron oxide	Columb-
Mapico Yellow 3100	Iron oxide	Columb-
Mearlin Fine-Pearl #139V	Powdered pearl pigment	Mearl
Medium Chrome Yellow X2075	Yellow	Ciba-
Medium Chrome Yellow X-3356	Yellow	Ciba-

RAW MATERIAL	CHEMICAL DESCRIPTION	SOURCE
Merbac 35	Bactericide	Calgon
Methocel HG-400	Hydroxypropyl methylcellulose	Dow
Methocel J-12-HS	Hydroxypropyl methylcellulose	Dow
Methyl Carbitol	Diethylene glycol monomethyl ether/UnCarb	
Methyl Propasol	Solvent	UnCarb
Mica C-3000	Lamina mica pigment	Smith
Mica 325WG	Mica	KMG
Michem Emulsion 39235	Wax emulsion	Michel-
Michem Emulsion 43040	Wax emulsion	Michel-
Michemlube 110	Wax additive	Michel-
Michemlube 155	Wax additive	Michel-
Michemlube 511	Wax additive	Michel-
Michemlube 743	Wax additive	Michel-
Minex 4	Nepheline syenite	Unimin
Minex 7	Nepheline syenite	Unimin
Minex 10	Nepheline syenite	Unimin
Min-U-Sil 10	Micron-sized silica	Unimin
Min-U-Sil 15	Micron-sized silica	Unimin
Min-U-Sil 40	Micron-sized silica	Unimin
Molybdate Orange UE-637-D	Orange	Heubach
Moly-white MZAP	Anti-corrosive pigment	SherWil
Monarch Blue X3367	Blue	Ciba-
Monastral Blue BT	Phthalocyanine blue	Ciba-

RAW MATERIAL	CHEMICAL DESCRIPTION	SOURCE
Monastral Red Y RT-759-D	Red	Ciba-
Monastral Violet R RT-201-D	Violet	Ciba-
M-P-A 1075	Anti-settling agent	NLChem
Nacorr 1552	Rust inhibitor	King
Nacure 155	Catalyst	King
Nacure 5225	Catalyst	King
Nalco 43J-36	Antifoam	Nalco
Nalco 65J-769	Antifoam	Nalco
Nalco 955-815	Antifoam	Nalco
Nalco #2300	Antifoam	Nalco
Nalco 2302	Antifoam	Nalco
Nalco 2309	Antifoam	Nalco
Nalzin 2	Zinc hydroxy phosphite	Rheox
Natrosol Plus, 330	Modified hydroxyethylcellulose	Aqualon
Natrosol Plus	Modified hydroxyethylcellulose	Aqualon
Natrosol 250	Hydroxycellulose	Aqualon
Natrosol 250HR	Hydroxyethylcellulose	Aqualon
Natrosol 250MR	Hydroxyethylcellulose	Aqualon
Neocryl A-640	Acrylic copolymer	ICIRes
Neosil A	Silica extender	Tammsco
Niax PCP-0301	Polyester polyol resin	UnCarb

RAW MATERIAL	CHEMICAL DESCRIPTION	SOURCE
Nickel HCT		Novamet
Nigrosine 128-B Conc.Pdr		
Nopco KFS	Defoamer	Henkel
Nopco NDW	Defoamer	Henkel
Nopco NXZ	Defoamer	Henkel
Nopco 1907	Defoamer	Henkel
Nopcocide N-40-D	Fungicide	Henkel
Nopcocide N-96	Mildewcide	Henkel
Nopcosperse 44	Dispersant	Henkel
Novacite L207A	Silica extender pigment	Malvern
Novacite L-337	Silica extender pigment	Malvern
Novacite S-325	Silica extender pigment	Malvern
Novaperm F3RK70		Hoechst
Novaperm Red F5RK		Hoechst
Nuocide 404-D	Mildewcide	Huls
Nuocure Manganese 6%	Catalyst	Huls
Nuocure Cobalt 10%	Catalyst	Huls
Nuocure CK-10	Catalyst	Huls
Nuocure Mn	Catalyst	Huls
Nuosept 95	Preservative	Huls
Nuosept 145	Preservative	Huls
Nuosperse 657	Dispersing agent	Huls
Nyad G	Wollastonite	Nyco

RAW MATERIAL	CHEMICAL DESCRIPTION	SOURCE
Nytal 300	Talc	Vander-
Nytal 400	Talc	Vander-
OK-412	Flatting agent	Degussa
OK-500	Flatting agent	Degussa
Oncor F-31	Anti-corrosive pigment	NLChem
Oncor M50	Anti-corrosive pigment	NLChem
Optiwhite	Anhydrous aluminum silicate	Burgess
Oswego Orange X-2065	Orange	Ciba-
PA-328	Antifoam	USMovid
PA-454	Antifoam	USMovid
PAG-188		
Paint Additive #14	Flow and leveling agent	DowCorn
Palomar Green G5406	Green	Miles
Paraplex WP-1	Plasticizer	Rohm&
Pasco 311	Zinc oxide	Pacific
Patcote 519	Defoamer	Patco
Patcote 531	Defoamer	Patco
Patcote 550	Defoamer	Patco
Patcote 577	Defoamer	Patco
Patcote 847	Defoamer	Patco

RAW MATERIAL	CHEMICAL DESCRIPTION	SOURCE
Patcote 888	Defoamer	Patco
Pentex 99	Anionic surfactant	Rhone-
Phthalo Blue BT-417D	Blue	duPont
Pliolite 7103	Styrene-butadiene copolymer	Goodye-
Pliolite 7104	Styrene-butadiene copolymer	Goodye-
Polyco 2161	Polyvinyl acetate emulsion	Rohm&
Polycompound W-953	Leveling aid	BASF
Polygloss 90	Clay	Huber
Polymon Blue	Pigment	ICI
Polyphase AF-1	Fungicide	Troy
Polyphobe 111		
Polywet ND-2	Dispersant/surfactant	Uniroy-
Propasol B	Industrial solvent	UnCarb
Propasol P	Industrial solvent	UnCarb
Propyl Carbitol	Industrial solvent	UnCarb
Propyl Cellosolve	Industrial solvent	UnCarb
Propyl Propasol	Coalescent	UnCarb
Proxel GXL	Chemical biocide	ICI
Pyrax WA	Pyrophyllite mineral	Vander-
QR-681-M	Rheology modifier	Rohm&

RAW MATERIALS	CHEMICAL DESCRIPTION	SOURCE
R-2899D	Red iron oxide	Pfizer
Raven Black 16	Carbon black	Columb-
Raven #14 Powder	Carbon black	Columb-
Raven 150	Carbon black	Columb-
Raven 420	Carbon black	Columb-
Raven 1020	Carbon black	Columb-
Raven 1035	Carbon black	Columb-
Raven 1250	Carbon black	Columb-
Raven 1255	Carbon black	Columb-
Raven 2000	Carbon black	Columb-
Raybo 62-HydroFlo	Additive	Raybo
Red Iron Oxide R-4098	Pure red iron oxide	Pfizer
Red Oxide RO-3097	Red iron oxide	Harcros
Red Oxide 130M	Red iron oxide	
Red Oxide #4098-D	Red iron oxide	Pfizer
Resimene I-720	Melamine resin	Monsan-
Resimene 740	Melamine resin	Monsan-
Resimene 745	Melamine resin	Monsan-
Resimene 797	Melamine resin	Monsan-
Resin BAL-389	Water borne polyester	Amoco
Resin WA-17-2T	Water-Reducible alkyd enamel	Eastman
Resin WA-17-6C	Polyester resin	Eastman
Resin WB-17-INS	Polyester resin	Eastman
Resin WB-151	Air drying alkyd	Amoco

RAW MATERIAL	CHEMICAL DESCRIPTION	SOURCE
Resin WB-300	Polyester	Amoco
Resin WB-408	Alkyd resin	Amoco
Resin WB-3823	Alkyd resin	Amoco
Resin WS-3-1C	Polyester resin	Eastman
Resin WS-3-2C	Polyester resin	Eastman
Resin WS-17-1T	Polyester resin	Eastman
Rex Orange X-2806	Molybdate. Red 104	Ciba-
Rheolate 255	Rheological additive	Rheox
Rheolate 278	Rheological additive	Rheox
Rheology Modifier QR-708	Thickener	Rohm&
Rheology Modifier QR-1001	Thickener	Rohm&
Rhodopol 23	Xanthan	Rhone-
Rhoplex AC-33	Acrylic emulsion. 44.5% solids	Rohm&
Rhoplex AC-64	Acrylic emulsion. 60.5% solids	Rohm&
Rhoplex AC-388	Acrylic latex. 50% solids	Rohm&
Rhoplex AC-417	Acrylic latex. 48% solids	Rohm&
Rhoplex AC-490	Acrylic latex. 46% solids	Rohm&
Rhoplex AC-507	Acrylic latex. 46.5% solids	Rohm&
Rhoplex AC-604	Acrylic latex. 46.0% solids	Rohm&
Rhoplex AC-658	Acrylic latex. 47.0% solids	Rohm&
Rhoplex AC-707	Acrylic latex	Rohm&
Rhoplex AC-829	Acrylic latex. 55.0% solids	Rohm&
Rhoplex AC-1024	Acrylic latex. 55.0% solids	Rohm&
Rhoplex AC-1230	Acrylic latex. 47.0% solids	Rohm&

RAW MATERIAL	CHEMICAL DESCRIPTION	SOURCE
Rhoplex AC-1822	Acrylic latex. 46.5% solids	Rohm&
Rhoplex AC-2045	Acrylic latex	Rohm&
Rhoplex CL-103	Acrylic latex	Rohm&
Rhoplex CL-104	Acrylic latex	Rohm&
Rhoplex MC-76	Acrylic latex	Rohm&
Rhoplex MV-2	Acrylic latex. 46.0% solids	Rohm&
Rhoplex MV-9	Acrylic latex. 45.5% solids	Rohm&
Rhoplex MV-23	Acrylic latex. 43% solids	Rohm&
Rhoplex SG-10	Acrylic latex	Rohm&
Rhoplex WL-71	Acrylic polymer. 41.5% solids	Rohm&
Rhoplex WL-96	Acrylic polymer. 42% solids	Rohm&
RM-825		Rohm&
Ropaque OP-62	Opaque polymer	Rohm&
Ross&Rowe 551	Modified lecithin	Ross&
Samhide 583		JMHuber
Santicizer 160	Plasticizer	Monsan-
Satintone #1	Calcined aluminum silicate	Engel-
Satintone #5	Calcined aluminum silicate	Engel-
Satintone W	Calcined aluminum silicate	Engel-
SCT 270	Associative thickener	UnCarb
SCT 275	Associative thickener	UnCarb
Shieldex	Inhibitive pigment	Grace

RAW MATERIAL	CHEMICAL DESCRIPTION	SOURCE
Sicor ZNP-M	Anti-corrosion pigment	BASF
Silica #19	Silica	WC&D
Silwet L-77	Surface active copolymer	UnCarb
Silwet L-7600	Surface active copolymer	UnCarb
Silwet 7605	Surface active copolymer	UnCarb
Skane M8	Mildewcide	Rohm&
Skino #1	Antiskinning agent	OMGroup
Slip Ayd 18	Surface conditioner	Daniel
Snowflake	Calcium carbonate	ECC
Snow*Tex 45	Calcined clay	USSili-
Spacerite S-11		
Special Black 4	Carbon black	Degussa
Special Black 4A	Carbon black	Degussa
Spensol L44	Urethane polyol	Reich-
Stapa Hydrolac W 60n.1	Aluminum pigment	Alumin-
Sterling R	Carbon black	Cabot
St. Joe #40 ZnO	Zinc oxide	St.Joe
Strodex MOK-70	Surfactant	Dexter
Strontium Chromate J1365	Strontium chromate	Mineral
Strontium Chromate 30-AC-3008	Strontium chromate	Hilton-
Sun Fast 264-8142	Pigment	SunChem
Super-Ad-It	Fungicide	Huls
Super Beckamine 27-568	Synthetic resin	Reich-

RAW MATERIAL	CHEMICAL DESCRIPTION	SOURCE
Supercoat		ECC
Surfynol DF-75	Surfactant	AirProd
Surfynol GA	Surfactant	AirProd
Surfynol PC	Surfactant	AirProd
Surfynol TG	Surfactant	AirProd
Surfynol 104	Surfactant	AirProd
Surfynol 104A	Surfactant	AirProd
Surfynol 104BC	Surfactant	AirProd
Surfynol 104E	Surfactant	AirProd
Surfynol 104H	Surfactant	AirProd
SWS 211	Antifoam	Wacker
Syloid 72	Silica flatting agent	Davison
Syloid 74	Silica flatting agent	Davison
Syloid 74x3500	Silica flatting agent	Davison
Syloid 83	Silica flatting agent	Davison
Syloid 161	Silica flatting agent	Davison
Syloid 166	Silica flatting agent	Davison
Syloid 169	Silica flatting agent	Davison
Syloid 234	Silica flatting agent	Davison
Syloid 308	Silica flatting agent	Davison
Syloid 978	Silica flatting agent	Davison
Syntex 3981	Accelerator	Rhone-
Synthemul 40-412	Acrylic polymer. % solids: 50	Reich-
Synthemul 40-422	Acrylic polymer. % solids: 49	Reich-

RAW MATERIAL	CHEMICAL DESCRIPTION	SOURCE
Synthemul 40-423	Acrylic polymer. % solids: 45	Reich-
Synthemul 40-424	Acrylic polymer. % solids: 40	Reich-
Synthemul 40-430	Acrylic polymer. % solids: 50	Reich-
Synthemul 40-431	Acrylic polymer. % solids: 50	Reich-
Synthemul 97-603	Acrylic polymer. % solids: 45	Reich-
Synthemul 97-799	Acrylic polymer. % solids: 50	Reich-
Synthetic Iron Oxide Yellow 1103		Mineral
Synthetic Red Iron Oxide RO-3097		Harcros
Synthetic Yellow Iron Oxide YLO-3288-D		Harcros
Talc 399	Extender pigment	WC&D
Talc 499	Extender pigment	WC&D
Talcron MP40-27	Talc	Pfizer
Tamol SG-1	Dispersant	Rohm&
Tamol 165	Dispersant	Rohm&
Tamol 681	Dispersant	Rohm&
Tamol 731	Dispersant	Rohm&
Tamol 850	Dispersant	Rohm&
Tamol 960	Dispersant	Rohm&
Tamol 1124	Dispersant	Rohm&
Tamol 1254	Dispersant	Rohm&
Tan 10A		Columb-
Tego 1488		Golds-

RAW MATERIAL	CHEMICAL DESCRIPTION	SOURCE
Tektamer 38	Preservative	Calgon
Tenneco 8800 Line	Lampblack 0.5/Raw Umber 0.2	Huls
Tergitol NP-40	Nonionic surfactant	UnCarb
Texanol	Coalescent	Eastman
Thickener LN		GAF
Tinuvin 440		Ciba-
Tinuvin 1130		Ciba-
Tiona RCL-9	Titanium dioxide	SCM
Tiona RCL-535	Titanium dioxide	SCM
Tiona RCL-628	Titanium dioxide	SCM
Tioxide TR92	Titanium dioxide	Tioxide
Ti-Pure R-700	High gloss titanium dioxide	DuPont
Ti-Pure R-702	Titanium dioxide	DuPont
Ti-Pure R-900	Titanium dioxide. Rutile	DuPont
Ti-Pure R-900HG	Titanium dioxide. Rutile	DuPont
Ti-Pure R-901	Titanium dioxide.Rutile.88% assay	DuPont
Ti-Pure R-902	Titanium dioxide.Rutile.91% assay	DuPont
Ti-Pure R-931	Titanium dioxide.Rutile.85% assay	DuPont
Ti-Pure R-960	Titanium dioxide.Rutile.89% assay	DuPont
Ti-Pure R-960 VHG	Titanium dioxide.Rutile	DuPont
Titanox AWD	Titanium dioxide	Kronos
Titanox 2020	Titanium dioxide.Rutile.94% assay	Kronos
Titanox 2101	Titanium dioxide.Rutile.92% assay	Kronos
Titanox 2160	Titanium dioxide.Rutile.90% assay	Kronos

RAW MATERIAL	CHEMICAL DESCRIPTION	SOURCE
Toluidine Red RD-0027		Uhlich
Tributoxy Ethyl Phosphate		FMC
Triton CA	Wetting agent	UnCarb
Triton CF-10	Wetting agent	UnCarb
Triton GR-7	Wetting agent	UnCarb
Triton GR-7M	Wetting agent	UnCarb
Triton N-57	Wetting agent	UnCarb
Triton N-101	Wetting agent	UnCarb
Triton X-100	Wetting agent	UnCarb
Triton X-102	Wetting agent	UnCarb
Triton X-114	Wetting agent	Uncarb
Triton X-207	Wetting agent	UnCarb
Triton X-405	Wetting agent	UnCarb
Tronox CR-800	Titanium dioxide.Rutile.95% assay	Kerr-
Tronox CR-808	Titanium dioxide	Kerr-
Tronox CR-812	Titanium dioxide	Kerr-
Tronox CR-813	Titanium dioxide	Kerr-
Tronox CR-820	Titanium dioxide.Rutile.87% assay	Kerr-
Tronox CR-821	Titanium dioxide	Kerr-
Tronox CR-828	Titanium dioxide	Kerr-
Troysan Polyphase AF-1	Fungicide	Troy
Troysan 174	Fungicide	Troy
Troysan 186	Fungicide	Troy
Troysan 364	Fungicide	Troy
TS-100	Flatting agent	Degussa

RAW MATERIAL	CHEMICAL DESCRIPTION	SOURCE
Ucar 367	Vinyl-acrylic polymer	UnCarb
Ucar Latex 376	Vinyl-acrylic polymer	UnCarb
Ucar 379	Vinyl-acrylic latex	UnCarb
Ucar Latex 430	Acrylic polymer	UnCarb
Ucar 522		UnCarb
Ucar 525	Modified acrylic polymer	UnCarb
Ucar Latex 624	Acrylic latex	UnCarb
Ucar Latex 625	Acrylic latex	UnCarb
Ucar Polyphobe 102	Thickener	UnCarb
Ucar Polyphobe 104	Rheology modifier	UnCarb
Ucar 4431X	Acrylic latex polymer	UnCarb
Ucar Filmer IBT		UnCarb
Ucar Latex 154		UnCarb
Ucar SCT 200	Thickener/rheology modifier	UnCarb
Ucar SCT 270	Thickener/rheology modifier	UnCarb
Ucar SCT 275	Thickener/rheology modifier	UnCarb
Ucar Solvent 2LM	Dipropylene glycol monoethyl	UnCarb
UCD 4820Q		Bee
Uformite MM83	Urea-formaldehyde resin	Reich-
Unitane OR-600	Titanium dioxide. Rutile.	Kemira
Unocal 3084		Rohm&
Uvinul N539	UV absorber	BASF

RAW MATERIAL	CHEMICAL DESCRIPTION	SOURCE
Vancide TH	Bactericide-fungicide	Vander-
Van Dyke Brown	Pigment	Landers
Van-Gel	Thixotrope	Vander-
Van-Gel B	Thixotrope	Vander-
Vansil W-10	Wollastonite	Vander-
Varsol No. 1	Aliphatic solvent	Exxon
Veegum T	Magnesium aluminum silicate	Vander-
Vera Blanc	Calcium carbonate	NLChems
Versaflow 102		Shamro-
Vicron 1515	Ground limestone	Pfizer
Vinac 885	Polyvinyl acetate homopolymer	AirProd

WA-8	Opaque stain	USCell-
Wallpol 40-136	Synthetic resin emulsion	Reich-
Water Reducible Modified Polyolefin 73-7358		Cargill
Wave 345	Wet adhesion terpolymer	AirProd
WB-138 Dispersion	Polyester resin	Amoco
WB-389 Dispersion	Polyester resin	Amoco
WD-2348 Tint Paste	Black	Daniels
Witcobond W234	Aqueous polyurethane dispersion	Witco
Witco 3056-A		Witco

RAW MATERIAL	CHEMICAL DESCRIPTION	SOURCE
X-1134 CP	Cr2O3 Green	Ciba-
X-1843 Madras Orange	Orange	Ciba-
X-2806 Red Orange	Orange	Ciba-
XAMA-7		Rohm&
XX-503R	Zinc Oxide	Zinc
Y-5775 Fast Yellow	Yellow	Harmon
Yellow Iron Oxide 2288D		Pfizer
Yellow Iron Oxide YLO-1888D		Pfizer
YLO-3288D	Ferric oxide	Pfizer
Zeolex 80	Sodium aluminosilicate	JMHuber
Zinc Oxide XX-503	Zinc oxide	NJZinc
Zinc Oxide XX-601	Zinc oxide	NJZinc
Zinc Phosphate J0852	Zinc phosphate	Mineral
Zinc Yellow Y-539D	Zinc yellow	Heubach
Zirconium Hydro-Cem	Zirconium drier, 12%	Mooney
Zirconium Hydrocure	Drier catalyst	OMGroup
Zopaque RCL-9	Titanium dioxide.Rutile.94% assay	SCM

RAW MATERIAL	CHEMICAL DECRIPTION	SOURCE
#1 White	Calcium carbonate	ECC
#10 White		
13-3060 Novoperm A20	Red	Hoechst
76 RES 4076	Polymer emulsion	Rohm&
106 Lo Micron Barytes	Extender pigment	WC&D
290 Lo Micron Barytes	Extender pigment	WC&D
325 Mica	Mica	English
399 Lo Micron Talc	Extender pigment	WC&D
499 Talc	Extender pigment	WC&D
1160 Silica	Silica	Illin-
3011		

Section XIV
Suppliers' Addresses

Air Products and Chemicals
7201 Hamilton Blvd.
Allentown, PA 18195
(215)-481-4911/(800)-345-3148

Akzo Chemicals, Inc.
300 S. Riverside Plaza
Chicago, IL 60175
(312)-906-7500/(800)-257-8292

Alcoa Industrial Chemicals
P.O. Box 67
Bauxite, AR 72011
(501)-776-4981/(800)-643-8771

American Cyanamid Co.
Chemicals Group
One Cyanamid Plaza
Wayne, NJ 07470
(201)-831-2000/(800)-443-0443

American Lecithin Co.
33 Turner Rd.
P.O. Box 1908
Danbury, CT 06813
(203)-790-2700

American Powdered Metals, Inc.
225 Broadway
New York, NY 10007
(212)-267-4900/(800)-322-0323

Amoco Chemical Co.
200 E. Randolph Dr.
Chicago, IL 60601
(312)-856-3200/(800)-621-4567

Angus Chemical Co.
1500 E. Lake Cook Rd.
Buffalo Grove, IL 60089
(708)-215-8600/(800)-362-2580

Aqualon
P.O. Box 15417
2711 Centrevile Rd.
Wilmington, DE 19850
(302)-996-2000/(800)-345-8104

ARCO Chemical Co.
3801 West Chester Pike
Newtown Square, PA 19073
(215)-359-2000

Asarco, Inc.
180 Maiden Lane
New York, NY 10038
(212)-510-2000

Ashland Chemical Co.
P.O. Box 2219
Columbus, OH 43216
(614)-889-3333/(800)-828-7659

BASF Corp.
100 Cherry Hill Road
Parsippany, NJ 07054
(201)-316-3000/(800)-526-1072

Buckman Laboratories Int'l. Inc.
1256 N. McLean Blvd.
Memphis, TN 38108
(901)-278-0330/(800)-BUCKMAN

Burgess Pigment Co.
P.O. Box 349
Sandersville, GA 31082
(912)-552-2544/(800)-841-8999

Byk-Chemie USA
524 S. Cherry St.
Wallingford, CT 06492
(203)-265-2086

Cabot Corp.
Rt. 36 West
Tuscola, IL 61953
(217)-253-3370/(800)-222-6745

Calgon Corp.
P.O. Box 1346
Pittsburgh, PA 15230
(412)-777-8000/(800)-4-CARBON

Cargill, Inc.
Box 5630
Minneapolis, MN 55440
(612)-475-6478/(800)-535-1443

Celite Corp.
P.O. Box 519
Lompoc, CA 93438
(805)-735-7791

Chemical Corp. of America
2 Carlton Ave.
E. Rutherford, NJ 07073
(201)-438-5800

Ciba-Geigy Corp.
7 Skyline Dr.
Hawthorne, NY 10532
(914)-347-4700/(800)-431-1900

Colloids, Inc.
1525 Church St.
Marietta, GA 30060

Columbian Chemicals Co.
1600 Parkwood Circle
Suite 400
Atlanta, GA 30339
(404)-951-5700/(800)-235-4003

Continental Chemical Co.
2686 Lisbon Rd.
Cleveland, OH 44104
(216)-721-4747/(800)-929-4633

Cosan Chemical Corp.
400 14th St.
Carlstadt, NJ 27072
(201)-460-9300

Crucible Chemical Co., Inc.
P.O. Box 6786
Greenville, SC 29606
(803)-277-1284/(800)-845-8873

Daniel Products
400 Claremont Ave.
Jersey City, NJ 07304
(201)-432-0800

Daubert Chemical Co.
S. Central Ave.
Chicago, IL 60638
(708)-496-7350

Davison Chemical Div.
W.R. Grace & Co.
P.O. Box 2117
Baltimore, MD 21203

Degussa Corp.
65 Challenger Rd.
Ridgefield Park, NJ 07305
(201)-641-6100

Dexter Chemical Corp.
845 Edgewater Road
Bronx, NY 10474

Dow Chemical USA
2020 Willard H Dow Center
Midland, MI 48674
(800)-447-4DOW

Dow Corning Corp.
Box 0994
Midland, MI 48686
(517)-496-4000

Drew Industrial Div.
Ashland Chemical Inc.
One Drew Plaza
Boonton, NJ 07005
(201)-263-7800/(800)-526-1015

Dry Branch Kaolin Co.
Route #1
Box 468-D
Dry Branch, GA 30120
(912)-7500-3500/(800)-DBK-CLAY

DuPont Co.
1007 Market St.
Wilmington, DE 19898
(800)-441-7515

Durkee/SCM Chemicals
2701 Broening Hwy
Baltimore, MD 21222
(301)-288-8884/(800)-638-3234

Eastman Chemical Co.
P.O. Box 431
Kingsport, TN 37662
(615)-229-2318/(800)-EASTMAN

ECC International
5775 Peachtree-Dunwoody Rd.
Atlanta, GA 30342
(404)-843-1551/(800)-843-3222

Engelhard Corp.
101 Wood Ave.
Iselin, NJ 08830
(908)-205-5000

English Mica Co.
Ridgeway Center Bldg.
Stamford, CT 06905
(203)-324-9531

Exxon Chemical Americas
13501 Katy Frwy
Houston, TX 77079
(713)-870-6000/(800)-231-6633

FMC Corp.
P.O. Box 4239 GCS
New York, NY 10163
(215)-299-6000/(800)-732-3278

GAF Chemicals Corp.
1361 Alps Rd.
Wayne, NJ 07470
(201)-628-3000

GE Silicones
260 Hudson River Rd.
Waterford, NY 12188
(518)-237-3330/(800)-255-8886

General Electric
21800 Tungsten Rd.
Euclid, OH 44117
(216)-266-2451/(800)-255-8886

Genstar Stone Products Co.
Exec Plaza IV
11350 McCormick Rd.
Hunt Valley, MD 21031
(410)-527-4000

Georgia Marble Co.
1201 Roberts Blvd., Bldg. 100
Kennesaw, GA 30144
(404)-421-6500

Gold Bond Bldg. Products Div.
National Gypsum Co.
2001 Rexford Rd.
Charlotte, NC 28211
(704)-365-7300

Goldschmidt Chemical Corp.
P.O. Box 1299
914 E. Randolph Rd.
Hopewell, VA 23860
(804)-541-8658/(800)-446-1809

Goodyear Tire & Rubber Co.
1144 E. Market St.
Akron, OH 44316
(216)-796-3010

W.R. Grace & Co.-Conn
55 Hayden Ave.
Lexington, MA 02173
(617)-861-6600

Hammond Lead Products, Inc.
P.O. Box 6408
5231 Hohman Ave.
Hammond, IN 46325
(219)-931-9360

Harcros Chemicals Inc.
5200 Speaker Rd.
Kansas City, KS 66106
(913)-321-3131

Harmon Colors Corp.
P.O. Box 419
Hawthorne, NJ 07507
(201)-274-3232

Harshaw/Filtrol Partnership
30100 Chagrin Blvd.
Cleveland, OH 44124

Henkel Corp.
300 Brookside Ave.
Ambler, PA 19002
(215)-628-1000/(800)-922-0605

Hercules, Inc.
Hercules Plaza
Wilmington, DE 19894
(800)-247-4372

Heucotech, Ltd.
99 Newbold Road
Fairless Hills, PA 19030
(215)-736-9533/(800)-HEUBACH

Hilton-Davis Co.
2235 Langdon Farm Rd.
Cincinnati, OH 45237
(513)-841-4000/(800)-477-1022

Hoechst Celanese Corp.
Specialty Chemicals Group
5200 77 Center Drive
Charlotte, NC 28217
(704)-559-6000/(800)-365-2436

Hoover Color Corp.
P.O. Box 218
Hiwassee, VA 24347
(703)-980-7233

J.M. Huber Corp.
Calcium Carbonate Div.
2029 Woodlands Pkwy
St. Louis, MO 63146
(217)-224-1100/(800)-637-8176

J.M. Huber Corp.
Clay Div.
One Huber Rd.
Macon, GA 31298
(912)-745-4751/(800)-637-8176

Huls America Inc.
80 Centennial Ave.
Piscataway, NJ 08855
(908)-980-6800/(800)-631-5275

ICI Resins US
730 Main St.
Wilmington, MA 01887
(508)-658-6600/(800)-225-0947

ICI Specialties
Concord Pike & New Murphy Rd.
Wilmington, DE 19897
(302)-886-3000/(800)-822-8215

Indusmin
365 Bloor St.-Suite 200
Toronto, ON CN M4W 3L4

International Minerals &
 Chemicals Corp.
421 E. Hawley St.
Mundelein, IL 60060

Interstab Chemicals, Inc.
500 Jersey Ave.
P.O. Box 638
New Brunswick, NJ 08903

S.C. Johnson Polymer
1525 Howe St.
Racine, WI 53403
(414)-631-4875/(800)-231-7868

Kaopolite, Inc.
2444 Morris Ave.
Union, NJ 07083
(908)-851-2974

Kemira, Inc.
Box 368
Savannah, GA 31402
(912)-236-6171/(800)-4-KEMIRA

Kerr-McGee Chemical Corp.
Kerr-McGee Ctr.
P.O. Box 25861
Oklahoma City, OK 73125
(405)-270-1313/(800)-654-3911

King Industries, Inc.
Science Rd.
Norwalk, CT 06852
(203)-866-5551/(800)-431-7900

KMG Minerals, Inc.
P.O. Box 729
1433 Grover Rd.
Kings Mountain, NC 28086
(704)-739-3616/(800)-443-MICA

Kronos, Inc.
P.O. Box 60087
3000 N. Sam Houston Pkwy
Houston, TX 77205
(713)-987-6300

Landers-Segal Color Co.
84 Dayton Ave.
Passaic, NJ 07055
(201)-779-8948

Lubrizol Corp.
29400 Lakeland Blvd.
Wickliffe, OH 44092
(216)-943-4200

Mallinckrodt Specialty Chemicals
16305 Swingley Ridge Dr.
St. Louis, MO 63017
(314)-539-1241/(800)-325-7155

Malvern Minerals Co.
P.O. Box 1238
Hot Springs National Park,
 AK 71902
(501)-623-8893

Manchem, Inc.
105 College Road East
Princeton Forrestal Center
Princeton, NJ 08540

McCloskey Varnish Co.
4155 N.W. Yern Ave.
Portland, OR 97210

McWhorter
400 East Cottage Place
Carpentersville, IL 60110

The Mearl Corp.
41 E. 42 St.
New York, NY 10017
(212)-573-8500

Michelman, Inc.
9089 Shell Road
Cincinnati, OH 45236
(513)-793-7766

Miles Inc.
Mobay Rd.
Pittsburgh, PA 15205
(412)-777-2000/(800)-662-2927

Mineral Pigments Corp.
Davis Colors
7011 Muirkirk Rd.
Beltsville, MD 20705

Monsanto Chemical Co.
800 N. Lindbergh Blvd.
St. Louis, MO 63167
(314)-694-1000/(800)-325-4330

Mooney Chemicals Inc.
2301 Scranton Rd.
Cleveland, OH 44113

Nalco Chemical Co.
One Nalco Ctr.
Naperville, IL 60563
(708)-305-1000

New Jersey Zinc Co.
Fourth St. & Franklin Ave.
Palmerton, PA 18071

NL Chemicals, Inc.
P.O. Box 700
Hightstown, NJ 08520

Novamet Specialty Products, Inc.
10 Lawlins Pk.
Wyckoff, MJ 07481
(201)-891-7976

Huls America Inc.
80 Centennial Ave.
Piscataway, NJ 08855
(908)-980-6800/(800)-631-5275

Nyco Minerals Inc.
P.O. Box 368
Willsboro, NY 12996
(518)-963-4262

OM Group Inc.
2301 Scranton Rd.
Cleveland, OH 44113
(216)-781-8383/(800)-321-9696

Pasco Zinc Products Corp.
P.O. Box 280998
3380 Fite Rd.
Memphis, TN 38186

Patco Polymer Additives
C.J. Patterson Co.
3947 Broadway
Kansas City, MO 64111
(816)-561-9050/(800)-669-2250

Pfizer, Inc.
235 E. 42 St.
New York, NY 10017
(212)-573-7217

Raybo Chemical Co.
P.O. Box 2155K
Huntington, WV 25722

Reichhold Chemicals, Inc.
P.O. Box 13582
Research Triangle Park, NC27709
(919)-990-7500/(800)-448-3482

Rheox, Inc.
P.O. Box 700
Hightstown, NJ 08520
(609)-443-2500

Rhone-Poulenc
Performance Resins & Coatings
9808 Bluegrass Pkwy
Jeffersontown, KY 40299
(502)-499-4011/(800)-626-2613

Rohm & Haas Co.
Independence Mall West
Philadelphia, PA 19105
(215)-592-3000/(710)-670-5335

Sandoz Chemicals Corp.
4000 Monroe Rd.
Charlotte, NC 28205
(704)-331-7078/(800)-631-8077

SCM Chemicals
2701 Broening Hwy
Baltimore, MD 21222
(301)-288-8884/(800)-638-3234

Shamrock Technologies, Inc.
Foot of Pacific St.
Newark, NJ 07114
(201)-242-2999

Sherwin Williams Co.
Chemicals Div.
P.O. Box 6520
Cleveland, OH 44101
(216)-566-2344

Silberline Mfg. Co., Inc.
P.O. Box A
Lansford, PA 18232
(717)-668-6050

Stepan Co.
22 W. Frontage Rd.
Northfield, IL 60093
(708)-446-7500/(800)-745-7837

Sun Chemical Corp.
4526 Chickering Ave.
Cincinnati, OH 45242
(513)-681-5950/(800)-543-2323

Tammsco, Inc.
P.O. Box J
Tamms, IL 62988

Thompson-Hayward Chemical Co.
5200 Speaker Rd.
Kansas City, KS 66106
(913)-321-3131

3M Corp.
Bldg. 225-3S-05
St. Paul, MN 55144
(612)-733-1110/(800)-362-3456

Troy Corp.
72 Eagle Rock Ave.
East Hanover, NJ 07936
(201)-884-4300

Ulrich Chemical Inc.
3111 N. Post Rd.
Indianapolis, IN 46226
(317)-898-8632

Ultra Additives, Inc.
460 Straight St.
Paterson, NJ 07501
(201)-279-1306/(800)-524-0055

Unimin Corp.
258 Elm St.
New Canaan, CT 06840
(203)-966-8880/(800)-243-9004

Union Carbide Chemicals and
  Plastics
39 Old Ridgebury Rd.
Danbury, CT 06817
(203)-794-5300

Uniroyal Chemical Co.
Benson Rd.
Middlebury, CT 06749
(203)-573-2000/(800)-243-3024

U.S. Silica Co.
P.O. Box 187
Berkeley Springs, WV 25411
(304)-258-2500/(800)-243-7500

R.T. Vanderbilt Co., Inc.
P.O. Box 5150
30 Winfield St.
Norwalk, CT 06856
(203)-853-1400/(800)-243-6064

Wacker Silicones Corp.
3301 Sutton Rd.
Adrian, MI 49221
(517)-264-8500/(800)-248-0063

Whittaker Corp.
Heico Chemical Div.
Delaware Watergap, PA 18327
(717)-476-0353

Whittaker, Clark & Daniels, Inc.
1000 Coolidge St.
S. Plainfield, NJ 07080
(908)-561-6100/(800)-732-0562

Witco Corp.
520 Madison Ave.
New York, NY 10022
(212)-605-3941/(800)-238-9150

Zinc Corp. of America
300 Frankfort Rd.
Monaca, PA 15061
(412)-774-1020/(800)-962-7500

Other Noyes Publications

INDUSTRIAL WATER-BASED
PAINT FORMULATIONS

by

Ernest W. Flick

This collection of 220 up-to-date industrial water-based paint formulations will be of value to technical and managerial personnel in paint manufacturing firms, and firms which supply raw materials or services to these companies, and those generally interested in less hazardous, environmentally safer formulations.

U.S. paint manufacturers ship over 950 million gallons of paints and coatings per year, for a total dollar volume approaching $10 billion. The industry seems to thrive on competition, and technical challenges are being met with new generation finishes with improved durability and drying time, and, of course, a water base.

The data in the book consist of selections of manufacturers' suggested formulations made at no cost to, nor influence from, the makers or distributors of these materials. The information given is presented as supplied. Only the most recent data supplied have been included. Any solvent contained is minimal.

The book is divided into the following Sections:

1. Air Dry Coatings (14)

2. Air Dry or Force Dry Coatings (10)

3. Anti-Skid or Non-Slip Coatings (2)

4. Bake Dry Coatings (47)

5. Clear Coatings (10)

6. Coil Coatings (7)

7. Concrete Coatings (6)

8. Dipping Enamels (5)

9. Lacquers (27)

10. Primers (44)

11. Protective Coatings (3)

12. Spray Enamels (3)

13. Topcoats (18)

14. Traffic and Airfield Paints (7)

15. Miscellaneous (17)

Parenthetic numbers indicate the number of formulations in each section. In addition to the above, there are two other Sections which will be helpful:

16. A chemical trade name section where trade-named raw materials are listed with a chemical description and the supplier's name. The specifications which the raw materials meet are included, if applicable.

17. Main office addresses of the suppliers of trade-named raw materials.

Included in the descriptive information for each formulation, where available, the following may be listed:

1. Type of paint or coating

2. End use

3. Ingredients, by weight and/or volume

4. Mixing suggestions

5. Properties such as viscosity, total solids (by weight and/or volume), gloss, pencil hardness, pH, nonvolatiles, pigment/binder ratio, density, flash point, adhesion

6. Formulation source

ISBN 0-8155-1146-9 (1988)

277 pages

Other Noyes Publications

WATER-BASED
TRADE PAINT FORMULATIONS

by

Ernest W. Flick

This collection of 562 up-to-date water-based trade paint formulations will be of value to technical and managerial personnel in paint manufacturing companies and firms which supply raw materials or services to these companies, and to those interested in less hazardous, environmentally safer formulations. The book will be useful to both those with extensive experience as well as those new to this ever expanding $10 billion/year industry.

The data in the book consist of selections of manufacturers' suggested formulations made at no cost to, nor influence from, the makers or distributors of these materials. The information given is presented as supplied; only the most recent data supplied have been included. Any solvent contained is minimal.

Included in the descriptive information for each formulation, where available, the following properties may be listed: viscosity, solids content, % nonvolatiles, pigment volume concentration, density, pH, spatter, leveling, sag resistance, scrub cycles to failure, contrast ratio, ease of dispersion, fineness of grind, heat stability, freeze-thaw stability, ease of application, gloss, foaming, cratering, brightness, opacity, water spotting, adhesion to chalk, brush clean-up, reflectance, and sheen.

The formulations described are divided into three major sections, which cover exterior, interior, and exterior and/or interior water-based paints, enamels, and coatings. Further subdivision into chapters is as indicated below. Parenthetic numbers indicate the number of formulations per section or chapter. Also included are a Trade-Named Raw Materials section and a list of Suppliers' Addresses.

ISBN 0-8155-1147-7 (1988)

697 pages

Printed and bound by CPI Group (UK) Ltd, Croydon, CR0 4YY

15/10/2024

01774421-0001